Multivariable Cal

A Geometric Approa

Multivariable Calculus
A Geometric Approach

Frank Beatrous
University of Pittsburgh

Caspar R. Curjel
University of Washington

Prentice Hall
Upper Saddle River, New Jersey 07458

Library of Congress Cataloging-in-Publication Data

Beatrous, F.
 Multivariable calculus: a geometric approach / Frank Beatrous and Caspar R. Curjel.
   p. cm.
 ISBN 0-13-030437-9
  1. Calculus. I. Curjel, C.R. II. Title

 QA303.2 .B43 2002
 515—dc21

2001040033

Acquisition Editor: George Lobell
Editor-in-Chief: Sally Yagan
Vice President/Director of Production and Manufacturing: David W. Riccardi
Executive Managing Editor: Kathleen Schiaparelli
Senior Managing Editor: Linda Mihatov Behrens
Production Manager: Bob Walters
Manufacturing Buyer: Alan Fischer
Manufacturing Manager: Trudy Pisciotti
Marketing Manager: Angella Battle
Marketing Assistant: Vince Jansen
Director of Marketing: John Tweeddale
Editorial Assistant: Melanie VanBenthuysen
Art Director: Jayne Conte
Cover Design: Bruce Kenselaar
Cover Photo: Organ Recital Room, Arch. Christian De Portzamparc, Paris Music School / Antonio Martinelli

 ©2002 by Prentice-Hall, Inc.
Upper Saddle River, New Jersey 07458

Printed in the United States of America

10 9 8 7 6 5 4 3 2 1

ISBN: 0-13-030437-9

Pearson Education LTD., *London*
Pearson Education Australia PTY, Limited, *Sydney*
Pearson Education, Singapore Pte. Ltd.
Pearson Education North Asia Ltd., *Hong Kong*
Pearson Education Canada, Ltd., *Toronto*
Pearson Educaciûn de Mexico S.A. de C.V.
Pearson Education — Japan, *Tokyo*
Pearson Education Malaysia, Pte. Ltd.

Contents

Preface

Origins, points of view

This book grew out of lecture notes written by one of us (CRC) during a sabbatical year in 1992–93. Subsequently, the lecture notes were the multivariable calculus text of the Department of Mathematics at the University of Washington Seattle, Wash., for several years. Thus, most of the book's material has been class-tested by different instructors over an extended period.

The book embodies distinct views on what should be taught in a standard multivariable/vector calculus course[1] and how it should be taught. The thrust of our views, as tempered by classroom experience and input from colleagues, may be summarized as follows.

- We emphasize geometric interpretations and geometric reasoning in 3-space. Such an emphasis is appropriate for various reasons. For one, much of the multivariable calculus in question reflects the physics of basic phenomena taking place in 3-space. Then, 3-space with its points, lines, planes, surfaces, and solids, is, after all, the stage of our everyday material lives. In short, working with these geometric objects and their analytic representations constitutes an authentic application of multivariable calculus.

- If it comes down to it, the substance of a course is defined by what we ask students to do *themselves*, by the mathematical activities we expect them to become able to perform *on their own*. In this vein, we have given much thought to the exercises. In addition to standard exercises, we present some that are different in form, content, and also location within the exposition of the material.

Our views surface at times as novel features, with the result that the book is a traditional text with some nontraditional traits. We now give some examples of the latter.

Geometric reasoning

To describe what we mean by geometric reasoning in the context of multivariable calculus, we first discuss a sample exercise and then go into the idea of geometric reasoning in general.

---

[1] That is, a course covering vectors, functions of several variables, vector fields, line and surface integrals, and the theorems of Gauss and Stokes aimed at students preparing for programs in the natural sciences and engineering.

For the statement of the exercise, it suffices to know how the picture of the xyz-system used in the sample exercise comes about: It is the result of mapping an xyz-system in space into the plane by parallel orthogonal projection in the direction of the vector $-\vec{\imath} - \vec{\jmath} - \vec{k}$. In informal language, it is a photograph of an xyz-system obtained while standing at a point very far away and aiming with the camera in the direction of $-\vec{\imath} - \vec{\jmath} - \vec{k}$.

The sample exercise deals with a straight line in space considered on the one hand as geometric object defined by two points and on then other hand as track of a particle moving at constant speed, as follows.

> Below are shown two points A, B in space together with their projections A_0, B_0 onto the xy-plane.

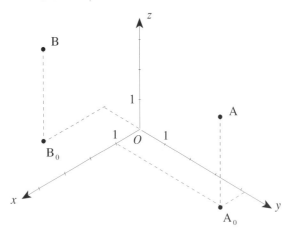

> The two points determine a straight line d. Draw d and its projection d_0 onto the xy-plane.
>
> (i) Find the points of intersections with the three coordinate planes by reasoning geometrically with d and d_0 in space and using only ruler and pencil.
>
> (ii) A particle moves on d at constant speed. At time $t = 0$, it is located at A, and at time $t = 1$ at B. Use formulas to find the times t at which the particle is located in the three coordinate planes.

This exercise takes place in front of the following background:

- By *geometric reasoning*, we mean the ability of working in one's mind with geometric objects such as points, lines, and planes as if they were tangible objects.

- One way to develop this ability is to make *drawings* of geometric objects. Working with such drawings, like working with blueprints or floor plans, requires understanding and visualizing the geometry.

- For the drawings to be as edifying as blueprints or floor plans, they must *look real* (i.e., they must be suggestive of reality). In passing from reality to drawing and back, the loss of information must be controllable. Given that experience and dexterity with ruler and pencil varies from person to person, our drawing procedures have to be realistically simple so that we can expect students to master them in a short time.

- The drawing procedures explained above before the example satisfy the requirements just stated. The technical name for the map of R^3 onto R^2 in question is "isometric axonometry."[2]

- By now, many students have done ruler and pencil work in single variable calculus. We use isometric axonometry to create pencil and ruler problems for multivariable calculus. We found that such problems work well in class, homework, and examinations. Among other things, they prepare students for the geometry of 3-space in which the main concepts of 3D vector analysis are staged.

Interspersed exercises

For most students, multivariable calculus is an order of difficulty higher than single variable calculus. This is notable especially when we introduce new concepts, many of which are much more "mathematical" than students expect. Over the years, we found it productive to intersperse one or several rounds of simple exercises in expository passages. Such exercises seem to loosen up the students' mind for the new material, help them ground the latter more solidly, present an opportunity to ask questions before misunderstandings are cemented, and provide satisfaction of having done something. For example:

- Shortly after the definition of flux, we consider a rectangle in the yz-plane and three vector fields and ask whether the flux is positive, or zero, or negative. Students have been introduced to flux only a few minutes ago, but most of them are able to work the questions.

- We begin the section on three-dimensional vector fields by examining the velocity field of a rigid rotation around the z-axis. Right at this point, we ask students to find the components of the vector field by using what they did earlier in the section on vector fields in the plane when working through the velocity field of rotating the xy-plane around the origin.

- The typical beginning multivariable calculus student finds the dot product of two vectors a strange concept. Shortly after defining the dot product geometrically, we provide four pairs of vectors that are either parallel, opposite, or at angles a multiple of $\frac{\pi}{4}$. Students are asked to use ruler, pencil, and calculator to find the dot products. Early hand-to-hand contact with a new concept seems to make the concept more palatable.

The interspersed problems may be used in different ways. The instructor may:

- Have students work on them in class;
- Work them out in full on the board;
- Outline the way they are solved;

---

[2]See, for example, *Computer Graphics*, R. A. Plastick and G. Kalley, Schaum's Outline Series, McGraw-Hill, 1986, p. 160.

- Present a slightly changed version so that the book's version can be assigned as homework.

Problems that require reading and writing

In addition to problems that have no solution or infinitely many, a number of them require reading and writing skills which we think are appropriate for multivariable calculus and to which students have to be introduced. For example:

- Students compare the scalar projections of $\vec{a}$ onto $\vec{b}$ and of $\vec{b}$ onto $\vec{a}$. Then they are asked to clear up the confusion underlying the question "Why are the two projections different? Did the professor not say in class that $\vec{a} \cdot \vec{b} = \vec{b} \cdot \vec{a}$ all the time?" Write out an answer that will make sense to the skeptical student.

- Students compute the circulation of a vector field around a circle centered at the origin. The circulation equals zero. A says: "It shows that the field is conservative." B disagrees. Who is right? Write down an answer that makes sense to both A and B.

- Students are presented with six statements such as "The triple integral of div $\vec{F}$ over the solid K equals the flux of $\vec{F}$ across ∂K" and "The line integral of grad f over the boundary of a surface S equals the flux of div f across S." Do all the expressions make sense? Which of the statements are true? Write out explanations for your answers.

Design conventions

As explained previously, each section contains within its expository parts several rounds of simple exercises. These problems are set off from the expository text by the heading **Problem(s)** and horizontal rules at the beginning and at the end. Additional problems are included at the end of each section.

Throughout the book, we distinguish between the *definition* of a concept and the mechanics of doing *calculations*. Headings of sections introducing *concepts* are flagged with the symbol ☞. Headings of sections concerning *calculation* are flagged with the symbol ✍.

The symbol ∎ is used to mark the end of an example.

Supplements

- The *Students' Solution Manual* contains written out solutions to the odd-numbered exercises. The text itself provides only the *answers* to the odd-numbered exercises.

- The *Instructor's Resource Manual* contains the written out solutions to *all* exercises. In addition, it contains a computer lab manual and instructional materials titled "Practicing Geometric Reasoning in 3-space," as follows.

  - The computer lab manual is part of a computer lab. We describe the lab further below.

  - "Practicing Geometric Reasoning in 3-space" begins with a detailed explanation of isometric axonometry, the map from 3-space to 2-space which students use to work ruler and pencil problems about geometric events in 3-space (see *Geometric Reasoning* at the beginning of this preface). Then it provides a number of practice problems which help students visualize operations with points, straight lines, and planes.

- The *computer lab* has students work through standard problems of multivariable calculus by geometric manipulation instead of computation. The lab works as follows. *First,* students download a free plug-in. *Second,* they log on to the Prentice-Hall web site (see below) to download the lab manual. *Third,* they go to special graphics files on the web site. These files can be modified by means of the plug-in. Students work problems by manipulating the graphics files according to instructions of the lab manual.

- The web site for the book is `www.prenhall.com/beatrous`. The web site contains the computer lab manual for downloading by students, and other material related to the book. It will be operational as of October 31, 2001.

Acknowledgments

We would like to extend our thanks to:

- G. S. Monk of the Mathematics Department at the University of Washington, with whom one of us (CRC) began working on instructional issues already in the 70s;

- The College of Arts and Sciences and the Department of Mathematics at the University of Washington, Seattle, Wash. for the support of a project that some consider to be outside the conventional view of a multivariable calculus text;

- Fellow faculty, teaching assistants, and the many students at the University of Washington who helped us with their comments and alerted us to all kinds of weak spots;

- Ken Plochinski, the author of the solutions manual, with whom CRC had many helpful conversations on the project during the time when he was a colleague at the University of Washington;

- Rose L. Pugh (Bellevue Community College, Bellevue, WA), the coauthor of *Practicing Geometric Reasoning in 3-space*, from whom we learned to look at the material with the eyes of a much wider student body;

- John Sylvester, a colleague at the University of Washington and the coauthor of the Computer Lab, with whom CRC had a quarter of intense discussions on what a such a lab should and should not be;

- Michael O'Connell, of Math Sciences Computing Center, College of Arts and Sciences, University of Washington, whose assistance made the long-distance collaboration of the authors possible.

At Prentice-Hall, we had the good fortune of working with George Lobell. His insights into, and tolerance of, the strange minds of mathematicians made it a pleasure to work with him—thank you, George. We also would like to thank the colleagues who reviewed the book, among them Robert Boyer (Drexel University),

Paul C. DuChateau (Colorado State University), Ashwani K. Kapila (Rensselaer Polytechnic Institute), and William J. Keane (Boston College).

Last, but by no means least, we are deeply grateful to our families for putting up with our extended mental and physical absences during the writing of this book, that would not have taken shape without their supporting us in so many ways.

Frank Beatrous (`beatrous@pitt.edu`)
Caspar R. Curjel (`curjel@math.washington.edu`)

Multivariable Calculus
A Geometric Approach

Chapter 1

Vectors and Curves

1   Vectors in the Plane

Shifting around objects.   In the plane of this sheet, we shift around objects in a particular way, as follows. An object located at point A is shifted to point B. An object located at K gets the same shift and ends up at L. To indicate the shift from A to B, we draw an arrow with initial point A and endpoint B, and similarly for K and L, as shown below left.   The arrow AB (or KL) shows the direction of the

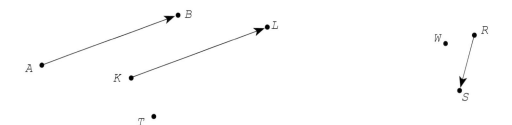

Figure 1.1

move, and the length of the arrow indicates its "magnitude" (i.e., how far we go). We use the symbol $\overrightarrow{AB}$ to denote the shift defined by AB. Similarly, we write $\overrightarrow{KL}$ for the shift KL. The shifts $\overrightarrow{AB}$ and $\overrightarrow{KL}$ are the same.

Now we consider a second shift $\overrightarrow{RS}$ shown above right. It moves every object the way R is shifted to S.

Problem _____

1.1. The questions a) – c) refer to Figure 1.1 above.

    a) We apply the shift $\overrightarrow{AB}$ to the point T and immediately afterwards the shift $\overrightarrow{RS}$. After the two shifts, the point T lands at a point U. Use ruler and pencil to draw point U.

    b) We apply the same two shifts to T, but in the other order: First $\overrightarrow{RS}$ and then $\overrightarrow{AB}$. Draw the point where T will land.

    c) Find a point V with the following property: When you apply the two shifts in a row, as in b), to the point V, the point V lands at W shown above right.

What does the discussion of shifts and Problem 1.1 tell us?

- A shift is determined by a direction and a real number. We call this number the *magnitude* or *length* of the shift.
- A shift can be represented as an arrow. The direction of the arrow is the direction of the shift, and its length is the magnitude of the shift.
- Given two shifts, we can combine them and obtain another shift. That is, we can "add" two shifts. This "addition" is different from the ordinary addition of numbers.

Vectors and Scalars.   In the sciences and engineering, one works with time, mass, temperature, force, velocity, acceleration, electric and magnetic fields, etc. Some of these concepts such as time, length, mass, and temperature are measured by a *number*. Others, such as shifts discussed in Problem 1.1 are characterized not just by a number but by a *number together with a direction*. Such a concept is called a *vector*.[1] Other examples of vector quantities are force, velocity, acceleration, and the electric and magnetic fields.

In this course, we study vectors as mathematical objects. We develop *one* mathematical tool that we can use for *many* different concepts in different areas.

Working with vectors is different from working with numbers. To stress the difference between vectors and numbers, it is customary to refer to numbers as *scalars*. The word "scalar" reminds us of numbers as being points on a time scale, temperature scale, etc.

The two questions.   When introducing a new concept, we will always ask the following two questions.

☞ What *is* the concept? How do we define and describe it?
✍ How does one *work* with it? How do we perform computations?

Correspondingly, there are "definition and description" problems and "computation" problems. In quizzes and tests, there will be both kinds.

☞ **What are vectors in the plane?**

By "plane," we mean this sheet of paper from which you read and on which you draw. A vector in the plane is a directed line segment in the plane. "Directed" means that the line segment carries an arrowhead that defines a direction, as shown in Figure 1.2.

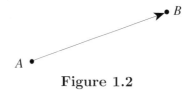

Figure 1.2

---
[1]Latin *vector,* "carrier." The word "vehicle" has the the same Latin root.

Notation and terminology. If the line segment has endpoints A and B, and if the arrow points from A to B, we write $\overrightarrow{AB}$ for the vector defined by A, B and the arrow. The point A is called the *initial point* and B the *endpoint* of $\overrightarrow{AB}$ (see Figure 1.2 on the preceding page).

Different books write vectors in different ways, e.g., by underlining, hat, or boldface:

$$\underline{v}, \quad \underline{AB}, \quad \ldots; \quad \hat{w}, \quad \widehat{PQ}, \quad \ldots; \quad \mathbf{R}, \quad \mathbf{T}.$$

We will use arrows:

$$\vec{a}, \ \vec{b}, \ \overrightarrow{PQ}, \ \ldots.$$

The *magnitude* or *length* of a vector is the length of the line segment which represents it. We will use the notation:

$$|\underline{v}|, \quad |\underline{AB}|, \ldots; \qquad |\hat{w}|, \quad |\widehat{PQ}|, \ldots; \qquad |\mathbf{R}|, \quad |\mathbf{T}|.$$

Note: The symbol $|\vec{a}|$ means "magnitude of the vector $\vec{a}$." It is a non-negative scalar. On the other hand, if c is a scalar, the symbol $|c|$ means "absolute value of the scalar c."

Problems

1.2. a) A is a given point shown below left. Draw four additional points Q_1, Q_2, Q_3, and Q_4 such that

$$|\overrightarrow{AP_1}| = |\overrightarrow{AP_2}| = |\overrightarrow{AP_3}| = |\overrightarrow{AP_4}| = 1.5 \text{ cm}.$$

b) Use one or more complete sentences to describe the collection of *all* points P for which $|\overrightarrow{AP}| = 1.5$ is true.

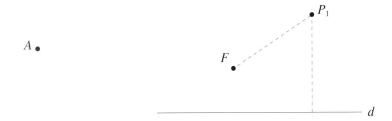

1.3. a) You are given the point F and the straight line d shown above right. Draw seven points P such that $|\overrightarrow{FP}|$ equals the distance from P to the line d. One point P_1 has already been drawn for you.

b) Make a guess: What will be the shape of the collection of *all* points P for which $|\overrightarrow{FP}|$ equals the distance of P from d?

What is meant by $\vec{a} = \vec{b}$? That is, when are two vectors equal? Go back to shifts: Two shifts $\vec{a}$ and $\vec{b}$ are the same when their representing arrows point in the same direction and have the same length. That is the definition of equality of vectors in general: *Two vectors $\vec{a}$ and $\vec{b}$ are equal, written $\vec{a} = \vec{b}$, if the following conditions 1 and 2 below are satisfied at the same time:*

1. The two vectors have the same magnitude, that is, $|\vec{a}| = |\vec{b}|$;
2. the line segments are parallel and the arrows point the same way. We will say, "$\vec{a}$ and $\vec{b}$ have the same direction."

This is illustrated in Figure 1.3.

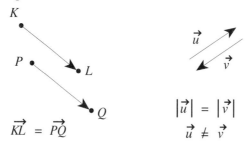

$$\overrightarrow{KL} = \overrightarrow{PQ}$$

$$|\vec{u}| = |\vec{v}|$$
$$\vec{u} \neq \vec{v}$$

Figure 1.3

Note: The example $\overrightarrow{KL} = \overrightarrow{PQ}$ shows that two different line segments may represent the same vector. Given any vector, you can move its initial point around unless you are told not to do it.

Addition of vectors. We add vectors as we add shifts, namely by attaching the second to the end of the first. In general: Given two vectors $\vec{u}$ and $\vec{v}$ (below left), we move $\vec{v}$ so that its initial point is at the endpoint of $\vec{u}$. Then $\vec{u} + \vec{v}$ is the vector that connects the initial point of $\vec{u}$ with the endpoint of $\vec{v}$, as shown below right.

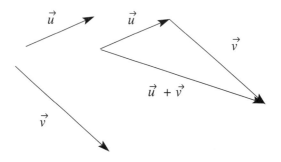

Problems _____

1.4. The vectors $\vec{u}$ and $\vec{v}$ are shown above. Into the figure above draw the vectors
    (a) $\vec{v} + \vec{u}$ (b) $(\vec{u} + \vec{v}) + \vec{u}$ (c) $\vec{u} + \vec{u} + \vec{u}$.

1.5. Below you find four statements. For each of them, answer the following questions: Does the statement make sense? If yes, is it true for all possible choices of $\vec{a}$ and $\vec{b}$? If it does not make sense, explain why.
    (a) $\vec{a} + \vec{b} = \vec{b} + \vec{a}$ (b) $\vec{a} + |\vec{b}| = |\vec{a} + \vec{b}|$
    (c) $|\vec{b} + \vec{a}| = |\vec{a} + \vec{b}|$ (d) $|\vec{a} + \vec{b}| = |\vec{a}| + |\vec{b}|$

Scalar multiples of a vector. We are given a vector $\overrightarrow{AB}$ and a scalar m (i.e., a real number m). Then the symbol $m\,\overrightarrow{AB}$ is read as "m times $\overrightarrow{AB}$." It denotes a vector which has magnitude $|m||\overrightarrow{AB}|$, and whose direction depends on m:

    if $m > 0$: same direction as $\overrightarrow{AB}$;   if $m < 0$: direction opposite to $\overrightarrow{AB}$.

The case $m = 0$ is discussed in (iii) below. Here are some examples:

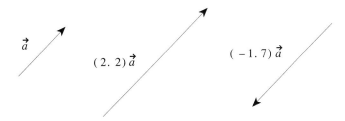

Note:

1. Forming a scalar multiple of a vector is also called "multiplying a vector with a scalar."

2. The vector $(-1)\vec{a}$ is written simply as $-\vec{a}$.

3. The vector $0\,\vec{a}$ (zero times $\vec{a}$) has magnitude zero. Its initial point is the same as its endpoint. It is called the *zero vector*, and it is written $\overrightarrow{0}$. Be sure to distinguish the scalar 0 (zero) from the zero vector $\overrightarrow{0}$.

4. "Taking a scalar multiple of a vector" or "multiplying a vector with a scalar" means "stretching or shrinking a vector, possibly also reversing its direction."

5. For the operations "taking the magnitude of a vector" and "shrinking (or stretching) a vector," the following relation holds:

$$|m\,\vec{a}| = |m|\,|\vec{a}|.$$

In this formula, $|m|$ is the absolute value of the scalar m, whereas $|m\,\vec{a}|$ and $|\vec{a}|$ are magnitudes of vectors.

6. Two nonzero vectors $\vec{a}$ and $\vec{b}$ are called *parallel* if their line segments are parallel. To say "$\vec{a}$ and $\vec{b}$ are parallel" means that $\vec{a} = c\,\vec{b}$ for some nonzero scalar c. The scalar c may be positive or negative.

Subtraction of vectors. Given $\vec{a}$ and $\vec{b}$ (below left), $\vec{a} - \vec{b}$ is defined as $\vec{a} + (-1)\,\vec{b}$ (below right).

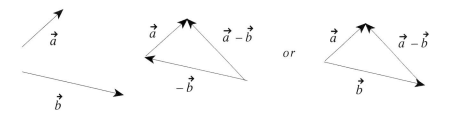

Note:

1. $\vec{a} - \vec{a}$ is the zero vector $\overrightarrow{0}$. If $\vec{b}$ is any vector, we clearly have $\vec{b} + \overrightarrow{0} = \vec{b}$.

2. The difference $\vec{a} - \vec{b}$ is easily remembered in the form "$\vec{a} - \vec{b}$ points from $\vec{b}$ to $\vec{a}$." This is a short way of saying *"If $\vec{a}$ and $\vec{b}$ have the same initial point, then $\vec{a} - \vec{b}$ points from the endpoint of $\vec{b}$ to the endpoint of $\vec{a}$."*

Problem _____

1.6. Below are shown seven vectors $\vec{a}, \ldots, \vec{g}$.

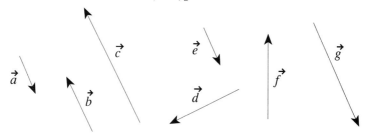

a) Are any of these two vectors equal? Write down all equal pairs.

b) Can you find a scalar x so that $\vec{a} = x\,\vec{g}$? In other words: Can you express $\vec{a}$ as a scalar multiple of $\vec{g}$? If yes, find such an x; if no, explain.

c) As b) but for the equation $\vec{b} = x\,\vec{d}$.

d) Can you solve the equation $\vec{f} = u\,\vec{c} + v\,\vec{d}$ for the scalars u and v? If yes, find u and v; if no, explain.

_____

Note: The operations "addition of vectors" and "taking a scalar multiple of a vector" interact the way we would expect:

$$m(\vec{a} + \vec{b}) = m\,\vec{a} + m\,\vec{b}, \qquad m(n\,\vec{a}) = (mn)\,\vec{a}.$$

Unit vectors. A *unit vector* is a vector of magnitude one. Given a vector $\vec{a}$, we want to find a vector which has the same direction as $\vec{a}$ but that has magnitude 1. In other words: *We want to find a unit vector in the direction of $\vec{a}$.* We reason as follows:

$|\vec{a}|$, the magnitude of $\vec{a}$, is a scalar. Therefore, $1/|\vec{a}|$ is also a scalar. Now look at the vector $(1/|\vec{a}|)\vec{a}$. It is of has magnitude 1 because

$$\left| \frac{1}{|\vec{a}|}\, \vec{a} \right| = \frac{1}{|\vec{a}|}|\vec{a}| = 1.$$

Also, it has the same direction as $\vec{a}$. Therefore it is a unit vector in the direction of $\vec{a}$.

Usually, the unit vector $(1/|\vec{a}|)\vec{a}$ is written in the form $\vec{a}/|\vec{a}|$. Note that $\vec{a}/|\vec{a}|$ does not mean that we divide by $\vec{a}$.

There is no standard abbreviation for unit vectors.

Problem _____

1.7. Figure 1.4 on the facing page shows five points $A, B, \ldots, E$.

a) At the beginning of the section *Answers to Odd Problems* on page 403, there is a statement in a box. Read the statement.

b) Draw the following two vectors $\vec{u}$ and $\vec{v}$:

(i) $\vec{u} = (2.5)\overrightarrow{AB} + (-0.8)\overrightarrow{CD}$ with initial point at A;

(ii) $\vec{v} = (2.5)\overrightarrow{BA} - (-0.8)\overrightarrow{CD}$ with initial point at D.

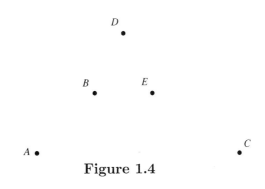

Figure 1.4

c) Draw

$$\vec{w} = 5\,\frac{\overrightarrow{AB}}{|\overrightarrow{AB}|}$$

with initial point at A. Use units of length 1 cm for b) and c).

d) Determine the scalar k so that $|k\,\overrightarrow{AC}| = 43.98\ m$.

✍ How do we compute with vectors in the plane?

The idea. If we want to do computations with points in the plane, we use a coordinate system. Usually we take the Cartesian xy-system. If we want to do computations with vectors in the plane, we do the same and use the same xy-system.

Working out the idea. Take the vector that points from the origin $(0,0)$ to the point $(1,0)$ and call it $\vec{\imath}$. Similarly, $\vec{\jmath}$ is the vector pointing from the origin to $(0,1)$. If $\vec{w}$ is any given vector, we move its initial point to the origin. Then the endpoint of $\vec{w}$ has xy-coordinates (a,b). If we now "stretch" $\vec{\imath}$ by the scalar a and $\vec{\jmath}$ by the scalar b then $\vec{w}$ will be the sum of $a\,\vec{\imath}$ and $b\,\vec{\jmath}$:

$$\vec{w} = a\,\vec{\imath} + b\,\vec{\jmath},$$

as shown in Figure 1.5. That is, *we can write any given vector $\vec{w}$ in the form $\vec{w} = a\,\vec{\imath} + b\,\vec{\jmath}$.*

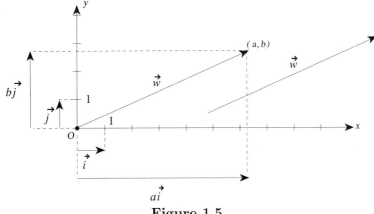

Figure 1.5

Terminology and notation

- The two vectors $\vec{\imath}$ and $\vec{\jmath}$ are often called *basis vectors* because any vector can be expressed in terms of them in the form $a\vec{\imath} + b\vec{\jmath}$.

- The two scalars a and b that we use to describe $\vec{w}$ are called the *components* of $\vec{w}$ with respect to our xy-system. Usually we omit the reference to the xy-system and say that (a, b) are the components of $\vec{w}$.

- The statement "$\vec{w}$ has components (a, b)" means "$\vec{w} = a\vec{\imath} + b\vec{\jmath}$." A shorter way of writing this equation is $\vec{w} = (a, b)$. This is the way we will be writing the components of a vector.

- In an xy-system, points have coordinates, and vectors have components. *What is the relation between "coordinates of a point" and "components of a vector"?*

  Answer: *Given any vector $\vec{w}$. If you move its initial point to the origin $(0, 0)$ of the xy-system, then the coordinates of its endpoint are the components of $\vec{w}$.*

- We distinguish coordinates and components by the way we write them. For example, we consider the point P with coordinates $(3, 1)$ and the vector $\vec{a}$ with components $(9, -2)$. We write

  $$P(3, 1) \text{ for "the point } P \text{ has coordinates } (3, 1),\text{"}$$

  and we write

  $$\vec{a} = (9, -2) \text{ for "the vector } \vec{a} \text{ has components } (9, -2).\text{"}$$

  Remember that $\vec{a} = (9, -2)$ means $\vec{a} = 9\vec{\imath} + (-2)\vec{\jmath}$.

Problem ─────────────────────────────

1.8.   a) Into the xy-system shown in Figure 1.6 draw (i) the vector $\vec{v}$ with components $(4, 1)$ and initial point at the origin, and (ii) $\vec{v} = (4, 1)$ with initial point $Q(3, 2)$.

       b) Find the components of the vector $\vec{w}$ shown in Figure 1.6.

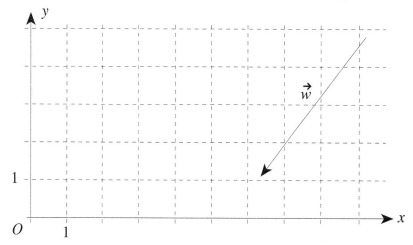

Figure 1.6

Questions. How do we compute the magnitude of a vector from its components? Given two vectors, how do we recognize from their components if they are equal or not? How do we add vectors? How do we take scalar multiples? Subtraction? Unit vector? Below we list the formulas for all the operations discussed under the heading *What are vectors in the plane?* on page 2. The formulas below are for vectors $\vec{u} = (a, b)$, $\vec{v} = (r, s)$.

- *How do we find the magnitude $|\vec{u}|$?*

  Answer: $|\vec{u}| = \sqrt{a^2 + b^2}$

- What does $\vec{u} = \vec{v}$ mean?

  Answer: The components must be the same: $a = r$, $b = s$.

- What are the components of $\vec{u} + \vec{v}$?

  Answer: $\vec{u} + \vec{v} = (a + r, b + s)$

- What are the components of $m\,\vec{u}$?

  Answer: $m\,\vec{u} = (ma, mb)$

- What are the components of $\vec{u} - \vec{v}$?

  Answer: $\vec{u} - \vec{v} = (a - r, b - s)$

- What are the components of the unit vector $\vec{u}_0$ in the direction of $\vec{u}$?

  Answer: $\vec{u}_0 = \dfrac{1}{|\vec{u}|}\,\vec{u} = \left(\dfrac{a}{\sqrt{a^2 + b^2}}, \dfrac{b}{\sqrt{a^2 + b^2}} \right)$

A formula for unit vectors. A convenient way to write a unit vector $\vec{p}$ in the plane is to use trigonometric functions, as follows. Take $\vec{p} = (\cos\theta, \sin\theta)$ with initial point at the origin. Then $|\vec{p}| = \cos^2\theta + \sin^2\theta = 1$. It means that the endpoint is on the circle of radius 1. That is, $\vec{p}$ is a unit vector, as shown in Figure 1.7.

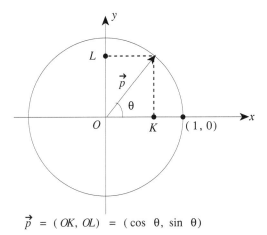

$$\vec{p} = (OK, OL) = (\cos\,\theta, \sin\,\theta)$$

Figure 1.7

Throughout the book, you will be asked many times to draw an xy-system yourself. Use 1 cm as unit on the x-axis and y-axis. To remind you of taking 1 cm as units, we will call such an xy-system a standard xy-system.

Problems

1.9.   a) With $K(7,1)$, $L(-4,5)$, find the components of $\overrightarrow{KL}$.

       b) R and S are the points with coordinates $R(a,b)$ and $S(u,v)$. Find the components of the vector $\overrightarrow{SR}$ from S to R.

1.10. In a standard xy-system draw the following points: $A(0,0)$, $B(2,2)$, $C(0,7)$, $D(3,4)$, $E(4,2)$. Label the points clearly.

       a) Find the components of the following vectors:
          (i) $\vec{u} = (2.5)\overrightarrow{AB} + (-0.8)\overrightarrow{CD}$;   (ii) $\vec{v} = (2.5)\overrightarrow{BA} - (-0.8)\overrightarrow{CD}$.

       b) Do the same for the vector

$$\vec{w} = 5\,\frac{\overrightarrow{AB}}{|\overrightarrow{AB}|}.$$

       c) Work with components to determine the scalar k so that $|k\,\overrightarrow{AC}| = 43.98\ m$.

       d) Find the components of a vector $\vec{p}$ that has the same direction as $\overrightarrow{EA}$ and whose magnitude equals 2 cm.

1.11. This is a problem about writing a unit vector with trig functions, as discussed just before Problem 1.9.

       a) Find the components of a vector that forms an angle of $14°$ with $\vec{\imath}$, and that has magnitude 147.49 cm.

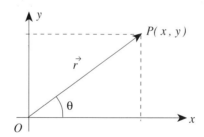

       b) Find the components (x, y) of a vector $\vec{r}$ that forms an angle of θ with $\vec{\imath}$ and that has magnitude r. Note that (r, θ) are the polar coordinates of the point $P(x, y)$. For a review of polar coordinates, see the beginning of Section 26.

Additional Problems

Definition and operations without components

1.12. K, L, M are points in the plane. The vector $\vec{p}$ is defined by

$$\vec{p} = -2\overrightarrow{KM} + 3\overrightarrow{LK} + \overrightarrow{MK} - \overrightarrow{LM} + (1.5)\overrightarrow{KL} - 2\overrightarrow{ML}.$$

Find scalars u, v, w so that you can write $\vec{p}$ in the form

$$\vec{p} = u\overrightarrow{KL} + v\overrightarrow{KM} + w\overrightarrow{LM}.$$

1.13. $\vec{u}$ and $\vec{v}$ are the two vectors shown in Figure 1.8.
  a) Find a scalar a such that $|a\vec{u}| = |\vec{v}|$.
  b) Find a scalar b such that $|b\vec{v}| = |\vec{u}|$.

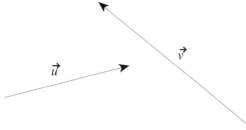

Figure 1.8

1.14. A, B, C, and D are four points in the plane, as shown in Figure 1.9. We consider two variable vectors $\vec{p}$ and $\vec{q}$ defined as follows:
  - $\vec{p} = t\overrightarrow{AB}$ where t is a variable scalar, with initial point always at A;
  - $\vec{q} = u\overrightarrow{CD}$ where u is a variable scalar, with initial point always at C.
  a) Draw $\vec{p}$ for $t = -0.5$ and $\vec{q}$ for $u = 1.5$.
  b) Find values t and u such that the endpoints of $\vec{p}$ and $\vec{q}$ are the same.

B •

• C

A •

• D

Figure 1.9

1.15. The four points P, Q, R, and S are given, as shown below. The three points Q, R and S are on a straight line.
  a) Use ruler and calculator to find the scalar

$$k = \frac{|\overrightarrow{QR}|}{|\overrightarrow{RS}|}.$$

  b) Find a scalar t such that $\overrightarrow{PR} = \overrightarrow{PQ} + t\,\overrightarrow{QS}$.

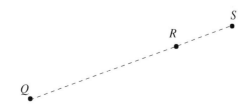

Computations with components

1.16. The line m in the xy-plane is shown below.

    a) Find the components of a vector $\vec{a}$ that is parallel to m and has magnitude 193 cm.

    b) Draw a line q perpendicular to m and find the components of a unit vector $\vec{b}$ that is parallel to q.

    c) Compute the slope of q from the slope of m. Use the slope of q to find the components of the unit vector $\vec{b}$ of part b.

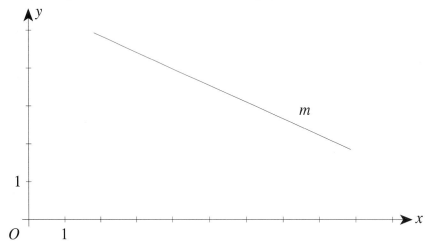

1.17. Let A, B, and C be the three points, as shown below.

    a) Draw a standard xy-coordinate system with origin at A, with horizontal x-axis and vertical y-axis.

        (i) Find the xy-components of $\overrightarrow{AB}$, $\overrightarrow{AC}$ and $\overrightarrow{AB} + \overrightarrow{AC}$.

        (ii) Use the xy-components you just found to draw $\overrightarrow{AB} + \overrightarrow{AC}$ with initial point at A.

    b) Use a colored pen to draw a standard uv-coordinate system as follows: Origin at C, u-axis horizontal and v-axis vertical.

        (i) Find the uv-components of $\overrightarrow{AB}$, $\overrightarrow{AC}$ and $\overrightarrow{AB} + \overrightarrow{AC}$.

        (ii) Use the uv-components you just found to draw $\overrightarrow{AB} + \overrightarrow{AC}$ with initial point at C.

    c) Compare the vectors obtained in a) and b).

C •

• B

A •

1.18. The vector $\vec{b}$ has magnitude 3.9 and forms the angle $\theta = \frac{\pi}{3}$ with the positive x-axis. Find the components of $\vec{b}$. *Hint:* Use the representation of unit vectors by means of trigonometric functions (page 9).

1.19. In a standard xy-system draw the straight line $y = -1$ and the point $F(0, 1)$. Label F clearly.

    a) On the straight line $y = 4$ find a point R such that $|\overrightarrow{RF}|$ equals the distance of R from the line $y = -1$.

    b) Now we consider *all* points $P(x, y)$ such that $|\overrightarrow{PF}|$ equals the distance of P from the line $y = -1$. Write down the equation the components (x, y) of $\overrightarrow{OP}$ satisfy. *Simplify the equation you get as much as possible.* What is the shape of the figure on which all our points P lie?

1.20. If t is a scalar, then the vector $\vec{r} = (4, 3) + t(2, -1)$ is a function of t, and we write, therefore, $\vec{r}(t)$ instead of $\vec{r}$. *We take the origin as the initial point of $\vec{r}(t)$, no matter what the the value of t is.*

    a) Into a standard xy-system, draw $\vec{r}(t)$ for $t = 0, 1, 2, 3, -1, -1.5$.

    b) Describe in words the pattern in which the endpoints of all vectors $\vec{r}(t)$ are arranged. Why is the pattern the way it is?

1.21. Figure 1.10 shows a curve C and a point P on C.

    a) Use ruler and pencil to find the components of a tangent vector $\vec{b}$ to the curve C at the point P.

    b) Find a point Q on the curve C at which $\vec{n} = (2, 3)$ is perpendicular to C (i.e., perpendicular to the tangent of C at Q).

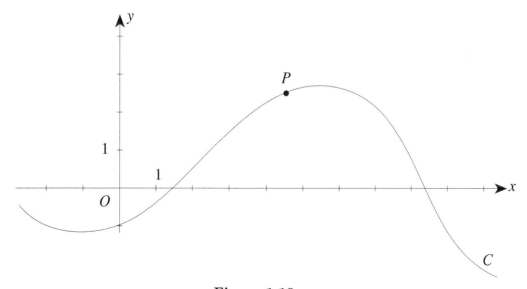

Figure 1.10

2   Lines and Curves in the Plane

Lines and their parametric representations

Point moving on a line.[1]   A point P moves *at constant speed* along the line m shown below. We write $P(t)$ for the position of P at time t sec, and we write (x, y) for the coordinates of $P(t)$. We are told: $P(0) = A$, $P(1) = B$. This is illustrated below. Note that the xy-system of Figure 2.1 is *not* a standard xy-system.   The

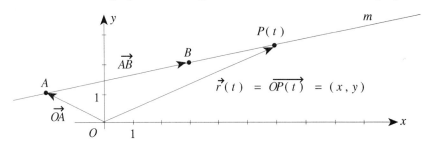

Figure 2.1

vector $\overrightarrow{OP(t)}$ gives the position of $P(t)$. We write $\vec{r}(t)$ for $\overrightarrow{OP(t)}$. As time passes, $\vec{r}(t)$ varies. That is, the components (x, y) of $\vec{r}(t)$ must be functions of t:

$$\vec{r}(t) = (x, y) = (a(t), b(t)).$$

What are the formulas for $a(t)$ and $b(t)$?
First, we observe that $\vec{r}(t) = \overrightarrow{OA} + \overrightarrow{AP(t)}$ (look at Figure 2.1). Then we observe that between $t = 0$ and $t = 1$ the point moves from A to B. That is, in one second it moves one length of $\overrightarrow{AB}$. Therefore, in the t seconds from time 0 to time t it moves the length of $t\overrightarrow{AB}$. But in these t seconds it moves from A to $P(t)$. Therefore, $\overrightarrow{AP(t)} = t\overrightarrow{AB}$, and we obtain

$$\vec{r}(t) = \overrightarrow{OA} + t\overrightarrow{AB}.$$

We read off coordinates: $A(-2, 1)$, $B(3, 2)$. We get: $\overrightarrow{OA} = (-2, 1)$, $\overrightarrow{AB} = (5, 1)$. If we go with these numbers into the equation for $\vec{r}(t)$ and write it out in components, we obtain

$$\begin{aligned} x &= a(t) = -2 + 5t \\ y &= b(t) = 1 + t. \end{aligned}$$

Velocity vector and speed.   The point moves at constant speed, and in one second, it moves the length of $\overrightarrow{AB}$. Therefore, $\overrightarrow{AB}$ is called the *velocity vector* of the moving point. The direction of $\overrightarrow{AB}$ shows the direction in which P moves. The magnitude $|\overrightarrow{AB}|$ is called the *speed* of the moving point. Note that "velocity" is a vector and "speed" is a scalar.

Parametric representation of a line.   The formulas $x = -2 + 5t$, $y = 1 + t$ describe the line m in the following sense: For every value of t, there is a corresponding point on the line, and for any point on the line, there is a corresponding value of t. The formulas $x = -2 + 5t$, $y = 1 + t$ constitute what is called *parametric*

---

[1] "Line" will always mean "straight line."

representation of the line m or a set of *parametric equations* for m, and t is called the *parameter*. The general form of a parametric representation of a line is as follows:

$$x = p + vt$$
$$y = q + wt$$

In this formula, (p, q) are the coordinates of a point of the line, and (v, w) are the components of a vector parallel to the line. We can look at the formulas $x = p + vt, y = q + wt$ in two ways:

(i) They describe a point P moving on a line m at constant speed.

(ii) They provide a parametric representation of the line m.

When we talk about a point moving on a line, we speak the language of physics. When we are working with a parametric representation of a line, we speak the language of geometry.

Problems

2.1. The line p is given by the parametric representation

$$x = 5 - 4t, \qquad y = 1 + 2t.$$

a) Draw the line into a standard xy-system.

b) Mark on p the points corresponding to $t = -1, 0, 1.5$.

c) Consider the given parametric representation as describing a point moving at constant speed on the line p, with t measured in seconds.

   (i) Find the velocity vector and speed of P.

   (ii) Find the time t at which P lies on the y-axis.

2.2. a) Find a parametric representation for the line plotted below. (Note that the xy-system shown is *not* a standard system.)

b) Find a parametric representation for the line through the origin that is parallel to the line below.

c) Find a parametric representation for the line through the point $P(2, 1)$ and perpendicular to the line below.

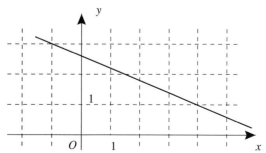

2.3. a) Find a parametric representation for the line $2x + 3y = 1$.

b) Find a parametric representation for the line through the origin which is perpendicular to the line $2x + 3y = 1$.

Curves and their parametric representations

Now our point P moves on a curve C in the plane. We use the same notation as for lines at the beginning of this section:

$P(t)$ is the position of P at time t sec, and $\vec{r}(t) = \overrightarrow{OP(t)}$ is the *position vector* of P.

This is illustrated in the figure below.

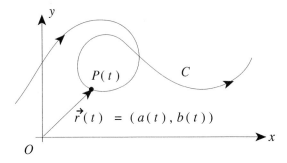

The components of the position vector $\vec{r}(t)$ are the coordinates (x, y) of the point $P(t)$ of C. As for straight lines, x and y must be functions of t:

$$\vec{r}(t) = (x, y) = (a(t), b(t)),$$

and we obtain a *parametric representation* of C:

$$x = a(t) \qquad y = b(t).$$

Example 1. In the discussion of unit vectors in Section 1, we looked at vectors of the form $\vec{p} = (\cos\theta, \sin\theta)$. In these formulas, θ is a variable. If we think of θ as time and write t instead of θ, we obtain a parametric representation $(x, y) = (\cos t, \sin t)$ of the circle of radius 1:

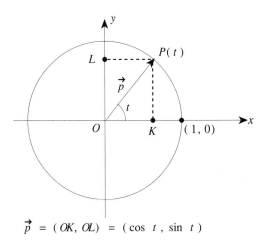

$$\vec{p} = (OK, OL) = (\cos t, \sin t)$$

As t varies, the point $P(t)$ moves around the circle of radius 1. ■

Problem

2.4. We consider the parametric representation $(x, y) = (\cos t, \sin t)$ of the "unit circle."

    a) Does P move clockwise or counterclockwise?

    b) On the circle, draw the point P that corresponds to the parameter value $t = \pi$. Label it $P(\pi)$. Draw also $P(\frac{7}{4}\pi)$.

Velocity vector and speed

Anything that moves has "velocity" or "speed." If a point moves on a straight line, then we know what is meant by the velocity vector and speed of the moving point. What do we mean by "velocity" and "speed" of a point moving on a curve? How do we define "velocity" and "speed"? How do we compute it?

☞ How are velocity and speed defined?

We consider a curve C given by $(x, y) = (a(t), b(t))$, as shown in Figure 2.2. We look at $P(t)$ and other points $P_1, P_2, P_3, \ldots$ close by, as shown in the figure.

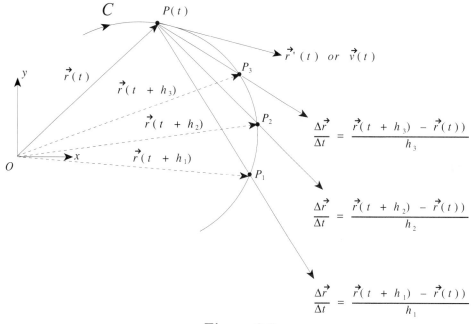

Figure 2.2

(i) We write $P_1 = P(t + h_1)$, where h_1 is a positive number. The vector $\overrightarrow{PP_1} = \vec{r}(t + h_1) - \vec{r}(t)$ represents the change of position of P between time t and $t + h_1$. We write $\Delta\vec{r}$ for $\overrightarrow{PP_1}$ and Δt for $(t + h_1) - t = h_1$ (the Greek letter Δ, read "delta," suggests "difference"). The vector $\frac{\Delta\vec{r}}{\Delta t}$ is the position change between times t and $t + h_1$ *per unit of time*. A "position change per time" is a velocity. However, $\frac{\Delta\vec{r}}{\Delta t}$ takes into account only P and P_1, and not what is happening in between. Therefore, $\frac{\Delta\vec{r}}{\Delta t}$ is the *average* velocity of the point between t and $t + h_1$.

(ii) We repeat (i) for $P_2 = P(t + h_2)$, where h_2 is a positive number less than h_1. Then $\frac{\Delta \vec{r}}{\Delta t}$ is the average velocity between two points that are closer together than in step (i).

(iii) We repeat (i) for $P_3 = P(t + h_3)$, where h_3 is a positive number less than h_2. Then $\frac{\Delta \vec{r}}{\Delta t}$ is the average velocity between two points that are closer together than in step (ii).

(iv) We continue with the process of taking average velocities for smaller and smaller numbers $h_4, h_5, h_6, \ldots, h_{4829}, h_{4830}, \ldots$. The corresponding average velocity vectors get closer and closer to a vector. We write $\vec{v}(t)$ or $\vec{r}'(t)$ for this vector. It is shown in Figure 2.2 on the page before. The vector $\vec{v}(t)$ or $\vec{r}'(t)$ is called the *velocity vector* of the point at P. The magnitude of the velocity vector at P is called the *speed* of the point at P.

The direction of $\vec{r}'(t)$ is tangent to the curve at C. Why? Look at Figure 2.2 on the preceding page. The average velocity vectors turn into $\vec{r}'(t)$ as the numbers h are getting closer and closer to zero. In summary:

> *The direction of the velocity vector is the direction in which the point moves at P. This direction is tangent to the curve at P. The magnitude of the velocity vector tells us how fast the point is moving at this moment.*

Note:

1. "Velocity" is a vector and "speed" is a scalar.

2. At $P(t)$ the curve C has many tangent vectors. They are all parallel to each other. The velocity vector is *one* of the many tangent vectors of the curve C at $P(t)$.

3. Another way of describing the average velocities getting closer and closer to the velocity at P is to say "the average velocity vectors stabilize at $\vec{r}'(t)$" or "the average velocity vectors vary less and less from $\vec{r}'(t)$." In technical language of mathematics, the vector $\vec{r}'(t)$ is the limit of the sequence of average velocities.

4. A tangent *vector* is not the same as a tangent *line*. The first is a vector, and the second is a straight line whose direction vector is parallel to a tangent vector.

✍ How do we compute velocity and speed?

To compute the velocity vector we translate the definition of the velocity vector into formula language.

Step 1: Formula for average velocity. We use the letter h for any of the numbers $h_1, h_2 \ldots$ which occur in the average velocities. The components of the position vector $\vec{r}(t)$ are $(a(t), b(t))$. First we find $\Delta \vec{r}$:

$$\begin{aligned} \Delta \vec{r} &= \vec{r}(t + h) - \vec{r}(t) \\ &= (a(t + h) - a(t), b(t + h) - b(t)). \end{aligned}$$

For the average velocity, we obtain

$$\frac{\Delta \vec{r}}{\Delta t} = \left(\frac{a(t + h) - a(t)}{h}, \frac{b(t + h) - b(t)}{h} \right).$$

Step 2: h tends to zero. If h tends to zero,

$$\frac{a(t+h) - a(t)}{h} \quad \text{and} \quad \frac{b(t+h) - b(t)}{h}$$

tend to $a'(t)$ and $b'(t)$, respectively, the ordinary derivatives of elementary calculus.

Result: *A point P moves on curve C. Its position at time t is given by $\vec{r} = (a(t), b(t))$. Its velocity vector at time t is given by*

$$\vec{r}'(t) = (a'(t), b'(t)).$$

Its speed at time t is given by

$$|\vec{r}'(t)| = \sqrt{(a'(t))^2 + (b'(t))^2}.$$

Problems _____

2.5. A point moves on the line m. At time t, the position of P is given by $(x, y) = (5, 1) + t(-4, 2)$. Use the derivative formulas to find the velocity vector and the speed.

2.6. This is about the way we define the velocity vector by means of average velocities (see "I. How are velocity and speed ...?").
The curve C is given by $(x, y) = (t^3, t^2)$. We look at the points that correspond to $t = 2$ and $t = 2 + h$. The figure below is *not* to scale.

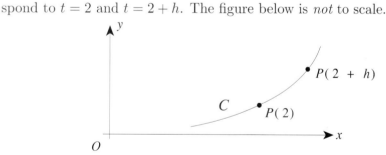

a) Find the components of the velocity vector at $t = 2$.

b) Find the average velocity between times 2 and $2 + h$. Your answer will be a vector whose components contain the letter h. Simplify your answer as much as possible.

c) Take $h = \dfrac{1}{249^{8117}}$. Compare the the velocity vector at $t = 2$ with the average velocity between 2 and $2 + \dfrac{1}{249^{8117}}$. What can you say?
   Note: The formulas for our curve C are very simple. Therefore, we can get the velocity vector at $t = 2$ by setting $h = 0$ in the formula for the average velocity from $t = 2$ to $t = 2 + h$. For more complicated curves, this easy way may not work.

2.7. The point P moves on a circle of radius 3 units in counterclockwise direction. The position of P at time t is defined as follows: At time t, the position vector $\vec{r}(t)$ of P forms the angle ωt with the positive x-axis. In the expression

$\omega\,t$ the symbol ω ("omega") is a constant scalar, called *angular velocity*. This is illustrated below. (Note that that the xy-system shown below is not a standard system.)

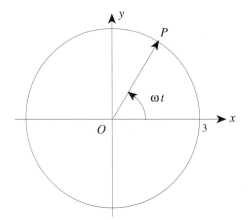

The position of P is given by $(x, y) = (3\cos\omega t, 3\sin\omega t)$. In questions a), b) (i), c) (i) below we take $\omega = 1.5$.

a) In the figure, above draw P and the velocity vector at P at time $t = 1.57\,(= \frac{\pi}{2})$. You have to use the derivative formula for the velocity vector.

b) How many units of time does it take the point to make one full revolution? Give the answer (i) with $\omega = 1.5$, and (ii) as an expression in terms of the symbol ω.

c) How many revolutions per unit of time does P make? Give the answer (i) with $\omega = 1.5$, and (ii) as an expression in terms of the symbol ω.

Note: The formulas $x = a(t), y = b(t)$ define a curve in the plane. As in the case of lines, we can look at the formulas in two ways:

(i) They describe a point P moving on a curve C.

(ii) They provide a parametric representation of the curve C.

When we talk about a point moving on a curve, we speak the language of physics. When we are working with a parametric representation of a curve, we speak the language of geometry.

Measuring arc length. You can think of the arc length of a curve as the length measured with a flexible tape measure laid along the curve. Thus, the arc length of the line segment from P to Q is just the distance from P to Q, and the arc length of a circle of radius a is $2\pi a$.

We now ask: *How can we use a parametric representation of a curve to measure its arc length?* It turns out that the answer to this question is hidden in the velocity calculation we have already done. Suppose you have a parametric representation for a curve C, and let $P(t)$ be the point on the curve corresponding to the parameter value t. Mark a point P_0 on the curve, corresponding to parameter value t_0, and for $t \geq t_0$, let $s(t)$ denote the arc length of the portion of the curve between P_0 and $P(t)$, as shown in the picture below.

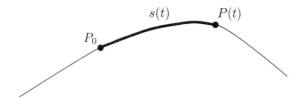

For $t < t_0$, we define $s(t)$ to be *minus* the arc length between $P(t)$ and P_0. The function $s(t)$ is called the *arc length function* for the curve.[2] If you knew the arc length function $s(t)$, then you could measure the arc length between any two points on the curve by simply calculating the change in $s(t)$ between those two points: For $a < b$,

$$\text{arc length from } P(a) \text{ to } P(b) = s(b) - s(a).$$

Thus, the arc length problem comes down to this: *How can we calculate the function* $s(t)$? Rather than address this question directly, we instead try to calculate the derivative ds/dt. If we can do this, then we can recover $s(t)$ by integration.

To get a handle on ds/dt, think of t as time, and $P(t)$ as the position of a moving particle at time t. Then $s(t)$ gives the distance that the particle has traveled since time t_0, measured along the curve, and its derivative ds/dt measures the speed of the particle. But the speed is just the magnitude of the velocity vector that we already know how to calculate. Letting $\vec{r}(t) = \overrightarrow{OP(t)}$ be the position vector of the particle, the velocity is $d\vec{r}/dt$, and so

$$\frac{ds}{dt} = \left| \frac{d\vec{r}}{dt} \right|.$$

You can now obtain the arc length function $s(t)$ by integration:

$$s(t) = \int_{t_0}^{t} |\vec{r}\,'(\tau)| \, d\tau.$$

The curve between the parameter values $t = a$ and $t = b$ (with $a < b$) by integration:

$$\text{arc length} = s(b) - s(a) = \int_a^b \left| \frac{d\vec{r}}{dt} \right| \, dt = \int_a^b \sqrt{\left(\frac{dx}{dt} \right)^2 + \left(\frac{dy}{dt} \right)^2} \, dt. \qquad (2.1)$$

The integrand in the above expression is sometimes called the *element of arc length* along the curve.

Problems

2.8. C is the circle of radius m given by $x = m \, \cos(2t), y = m \, \sin(2t)$.

   a) Write down and evaluate the integral (2.1) to find the arc length from $t = 0$ to $t = 2\pi$.

   b) Your colleague says: "There must be something wrong because I *know* that the circumference of a circle of radius m equals $2\pi \, m$." Write down a response.

---

[2]The use of the word "the" in referring to the arc length function is a little misleading, because there are different arc length functions for different choices of the initial parameter value t_0, which specifies where to position the arc length tape measure. However, any two arc length functions will differ by a constant, so the *change* in $s(t)$ between two parameter values doesn't depend on the choice of t_0.

2.9. The straight line which passes through $P(x_0, y_0)$ and $P(x_1, y_1)$ has a parametric representation

$$x = x_0 + t(x_1 - x_0), \qquad y = y_0 + t(y_1 - y_0).$$

  a) Set up, but do not yet evaluate, the integral (2.1) for the arc length from $t = 0$ and $t = 1$.

  b) Without evaluating the integral of a), you can predict what you get when evaluate it. What is your prediction? Why?

  c) Evaluate the integral.

Additional Problems

Lines

2.10. The point P moves at constant speed on the line m shown below. The velocity vector is $\vec{v}$. At time t sec, the position of P is written $P(t)$.

  a) $P(0) = A$, with A as shown. Draw $P(2)$.

  b) At which time t is $P(t) = K$?

  c) Another point Q moves on the same line m with constant velocity vector $\vec{w}$. We have $Q(0) = B$. Clearly P moves faster than Q. Find the time t at which P catches up with Q. *Hint:* Use ruler and pencil.

2.11. A line has parametric representation

$$x = 1 + 3t, \quad y = 2 - 2t.$$

  a) Sketch the line in a standard xy-system, and mark the points corresponding to $t = -1, 0, 1, 2$.

  b) If the above equations give the coordinates of a moving particle at time t, at what speed is the particle traveling?

  c) Find an equation for the line in the form $Ax + By = C$.

2.12. A particle moves along a straight line with constant speed of 2 distance units per second. At time $t = 0$, it is at the point $P(3, 4)$ and moving toward the origin.

  a) Find the velocity of the particle.

  b) Find its position as a function of time t.

  c) How long will it take for the particle to travel from the point $P(3, 4)$ to the origin?

2.13. The line m is given by the equation $3x - 2y + 4 = 0$.

  a) Find a parametric representation for m using x as parameter.

b) Find a parametric representation for m using y as parameter.

c) For any point P on the line m, let s denote the distance from P to the point $(0, 2)$ if P is to the right of the y-axis, and minus that distance if P is to the left of the y-axis, as shown below. Find a parametric representation of m using s as a parameter. *Hint:* How are s and x related?

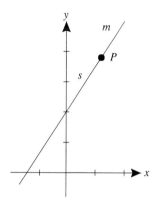

2.14. A general point $P = P(t)$ on the line below is determined by the parametric representation

$$\overrightarrow{OP} = \overrightarrow{OQ} + t\overrightarrow{QR}.$$

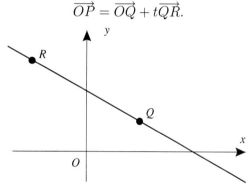

a) Estimate the values of t when $P(t)$ is on each of the coordinate axes.

b) For which values of t is $P(t)$ between the points Q and R?

2.15. The line p is given by $(x, y) = (4 + 2t, 10 + 6t)$, and the line q by $(x, y) = (6 + 3u, 9 + 2u)$ (in q the parameter is u). Find the coordinates of the point D of intersection of the two lines.

Curves

2.16. The curve C is given by $(x, y) = (t^3 + 1, 4t^2)$.

a) Find the coordinates of all points E on C which are at distance 16 from the x-axis.

b) Give a parametric representation of the tangent line to C at the point corresponding to $t = 1$.

c) Find the components of a unit vector (vector of length 1) which is perpendicular to the tangent line of b).

2.17. Find a parametric representation for the tangent line to each of the following curves at the given point.

$$C_1 \quad : \quad y = e^{\cos x} \quad \text{at} \quad x = \frac{\pi}{2}$$
$$C_2 \quad : \quad x = e^t \cos t, \quad y = e^t \sin t \quad \text{at} \quad t = 0.$$

2.18. The curve C is given by $(x, y) = (a(t), b(t))$, where the graphs of the functions $a(t)$ and $b(t)$ are shown below.

    a) Find the coordinates of the point Q on C that corresponds to $t = 1$.

    b) Find the components of a vector $\vec{b}$ that is parallel to the tangent line of C at the same point Q.

    c) Imagine that C describes the movement of a point P in the plane. Find the time t at which P lies on the line $x = y$.

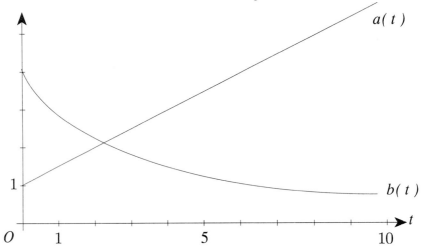

2.19. A *logarithmic spiral* has parametric representation

$$x = e^t \cos t, \quad y = e^t \sin t.$$

    a) Find the arc length from the point where $t = 0$ to the point where $t = 1$.

    b) Let $s(t)$ be the arc length function for the curve, starting at parameter value $t = 0$ (see page 21). Find a formula for $s(t)$.

    c) Use the formula you worked out in part b to express t in terms of the arc length s, and then to express x and y as functions of s along the curve. This process is called *parametrizing by arc length*.

2.20. The ellipse $(x/a)^2 + (y/b)^2 = 1$ has parametric representation

$$x = a \cos t, \quad y = b \sin t \quad \text{with} \quad 0 \le t \le 2\pi.$$

    a) Write down an integral that would give the arc length of the ellipse $\frac{x^2}{4} + y^2 = 1$. Do not attempt to evaluate the integral.

    b) Use a calculator or computer to obtain a numerical estimate for the arc length of the ellipse in part a.

3 Acceleration

As in the case of one-dimensional motion studied in elementary calculus, acceleration measures the rate of change of velocity with respect to time. However, for a particle moving in a *plane*, velocity can change in direction as well as magnitude, so a *vector* quantity is required to encode its rate of change.

☞ What is acceleration?

Imagine a particle moving along a plane curve, with position at time t given by $P(t)$. As we have done previously, we let $\vec{r}(t) = \overrightarrow{OP(t)}$ denote its position vector. The *velocity* at time time t is then

$$\vec{v}(t) = \vec{r}'(t).$$

Thus, $\vec{v}(t)$ is tangent to the path of the particle and points in the direction the particle moves, with magnitude equal to the particle's speed. The *acceleration* $\vec{a}(t)$ is the time derivative of the velocity:

$$\vec{a}(t) = \vec{v}'(t) = \vec{r}''(t).$$

This means that for a small time increment Δt, the acceleration is approximately equal to the ratio $\Delta \vec{v}/\Delta t$, where

$$\Delta \vec{v} = \vec{v}(t + \Delta t) - \vec{v}(t).$$

This is illustrated in the two figures below. The left figure shows the position and velocity vectors at time t and $t + \Delta t$. On the right, we have three vectors: $\Delta \vec{v}$, the change in velocity; $\Delta \vec{v}/\Delta t$, the approximate acceleration; $\vec{a}(t)$, the acceleration at time t.

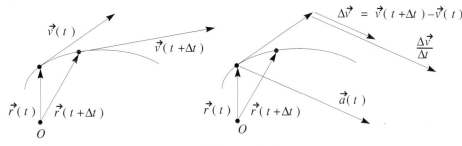

Figure 3.1

Problems

3.1.   a) The position of a particle at time t is given by

$$\vec{r}(t) = \overrightarrow{OP} + t\overrightarrow{PQ}.$$

Describe in words the direction and magnitude of the acceleration vector.

b) Another particle moves along a straight line, but, in contrast to part a), its velocity is *not* constant. Can you say anything about the direction of the acceleration vector? What about its magnitude?

3.2. The path of a moving particle is shown below. One of the vectors $\vec{b}$ and $-\vec{b}$ is the acceleration vector at the marked point. Which one is it? Explain your answer. *Hint:* As the particle moves from left to right along the curve, does the velocity vector rotate clockwise, or counterclockwise? What does that reveal about the direction of the acceleration vector?

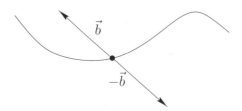

✎ Working with acceleration

We work out the formulas for calculating acceleration when the x and y coordinates are given as functions of time. Let $x(t)$ and $y(t)$ denote the coordinates of a particle's position at time t. Thus the position vector at time t is

$$\vec{r}(t) = x(t)\,\vec{\imath} + y(t)\,\vec{\jmath}.$$

We have already seen that the velocity is given by

$$\vec{v}(t) = \frac{d\vec{r}}{dt} = \frac{dx}{dt}\,\vec{\imath} + \frac{dy}{dt}\,\vec{\jmath}.$$

A second differentiation gives the acceleration:

$$\vec{a}(t) = \frac{d\vec{v}}{dt} = \frac{d^2x}{dt^2}\,\vec{\imath} + \frac{d^2y}{dt^2}\,\vec{\jmath}.$$

We work this out in the case of uniform motion about a circle of radius ρ with angular velocity ω (see Problem 2.7). In this case, the position at time t is

$$\vec{r} = \rho\cos\omega t\,\vec{\imath} + \rho\sin\omega t\,\vec{\jmath}.$$

The velocity is

$$\vec{v} = -\rho\omega\sin\omega t\,\vec{\imath} + \rho\omega\cos\omega t\,\vec{\jmath}$$

and the acceleration is

$$\vec{a} = -\rho\omega^2\cos\omega t\,\vec{\imath} - \rho\omega^2\sin\omega t\,\vec{\jmath}.$$

Problems

3.3. Here we continue with the circular motion described above. In a standard xy-system, draw a circle of radius 3 cm. The particle moves on the circle at constant speed. It takes four seconds to make one revolution.

    a) What is ω in our case? Write down the formulas that give the the position of the particle at time t.

    b) Draw the position, velocity vector $\vec{v}$, and acceleration vector $\vec{a}$ for $t = 0, 1, 2.25$.

c) Find the speed $v = |\vec{r}'(t)|$ for $t = 0, 1, 2.25$. What is the speed in general, that is, for generic values ρ, ω, t? Your answer to the latter question will contain some or all of the letters ρ, ω, t. How does v depend on t?

d) Find the magnitude of the acceleration vector $\vec{a}$ for our specific values $\rho = 3$, $\omega = \ldots$ at time $t = 0, 1, 2.25$.

e) Now we look at the general case, as we did in c). Express $|\vec{a}|$, the magnitude of the acceleration vector, in terms of the speed $v = |\vec{r}'(t)|$ and the radius ρ.

f) Describe in words the *direction* of the acceleration vector. *Hint:* Compare the formula for the acceleration vector to the one for the position vector $\vec{r}'(t)$.

3.4. In a suitably chosen coordinate system, the position of a baseball is given by

$$x = 30t, \qquad y = 25t - 16t^2.$$

Here x and y are measured in feet, and time t is measured in seconds.

a) Calculate the velocity $\vec{v}(t)$ and the acceleration $\vec{a}(t)$.

b) Write a sentence (no formulas) describing the physical meaning of the acceleration formula you worked out in part a.

Reversing the process: Recovering velocity and position from acceleration. It is common in mechanics to know the acceleration of a particle, and to want to calculate its velocity and position as functions of time. For example, when analyzing the motion of a satellite, the gravitational force acting on the satellite, and hence also its acceleration, can be calculated from Newton's Law of Universal Gravitation (see page 286).

Since acceleration is a derivative, going backward from the acceleration to get the velocity and position involves solving differential equations. In the special case when the acceleration is known explicitly as a function of time, the velocity and acceleration can be recovered by direct integration, once to get the velocity, and a second time to get the position. We will now carry this out in the simplest special case.

The case of constant acceleration. The simplest case is when the acceleration is known to be constant. Letting $\vec{a}$ denote the (constant) acceleration, you have

$$\frac{d\vec{v}}{dt} = \vec{a} = a_1\vec{\imath} + a_2\vec{\jmath},$$

so integration gives

$$
\begin{aligned}
\vec{v}(t) &= (a_1 t + c_1)\vec{\imath} + (a_2 t + c_2)\vec{\jmath} \\
&= (a_1\vec{\imath} + a_2\vec{\jmath})t + (c_1\vec{\imath} + c_2\vec{\jmath}) \\
&= \vec{a}t + \vec{c}.
\end{aligned}
$$

Notice that we have combined the two constants of integration c_1 and c_2 into a single vector constant $\vec{c} = c_1\vec{\imath} + c_2\vec{\jmath}$. To express this constant in a physically meaningful way, plug $t = 0$ into the above velocity formula to get $\vec{v}(0) = \vec{c}$. In other words, $\vec{c}$ is the velocity at time $t = 0$. We will call $\vec{v}(0)$ the *initial velocity*, and abbreviate it as $\vec{v}_0$. Thus, the velocity at time t is

$$\vec{v}(t) = \vec{a}t + \vec{v}_0.$$

To get the position at time t, just integrate a second time. You have

$$\frac{d\vec{r}}{dt} = \vec{v} = \vec{a}t + \vec{v}_0,$$

so

$$\vec{r}(t) = \frac{\vec{a}}{2}t^2 + \vec{v}_0 t + \vec{c},$$

where $\vec{c}$ is another (vector) constant of integration. Plugging in $t = 0$ gives $\vec{r}(0) = \vec{c}$, so $\vec{c}$ is just the initial position $\vec{r}(0)$, which we will denote by $\vec{r}_0$. Inserting this in the above formula for $\vec{r}(t)$ gives

$$\vec{r}(t) = \frac{\vec{a}}{2}t^2 + \vec{v}_0 t + \vec{r}_0.$$

Problem _____

    3.5. A particle has constant acceleration $\vec{a} = -\vec{j}$. At time $t = 0$ it is at the origin, with initial velocity $\vec{v}_0 = \vec{i} + 2\vec{j}$.

      a) Find a formula for its position as a function of time.

      b) Plot the position of the particle at times $t = 0, 1, 2, 3, 4$.

      c) What geometric shape describes the path of the particle? *Hint:* Part a gives you x and y as functions of time t. Eliminate t between the two equations to get a relation between x and y.

      d) Repeat parts a to c with initial velocity $\vec{v}_0 = 2\vec{j}$.

_____

    You found in part c of Problem 3.5 that, for the given values of $\vec{a}$, $\vec{v}_0$, and $\vec{r}_0$, the particle moves along a parabolic path, with the axis of the parabola parallel to the acceleration vector. The method you used there can be applied to *any* constant acceleration vector that is parallel to the y-axis. The result is that (aside from the special case when the initial velocity is either $\vec{0}$ or parallel to the acceleration) the particle follows a parabolic path, with the axis of the parabola parallel to the acceleration. We now ask: *What path does the particle follow if the acceleration is constant, but not parallel to the y axis?* The answer is exactly the same! The reason is that acceleration is defined in geometric terms *without reference to any particular coordinate system.* That means that you are free to *choose* an xy coordinate system so as to make your work as simple as possible.[1]

_____

[1]There is an important lesson to be learned here. For problems involving quantities that can be described in purely physical or geometric terms, without reference to a coordinate system, you are free to *choose* a coordinate system in a convenient way. You can often drastically simplify a problem by a suitable choice of coordinates. For example, it is much easier to describe the motion of the planets in a coordinate system with origin attached to the Sun than in one with origin attached to Mars!

Summary: *For a particle with nonzero constant acceleration $\vec{a}$:*

(i) If the initial velocity v_0 is a multiple of the acceleration $\vec{a}$, then the particle moves along a straight line parallel to $\vec{a}$.

(ii) If the initial velocity is not *a scalar multiple of the acceleration, the particle moves along a parabolic path, with the axis of the parabola parallel to $\vec{a}$.*

Motion of projectiles. A particle moving near the surface of the earth (e.g., a baseball or artillery shell) is subject to constant downward acceleration due to the earth's gravity.[2] We will call such a particle a *projectile*. From the general discussion of motion with constant acceleration, it follows that a projectile moves along either a parabolic path or a vertical line. The magnitude of the acceleration due to gravity is usually denoted by g. In the metric system, $g \approx 9.8 \text{m/sec}^2$, and in the English system, $g \approx 32 \text{ft/sec}^2$.

If an xy-system is chosen in the plane of the projectile path, with the y axis pointing straight up, the acceleration is given by

$$\vec{a} = -g\vec{j},$$

so the position is given by

$$\vec{r}(t) = -\frac{g}{2}t^2\vec{j} + t\vec{v}_0 + \vec{r}_0.$$

In terms of coordinates, this becomes

$$
\begin{aligned}
x &= \alpha t + x_0 \\
y &= -\frac{g}{2}t^2 + \beta t + y_0
\end{aligned}
$$

where $P(x_0, y_0)$ is the initial position, and α and β are the $\vec{i}$ and $\vec{j}$ components of the initial velocity $\vec{v}_0$.

Problem _____

3.6. A projectile is fired from ground level at an angle of 60 degrees from the horizontal, with an initial speed of 50 m/sec.

a) In a suitably chosen coordinate system, find the coordinates of the projectile as a function of time. *Tip:* It would be a good idea to place the origin at the position from which the projectile is fired.

b) How long will the projectile remain airborne? *Hint:* What is the y coordinate when the projectile strikes the ground?

c) How far is the point of impact from the point from which the projectile was fired?

_____

[2]By Newton's Second Law of Motion, the acceleration is the product of the mass of the particle and the net force that acts on it. In our treatment of projectile motion, we ignore the effects of any forces other than gravity, such as air resistance, electric, or magnetic forces.

Additional Problems

3.7. It is the bottom of the ninth inning, with two outs, and the Cubs are down by two runs, with two men on base. Sammy Sosa steps up to the plate, and hits the first pitch hard toward center. The ball's initial trajectory is at an angle of 40 degrees from horizontal, with an initial speed of 122 ft/sec. The center field fence in Wrigley Field is 440 ft away, and 11.3 ft high. Will the ball clear the fence for a home run?

3.8. A recreational bicyclist will typically use a gear ratio which propels her forward a distance of 4 meters for each revolution of the pedals. Relative to the bicycle, the pedals rotate around a circle of radius 170 mm = .170 m.

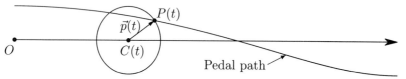

a) The cyclist steers in a straight line and turns the pedals at a rate of one revolution per second. A reference point $C(t)$ attached to the center of the pedal crank moves along a horizontal line with speed 4 m/second. Find the coordinates of $C(t)$ as a function of time t in a stationary coordinate system with the x-axis running along this line.

b) If $P(t)$ gives the location of the pedal at time t, then the vector $\vec{p}(t) = \overrightarrow{C(t)P(t)}$ describes a circle of radius .170 m at a rate of one revolution per second. Give a formula for this vector as a function of t. Assume that the pedal is at its highest point at time $t = 0$.

c) Give a formula for the pedal position in the stationary coordinate system you used in part a.

d) Calculate the acceleration of the moving pedal, and verify that the acceleration is always directed toward the crank center $C(t)$.

4 The Dot Product

Given two vectors $\vec{a}$ and $\vec{b}$, what is their angle? In particular, are they perpendicular? Questions like this come up frequently in many areas of the sciences and engineering. For such questions one uses an operation called the "dot product" of vectors. Some books use "scalar product" or "inner product" for what we call the dot product.

☞ How is the dot product of two vectors defined?

Two vectors $\vec{a}$ and $\vec{b}$ are given. We move them so that their initial points are the same. We write θ for the smaller of the two angles between $\vec{a}$ and $\vec{b}$.

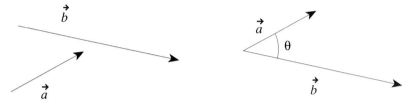

Definition. *The dot product of $\vec{a}$ and $\vec{b}$*
- *is defined as the* scalar $|\vec{a}||\vec{b}|\cos\theta$,
- *is written $\vec{a} \cdot \vec{b}$,*
- *is read "$\vec{a}$ dot $\vec{b}$."*

Note: If c is a scalar and $\vec{a}$, $\vec{b}$ are vectors, then $c\,\vec{a}$ is a vector, but $\vec{a} \cdot \vec{b}$ is a scalar.

Problem _____

4.1. a) Use the definition to evaluate the following dot products:

$$\text{(i) } \vec{\imath} \cdot \vec{\imath} \qquad \text{(ii) } \vec{\imath} \cdot \vec{\jmath} \qquad \text{(iii) } (5\vec{\jmath}) \cdot \vec{\imath} \qquad \text{(iv) } (-\vec{\jmath}) \cdot \vec{\jmath}$$

b) Use ruler and pencil to find the dot product of the following four pairs of vectors $\vec{a}$, $\vec{b}$ (take 1 cm as unit of measurement):

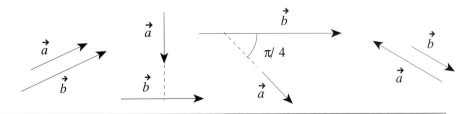

From the definition of the dot product we, conclude:

(i) $\vec{0} \cdot \vec{a} = \vec{a} \cdot \vec{0} = 0$

(ii) $\vec{a} \cdot \vec{a} = |\vec{a}||\vec{a}|\,1 = |\vec{a}|^2$

(iii) *If $\vec{a}$ is perpendicular to $\vec{b}$, then their dot product $\vec{a} \cdot \vec{b}$ is zero. If the dot product of two nonzero vectors is zero, then they are perpendicular to each other.*

Terminology. Two vectors which are perpendicular to each other are also called *normal* to each other or *orthogonal* to each other. The expressions "perpendicular to," "normal to," "orthogonal to" and "at a right angle to" mean the same thing.

Rules for the dot product.   There are three basic rules. The letter k stands for a scalar.

1. $\vec{a} \cdot \vec{b} = \vec{b} \cdot \vec{a}$
2. $(k\vec{a}) \cdot \vec{b} = \vec{a} \cdot (k\vec{b}) = k(\vec{a} \cdot \vec{b})$
3. $\vec{a} \cdot (\vec{b} + \vec{c}) = \vec{a} \cdot \vec{b} + \vec{a} \cdot \vec{c}$

We do not prove the rules. Rules (1) and (2) follow directly from the definition of the dot product. For a proof of rule (3) see, for example, R. Courant, *Differential and Integral Calculus, Volume II*, pages 3 to 8.

✍ How do we compute the dot product?

We consider the vectors $\vec{a}$ and $\vec{b}$:

$$\begin{aligned} \vec{a} &= (a_1, a_2) = a_1\vec{i} + a_2\vec{j}, \\ \vec{b} &= (b_1, b_2) = b_1\vec{i} + b_2\vec{j}. \end{aligned}$$

Now we work out the dot product, using the three rules above:

$$\begin{aligned} \vec{a} \cdot \vec{b} &= (a_1\vec{i} + a_2\vec{j}) \cdot (b_1\vec{i} + b_2\vec{j}) \\ &= a_1\vec{i} \cdot b_1\vec{i} + a_1\vec{i} \cdot b_2\vec{j} + a_2\vec{j} \cdot b_1\vec{i} + a_2\vec{j} \cdot b_2\vec{j} \quad \text{(by rule (3))} \\ &= a_1b_1\vec{i} \cdot \vec{i} + a_1b_2\vec{i} \cdot \vec{j} + a_2b_1\vec{j} \cdot \vec{i} + a_2b_2\vec{j} \cdot \vec{j} \quad \text{(by rule (2))} \\ &= (a_1b_1)1 + (a_1b_2)0 + (a_2b_1)0 + (a_2b_2)1 \\ &= a_1b_1 + a_2b_2. \end{aligned}$$

Result: *The dot product of $\vec{a} = (a_1, a_2)$ and $\vec{b} = (b_1, b_2)$ is computed by the formula*

$$\vec{a} \cdot \vec{b} = a_1b_1 + a_2b_2.$$

Problem

4.2. Evaluate the following dot products:
  (a) $(3, 2) \cdot (4, 5)$ (b) $(3\vec{i} + \vec{j}) \cdot (-3\vec{j})$ (c) $(7, 4) \cdot (-4, 7)$
  (d) $(7, 4) \cdot (-7, -4)$ (e) $((-3, 5) + (1, 1)) \cdot ((0, 6) + (-2, -1))$

The angle between vectors, orthogonal vectors.   We consider two vectors $\vec{a} = (a_1, a_2)$ and $\vec{b} = (b_1, b_2)$ and their angle θ between them, as we did at the beginning:

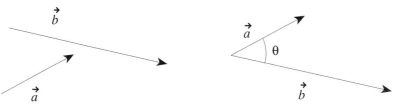

How do we compute the angle θ? The definition of $\vec{a} \cdot \vec{b}$ tells us $\vec{a} \cdot \vec{b} = |\vec{a}||\vec{b}| \cos\theta$. In this expression, we know how to compute the quantities $\vec{a} \cdot \vec{b}$, $|\vec{a}|$, and $|\vec{b}|$:

$$\vec{a} \cdot \vec{b} = a_1b_1 + a_2b_2, \quad |\vec{a}| = \sqrt{a_1^2 + a_2^2}, \quad |\vec{b}| = \sqrt{b_1^2 + b_2^2}.$$

Therefore, we can solve the equation $\vec{a} \cdot \vec{b} = |\vec{a}||\vec{b}| \cos \theta$ for $\cos \theta$ and obtain

$$\cos \theta = \frac{\vec{a} \cdot \vec{b}}{|\vec{a}||\vec{b}|} = \frac{a_1 b_1 + a_2 b_2}{\sqrt{a_1^2 + a_2^2} \sqrt{b_1^2 + b_2^2}}$$

Problem

4.3. In a standard xy-system, draw the vector $\vec{a} = (1, 6)$ with initial point at $(-2, -1)$ and $\vec{b} = (4, 3)$ with initial point at $(1, 1)$.

   a) Find the components of a vector $\vec{u} \neq \vec{0}$ that is normal (or orthogonal or perpendicular) to $\vec{a}$. *Hint:* Write $\vec{u} = (m, n)$ for the wanted vector. The dot product $\vec{a} \cdot \vec{u}$ must be equal to zero. Write out the dot product in components and use the equation to find m and n.

   b) Use formulas to find the angle of the vectors $\vec{a}$ and $\vec{b}$.

   c) We consider the collection N of all points P such that $\overrightarrow{OP}$ is perpendicular to the given vector $\vec{b} = (4, 3)$.

   (i) What does N look like? Describe N in words.

   (ii) Find the equation that the coordinates (x, y) of all points of N must satisfy. In other words, find the equation of N.

Projections of one vector onto another

We discuss two projections: "scalar projection" and "vector projection." They are closely related to each other.

The scalar projection of $\vec{a}$ onto $\vec{w}$. We are given two vectors $\vec{a}$ and $\vec{w}$. We move them so that their initial points coincide. Now we can describe the *scalar projection* of $\vec{a}$ onto $\vec{w}$. It is a scalar. It is defined as $\pm d$ where d is the projection of $\vec{a}$ onto the line of $\vec{w}$, as shown below. We take the plus sign if the angle θ is less than $\frac{\pi}{2}$, and the minus sign if θ is greater than $\frac{\pi}{2}$.

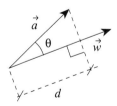
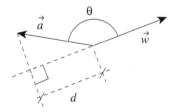

scalar projection of $\vec{a}$ onto $\vec{w}$: d scalar projection of $\vec{a}$ onto $\vec{w}$: $-d$

Note the following special cases:

- *If $\theta = 0$, then the scalar projection of $\vec{a}$ onto $\vec{w}$ equals $|\vec{a}|$ (why?).*
- *If $\theta = \frac{\pi}{2}$, then the scalar projection of $\vec{a}$ onto $\vec{w}$ equals zero (why?).*
- *If $\theta = \pi$, then the scalar projection of $\vec{a}$ onto $\vec{w}$ equals $-|\vec{a}|$ (why?).*

How do we compute the scalar projection of $\vec{a}$ onto $\vec{w}$? We use three facts:

Fact 1: *The scalar projection of $\vec{a}$ onto $\vec{w}$ equals $|\vec{a}| \cos \theta$. This is true by definition of the cosine in a right triangle, no matter if θ is less or greater then $\frac{\pi}{2}$.*

Fact 2: *The unit vector in the direction of $\vec{w}$ is given by $\frac{\vec{w}}{|\vec{w}|}$.*

Fact 3: *The dot product of $\vec{a}$ with $\frac{\vec{w}}{|\vec{w}|}$ equals $|\vec{a}| \cdot 1 \cdot \cos\theta$, and this is exactly the scalar projection.*

Result: *The scalar projection of $\vec{a}$ onto $\vec{w}$ is computed as $\vec{a} \cdot \vec{w}_0$ where $\vec{w}_0$ is the unit vector in direction of $\vec{w}$.*

Note:

1. The scalar projection of $\vec{a}$ onto $\vec{w}$ is also called "the scalar component of $\vec{a}$ in direction of $\vec{w}$" or "the component of $\vec{a}$ in direction of $\vec{w}$."

2. Some books use the symbol $\text{Proj}_{\vec{w}}\,\vec{a}$ for the scalar projection of $\vec{a}$ onto $\vec{w}$. We will not use this notation but will always spell out "scalar projection of . . . ."

3. For the scalar projection, it matters which vector we project and which is the vector onto which we project. The scalar projection of $\vec{a}$ onto $\vec{m}$ is different from the scalar projection of $\vec{m}$ onto $\vec{a}$.

Problems _____

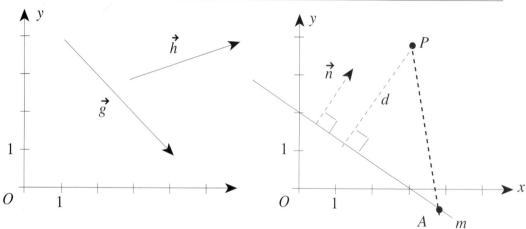

4.4. We consider the two vectors $\vec{g}$ and $\vec{h}$ shown above left.

a) Use ruler and pencil to find the scalar projection of $\vec{g}$ onto $\vec{h}$. For this, you have to make their initial points coincide.

b) Read off the components of $\vec{g}$ and $\vec{h}$. Then find the scalar projection of $\vec{g}$ onto (i) $\vec{\imath}$ and (ii) $\vec{\jmath}$.

c) Find the scalar projection of $\vec{g}$ onto $\vec{h}$ by computation.

4.5. We are given a line m and a point P as shown above right. We want to find the distance d of P from m. We do it as follows: Pick an arbitrary point A on m, and pick a vector $\vec{n}$ normal to m. Then d is the scalar projection of $\overrightarrow{AP}$ onto $\vec{n}$. In case $\vec{n}$ points to the other side, the scalar projection will be negative, and you take the absolute value. From the figure, find the equation of m and the components of a vector $\vec{n}$ normal to m. Then find the distance of the point $Q(20, 12)$ from our m. Pick for A an "easy" point of m.

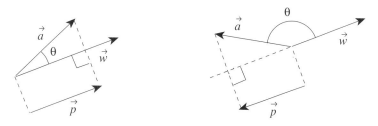

$\vec{p}$: Vector projection of $\vec{a}$ on $\vec{w}$

Figure 4.1

Vector projection of $\vec{a}$ onto $\vec{w}$. The *vector projection* of $\vec{a}$ onto $\vec{w}$ is the vector $\vec{p}$ defined in words by

"(scalar projection of $\vec{a}$ onto $\vec{w}$) times (unit vector in the direction of $\vec{w}$)"

This is illustrated in Figure 4.1. To find the formula for computing the vector projection we translate the verbal definition into formulas.

We obtain: *The vector projection of $\vec{a}$ onto $\vec{w}$ is given by*

$$\vec{p} = \left(\vec{a} \cdot \frac{\vec{w}}{|\vec{w}|} \right) \frac{\vec{w}}{|\vec{w}|}.$$

Note that the vector projection of $\vec{a}$ onto $\vec{w}$ is a scalar multiple of $\vec{w}$.

Problems _____

4.6. With $\vec{b} = (7, 2)$.

    a) Find the vector projection of $\vec{b}$ onto $\vec{\imath}$.

    b) Find the vector projection of $\vec{b}$ onto $\vec{\jmath}$.

    c) Write $\vec{b}$ as a sum $\vec{u} + \vec{v}$ so that $\vec{u}$ is parallel to $\vec{\imath}$ and $\vec{v}$ is parallel to $\vec{\jmath}$.

4.7. We are given two vectors $\vec{m}$ and $\vec{a}$. We want to write $\vec{m}$ as a sum

$$\vec{m} = \vec{u} + \vec{v}$$

so that $\vec{u}$ is parallel to $\vec{a}$ and $\vec{v}$ is perpendicular to $\vec{a}$, as shown in Figure 4.2.

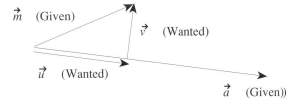

Figure 4.2

The diagram tells us: The vector $\vec{u}$ is the vector projection of $\vec{m}$ onto $\vec{a}$, and the vector $\vec{v}$ is given by $\vec{m} - \vec{u}$. Use this method to find $\vec{u}$ and $\vec{v}$ for $\vec{m} = (2, 1)$ and $\vec{a} = (3, -1)$.

Additional Problems

Dot product

4.8. We consider three points in the plane: $A(1,1)$, $B(5,2)$, $C(7,6)$.

    a) Can you find a *nonzero* vector $\vec{w} = (u,v)$ that is perpendicular to $\overrightarrow{AB}$ and $\overrightarrow{AC}$ at the same time? If yes, find such a vector $\vec{w}$; if no, explain.

    b) Evaluate the dot products $\overrightarrow{AB} \cdot \overrightarrow{AC}$ and $\overrightarrow{AB} \cdot \overrightarrow{BC}$.

    c) Think of the triangle ABC. Are all angles of ABC less than $\frac{\pi}{2}$? *Hint:* Use b).

    d) K is the collection of all points $P(x,y)$ such that $\overrightarrow{OP} \cdot \overrightarrow{BC} = 10$. Draw K in a standard xy-system.

4.9. Remember the notation for the dot product and for the scalar multiple of a vector:

- $\vec{a} \cdot \vec{b}$ is the dot product of the two vectors $\vec{a}$ and $\vec{b}$, always written with a dot between them. $\vec{a} \cdot \vec{b}$ *is a scalar.*

- If we write $k\,\vec{u}$, then k is a scalar and $k\,\vec{u}$ is the vector obtained by multiplying $\vec{u}$ with the scalar k. *Thus, $k\,\vec{u}$ is a vector.*

Look at each of the four expressions (i)–(iv). Does it make sense? If yes, is it a vector or a scalar? If not, explain why.

    (i) $(\vec{a} \cdot \vec{b})\,\vec{c}$ (ii) $(\vec{a} \cdot \vec{b}) \cdot \vec{c}$ (iii) $(\vec{a} \cdot \vec{b})(\vec{c} + \vec{d})$ (iv) $\vec{a} \cdot ((\vec{b} \cdot \vec{c})\,\vec{d})$

4.10. The two vectors $\vec{a}$ and $\vec{b}$ are shown at left in Figure 4.3. We want to find $\vec{a} \cdot \vec{b}$ with ruler and pencil, *but without measuring the angle between $\vec{a}$ and $\vec{b}$.* There are two steps.

    Step 1: Use ruler and pencil to find $\vec{a} \cdot \dfrac{\vec{b}}{|\vec{b}|}$. *Hint:* The second factor $\dfrac{\vec{b}}{|\vec{b}|}$ is a unit vector. Interpret $\vec{a} \cdot \dfrac{\vec{b}}{|\vec{b}|}$ as the scalar projection of a certain vector (which?) onto another vector (which?). Use ruler and pencil to measure the scalar projection (take 1 cm as unit of measurement).

    Step 2: The scalar product $\vec{a} \cdot \vec{b}$ can be written in the form $|\vec{b}| \left(\vec{a} \cdot \dfrac{\vec{b}}{|\vec{b}|} \right)$. Measure $|\vec{b}|$ and use Step 1 to find the value of $\vec{a} \cdot \vec{b}$.

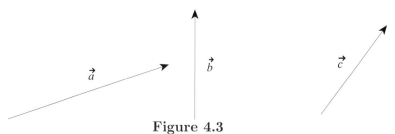

Figure 4.3

4.11. We work with the vector $\vec{c}$ on the right in Figure 4.3. We consider different vectors $\vec{w}$ *which are all of magnitude* 4 cm. In a) – c) we ask: For which choice of $\vec{w}$ is $\vec{c} \cdot \vec{w}$ zero? Highest possible? Lowest possible?

a) Draw $\overrightarrow{w}$, with initial point at the initial point of $\vec{c}$, so that $\vec{c} \cdot \overrightarrow{w} = 0$. Remember that $|\overrightarrow{w}| = 4$ cm.

b) As a) but so that $\vec{c} \cdot \overrightarrow{w}$ is highest possible.

c) As a) but so that $\vec{c} \cdot \overrightarrow{w}$ is lowest possible.

4.12. We are told that $\vec{a} \cdot \vec{a} = 27$. Find $|\vec{a}|$ if this is possible on the basis of the given information. If it is impossible, explain.

Projections

4.13. Find the scalar projection of $\vec{b} = (12, -4)$ onto: (i) $\vec{\imath}$, (ii) $\vec{\jmath}$, and (iii) $\vec{\imath} + \vec{\jmath}$.

4.14. $\overrightarrow{w}$ is shown below left. Draw the following vectors, all with their initial point at the initial point of $\overrightarrow{w}$.

a) $\vec{a}$ such that the scalar projection of $\vec{a}$ onto $\overrightarrow{w}$ equals -2 cm.

b) $\vec{b}$ such that the vector projection of $\vec{b}$ onto $\overrightarrow{w}$ equals $-(1/3)\overrightarrow{w}$.

c) $\vec{c}$ such that the vector projection of $\vec{c}$ onto $\overrightarrow{w}$ is the zero vector.

4.15. The vectors $\vec{u}$ and $\vec{a}$ are shown above right. Draw vectors $\vec{p}$ and $\vec{q}$ with the following properties:

- The vector $\vec{p}$ is parallel to $\vec{a}$ and $\vec{q}$ is perpendicular to $\vec{a}$.
- The given vector $\vec{u}$ can be written as $\vec{u} = \vec{p} + \vec{q}$.

4.16. With $\vec{v} = (5, 5)$ and $\vec{b} = (4, 2)$, write $\vec{v}$ as a sum $\vec{v} = \vec{r} + \vec{s}$ so that $\vec{r}$ is parallel to $\vec{b}$ and $\vec{s}$ is perpendicular to $\vec{b}$.

5   Coordinates and Vectors in Space

Construction of the xyz-coordinate system in space. When we work with points and vectors in the plane, we need *two* coordinates x, y. From now on, we want to work with points and vectors in space. In space, we need *three* coordinates x, y, z. We build an xyz-coordinate system by adding the third coordinate z to the xy-coordinate system as follows.

Below is an xy-coordinate system in the plane. Take a pencil. Put its eraser on the point O of the xy-system, and let the pencil stick out from the plane of this sheet at a right angle. Your pencil is the z-axis, and the tip of the pencil indicates the positive direction on the z-axis. The z-axis extends also below the xy-plane, and so your pencil is only a small part of the z-axis. The xy-coordinate system in the

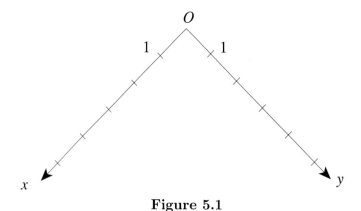

Figure 5.1

plane together with the z-axis sticking out of the plane is called an *xyz-coordinate system*. The figure below is a photograph-like picture of such an xyz-coordinate system.

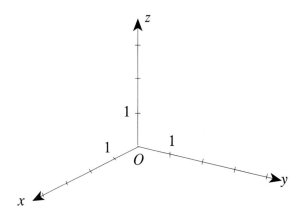

If P lies above the xy-plane, we draw a line that passes through P and that is parallel to the z-axis. This line intersects the xy-plane at a point $P_0(a, b)$, and we write c for the length of the segment PP_0, as shown below. Then we say that (a, b, c) are the xyz-coordinates of the point P in space. This is shown below.

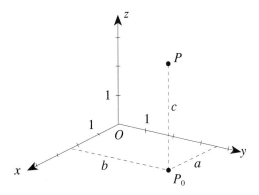

Note:

1. If P lies in the xy-plane, then its coordinates are $(a, b, 0)$.

2. If P is below the xy-plane at distance c from P_0, then the coordinates of P are $(a, b, -c)$.

If P is above P_0 and the sun shines down on P in the direction opposite to the z-axis, then P casts the shadow P_0. Therefore, we call P_0 the *shadow* of P. If P is below P_0 we still call P_0 the shadow of P. If P lies in the xy-plane, then it is its own shadow.

Terminology

(i) **The standard xyz-system.** The xyz-coordinate system you constructed by holding up the pencil will be called the *standard xyz-system in space.* In a standard system, the three axes are always arranged in the same way, in the same relative position to each other. For example, if you switched the x-axis with the y-axis and left the z-axis as it is, the system would no longer be a standard system. We use only standard systems.

(ii) **The coordinate planes.** Any of the planes defined by two of the three coordinate axes is called a "coordinate plane." The plane defined by the x-axis and the y-axis is called the *xy-plane* of the xyz-system. Similarly, the plane determined by the x-axis and the z-axis is called the *xz-plane*, and the plane determined by the y-axis and the z-axis is called the *yz-plane*.

Note:

1. Each coordinate plane extends beyond the two axes which determine it.

2. Each coordinate plane is perpendicular to the other two.

3. The x-axis is normal (i.e., perpendicular) to the yz-plane, the y-axis is normal to the xz-plane, and the z-axis is normal to the xy-plane.

Problems _____

5.1.   a) T is the point with coordinates (p, q, r). How can you tell from p, q, r whether T is in the xy-plane? In the xz-plane? In the yz-plane?

   b) Write down the coordinates of two points whose distance from the xz-plane is 4 units of length.

5.2. Imagine you look straight down onto the xy-plane. You see what is shown in Figure 5.1 on page 38.

    a) Consider the point Q with coordinates $(x, y, z) = (5, 2, 6)$. Take the plane which passes through Q and which is parallel to the xz-plane. What would you see of this plane when looking straight down onto the xy-plane? Draw into Figure 5.1 what you see.

    b) Consider the point $S(1, 4, -2)$. Take the plane which passes through S and which is parallel to the yz-plane. What would you see of this plane when looking straight down onto the xy-plane? Draw into Figure 5.1 what you see.

How to draw the xyz-system. Anything we look at looks differently from different viewpoints. The same is true for an xyz-system. The way it looks depends on where we stand when we look at it. Figure 5.2 shows three views of an xyz-system as seen from different points of view.

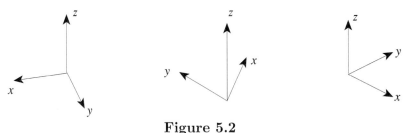

Figure 5.2

Problem

5.3. For each of the three views of an xyz-system shown in Figure 5.2, answer the following questions: Would the x-coordinate of the point from which we look have to be positive, or zero, or negative? What about the y-coordinate and the z-coordinate?

From now on, we draw the xyz-system as follows. From the point $(x, y, z) = (10^6, 10^6, 10^6)$ we aim a camera at the origin of the xyz-system, take a picture and magnify it. The picture we get is shown in Figure 5.3 on the next page.

Note:

1. In the picture, the angles between the coordinate axes are $120°$.

2. The units on the axes appear in the same length.

We choose the units on the axes in such a way that in the picture of the xyz-system they are 1 cm.[1]

Throughout the book, you will be asked many times to draw such an xyz-system yourself. Always use 1 cm as unit on the three axes, and have the axes form an angle of $120°$. To remind you of the 1 cm units and the $120°$ angle, we call such an xyz-system a standard xyz-system. At times, we use an xyz-system where the axes are at an angle of $120°$, but where the units are less than 1 cm. In such systems, the units will still be the same on the three axes.

---

[1]In reality, the units on the axes are greater than 1 cm because we look at the coordinate axes at an angle.

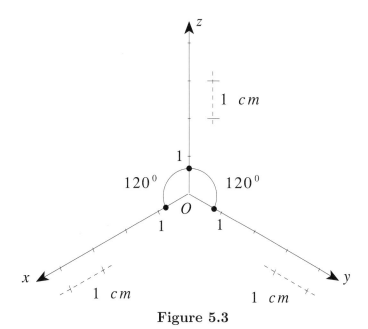

Figure 5.3

Note:

1. In a standard xyz-system, the length of a line segment parallel to a coordinate axis can be measured directly with the ruler. The length of a segment that is not parallel to a coordinate axis has to be computed from the coordinates of its endpoints.

2. We will draw points, planes, and lines. What we get are photographs taken by a camera extremely far away from the object. Then lines that are parallel in reality show up as parallel lines in the picture. In ordinary photography, we hold the camera closer to the object. If we do that, we get perspective effects in the following sense: Lines that are parallel in reality are not parallel in the picture. Figure 5.4 shows two pictures of a rectangular box. On the left, the camera is extremely far away, and parallel lines remain parallel. On the right, the camera is closer and parallel lines do not remain parallel.

Camera infinitely far away:
parallel lines look parallel

Ordinary photography:
parallel lines do not look parallel

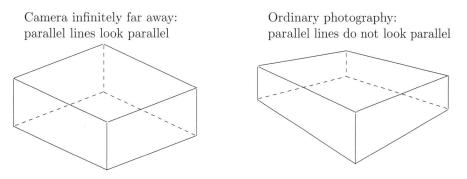

Figure 5.4

How to read off coordinates and draw points. We are working with a point $P(a, b, c)$ in space and its shadow $P_0(a, b, 0)$ in the xy-plane vertically below P as shown in Figure 5.5. (Note that the axes are at $120°$, but the units are less than 1 cm.) There are two questions:

1. *Given the picture of a point P and the picture of its shadow P_0, how do we read off the coordinates of P?*

   Answer: *The line connecting P and P_0 is parallel to the z-axis. Therefore we can read off the z-coordinate c by measuring the distance between P_0 and P. Then we find the coordinates $(a, b, 0)$ of the shadow P_0 by drawing parallels to the x-axis and y-axis (see Figure 5.5).*

2. *How do we draw the picture of a point P whose coordinates (a, b, c) are given?*

   Answer: *We reverse process (1): First we draw the shadow P_0 by using parallels to the x-axis and y-axis, and then we draw P by going up from P_0 by c units.*

Lines parallel to one of the coordinate axes will always be shown broken.

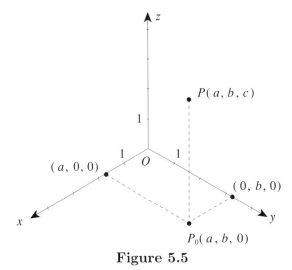

Figure 5.5

In problem 5.4, and in many more to come, you have to draw *points, lines, etc. These points, lines, etc. have letter names. For example, a point is called Q, and a line is called m. Write the letter name next to the object which you drew. Clear labels reduce the risk of confusion.*

Problem ———————————————————————————————

5.4. The questions refer to the xyz-system shown in Figure 5.6 on the facing page.

    a) Read the text in the box above.

    b) Into the system draw the points $Q(-2, -3, -5)$, $R(4, 0, 2)$, $S(2, 5, 9)$ and their shadows Q_0, R_0, S_0. Label the points you draw.

    c) The points K, L and their shadows are shown. Find the coordinates of K and L.

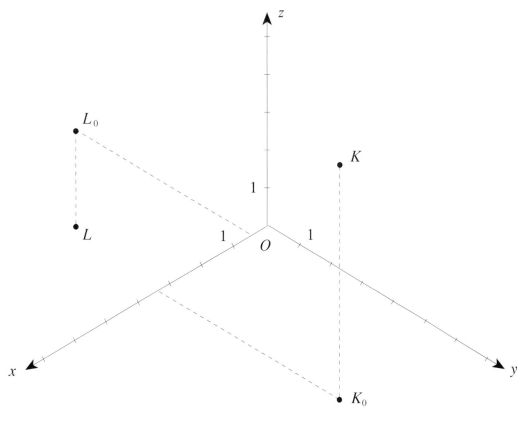

Figure 5.6

Vectors in space. Vectors in space work very much the same as vectors in the plane, except that a vector in space has three components instead of two. We can practically copy the text of Section 1.

☞ What are vectors in space? How are they defined?

A vector in space is represented by a directed line segment in space.

- Vectors in space are written using arrows, or hats, or underlines, or boldface letters. We use arrows: $\vec{u}$, $\overrightarrow{PQ}$, etc.
- For the vector $\overrightarrow{PQ}$, the point P is called the initial point and Q the endpoint.
- The length (or magnitude) $|\vec{u}|$ of $\vec{u}$ is the length of the line segment that constitutes $\vec{u}$.
- Two vectors $\vec{a}$ and $\vec{b}$ in space are the same if (1) they have the same length, and (2) if they have the same direction.
- If m is a scalar (i.e., a real number), then the scalar multiple $m\,\vec{u}$ is defined as in the plane.
- $\vec{a} \pm \vec{b}$ is defined as for vectors in the plane.
- If $\vec{a}$ is a given vector in space, then $\frac{\vec{a}}{|\vec{a}|}$ is a unit vector (i.e., a vector of length one) of the same direction as $\vec{a}$.

- Two nonzero vectors $\vec{u}$ and $\vec{v}$ are called parallel if $\vec{u} = c\vec{v}$ for a nonzero scalar c.

✐ How do we compute with vectors in space?

Now we have three coordinate axes. Therefore, we need a third basis vector. As third basis vector, we take a vector that points from $(0,0,0)$ to the point $(0,0,1)$, and we call it $\vec{k}$. Given any vector $\vec{u}$, we move its initial point to $(0,0,0)$. Then

$$\vec{u} = a\vec{\imath} + b\vec{\jmath} + c\vec{k},$$

and (a, b, c) are called the *components* of $\vec{u}$ (see below). We write $\vec{u} = (a, b, c)$ for $\vec{u} = a\vec{\imath} + b\vec{\jmath} + c\vec{k}$.

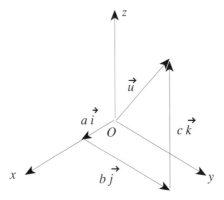

What is the relation between components of a vector and coordinates of a point? Given a vector $\vec{u}$, we move its initial point to the origin $(0,0,0)$. Then the coordinates of the endpoint of $\vec{u}$ are the components of $\vec{u}$.

The formulas for operating with vectors in space look like those for vectors in the plane. Below we give them for $\vec{u} = (a, b, c)$ and $\vec{v} = (r, s, t)$. By $\vec{u}_0$, we mean a unit vector in the direction of $\vec{u}$.

$$|\vec{u}| = \sqrt{a^2 + b^2 + c^2} \qquad \vec{u} = \vec{v} \text{ means } (a, b, c) = (r, s, t)$$
$$\vec{u} \pm \vec{v} = (a \pm r, b \pm s, c \pm t) \quad \vec{u}_0 = \left(\frac{a}{\sqrt{a^2 + b^2 + c^2}}, \frac{b}{\sqrt{a^2 + b^2 + c^2}}, \frac{c}{\sqrt{a^2 + b^2 + c^2}} \right)$$

Additional Problems

Shadows

We consider lines in xyz-space. The shadow of such a line m is a line in the xy-plane. We write m_0 for the shadow of m. How do we find m_0? We pick two points A, B of m and draw their shadows A_0, B_0. Then m_0 is the line determined by A_0, B_0.

5.5.   a) In a standard xyz-system, draw the point $A(1, 3, 4)$ and its shadow A_0. Then draw a point B and its shadow B_0 with the following two properties:

(i) The line segment AB is parallel to the y-axis.

(ii) The length of AB is 3.5 units.

How many answers are there?

b) In a second standard xyz-system, draw two lines p, q and their shadows p_0, q_0 as follows:

(i) The line p is parallel to the xz-plane but does not lie in the xz-plane. Also, p is neither parallel to the x-axis nor to the z-axis.

(ii) The line q is parallel to the xy-plane, but does not lie in the xy-plane. Also, q is neither parallel to the x-axis nor to the y-axis.

Note that there are many answers to part b.

5.6. The line u is parallel to the z-axis, but different from the z-axis. Use a complete sentence to describe the shadow u_0.

Vectors and lines

5.7. We are given five vectors $\vec{p}, \ldots \vec{t}$:

$$\vec{p} = (5, 0, 3), \quad \vec{q} = (3, 5, 0), \quad \vec{r} = (-26, 0, 0), \quad \vec{s} = (1, 1, 1), \quad \vec{t} = (0, -2, 0).$$

Which of these vectors are

a) Parallel to the xy-plane?

b) Perpendicular to the yz-plane?

Hints: (a) Given a vector $\vec{a} = (a_1, a_2, a_3)$, move its initial point to the origin. If $\vec{a}$ is parallel to the xy-plane to begin with, then the endpoint of the moved vector must lie in the xy-plane. (b) Any vector perpendicular to the yz-plane must be parallel to $\vec{\imath} = (1, 0, 0)$.

5.8. Let $\vec{a} = (2, 1, 3)$, $\vec{b} = (0, -1, 4)$, and $\vec{c} = (1, 5, 0)$.

a) Find the components of $\vec{u} = 2\vec{a} + \vec{b} - 3\vec{c}$.

b) Find a vector $\vec{v}$ of length 23.61 in the direction of $\vec{a} + \vec{b}$.

c) The vector $\vec{w} = \vec{a} + t\vec{b}$ depends on the scalar t. Find values t with the following properties (there may or may not be such values of t):

(i) $\vec{w}$ is parallel to the xz-plane.

(ii) $\vec{w}$ is parallel to the yz-plane.

(iii) $\vec{w}$ is perpendicular to the xy-plane.

(iv) $\vec{w}$ is of length 4.

(v) $\vec{w}$ is of length 2.

5.9. *Remember:* Lines drawn to locate points and vectors are parallel to the axes and are shown *broken*.

    a) Draw the following vectors:
       $\vec{a} = (3, -2, 4)$ with initial point at $P(0, 2, 1)$;
       $\vec{b} = (-4, 0, 5)$ with initial point at $Q(5, 5, -1)$.

    b) Use ruler and pencil to find the components of the vectors $\vec{c}$ and $\vec{d}$ shown in Figure 5.7.

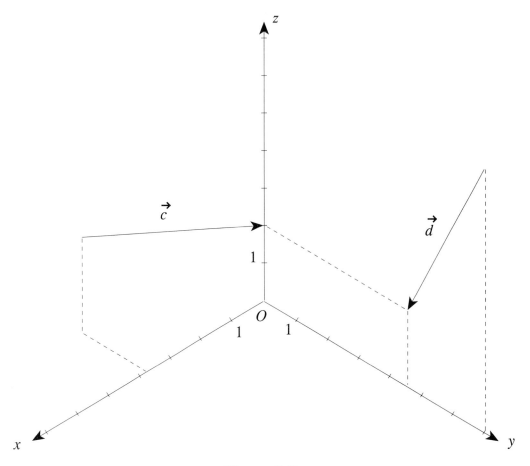

Figure 5.7

5.10. The points R, S, T are shown in Figure 5.8.

    a) Draw the vector $\overrightarrow{RT} + \overrightarrow{RS}$ with initial point at R.

    b) Find the components of a vector $\vec{b}$ which is parallel to the vector $\overrightarrow{RT} + \overrightarrow{RS}$ and that is of length 0.07 units.

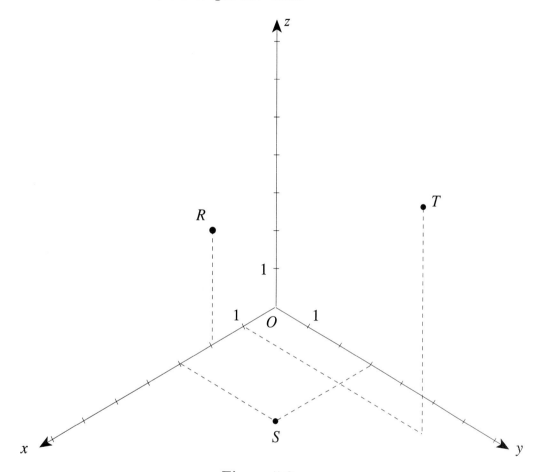

Figure 5.8

5.11. The line m and its shadow m_0 are shown below.

    a) A and B are two points on m, as shown. Find their coordinates.

    b) The line m intersects the xy-plane in the point E, the xz-plane in F, and the yz-plane in G. Draw the points E, F, G. *Hint for E:* Pick a point on m. Move it up and down the line while watching what its shadow does. At the moment the point hits the xy-plane, what can you say about its z-coordinate?

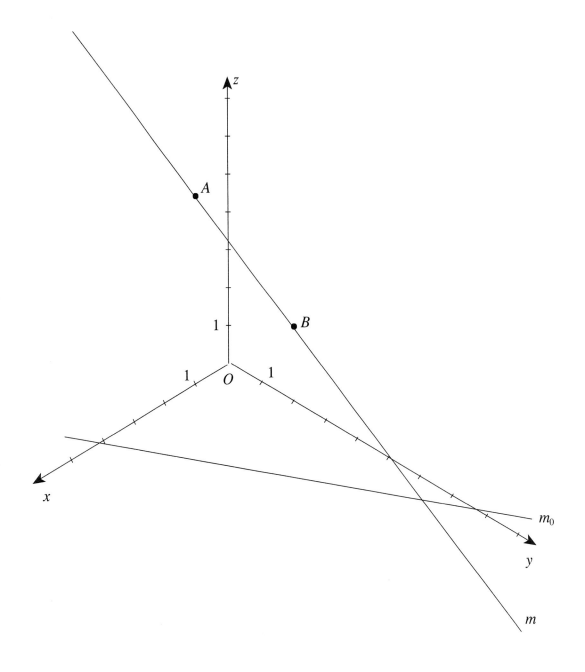

6   Lines in Space

Parametric representations.   Parametric representations for lines in space work just like parametric representations for lines in the plane.

The line m shown below is given by two of its points A and B. $P(x, y, z)$ is the general (or "generic") point of m. The vector $\overrightarrow{OP}$ is the *position vector* of P. We write $\vec{r}$ for $\overrightarrow{OP}$. Then

$$\vec{r} = \overrightarrow{OA} + t\overrightarrow{AB} \tag{6.1}$$

is a parametric representation of m. The position of P depends on t. Therefore, we write $P(t)$ for P and $\vec{r}(t)$ for $\vec{r}$.

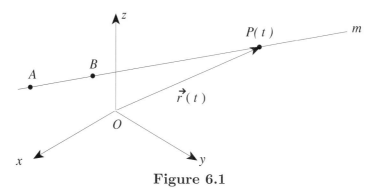

Figure 6.1

Problem _____

6.1. The line d passes through the points $K(2, 0, 3)$ and $L(1, 4, 3)$.
   a) Draw the points K, L and their shadows in a standard xyz-system.
   b) Write down a parametric representation of d.
   c) The line intersects the xy-plane at the point U, the yz-plane at V, and the xz-plane at W. Find the coordinates of these points.

Note:

1. A parametric representation of a line space can also be considered as a description of a point moving on a line at constant speed. For the line m shown in Figure 6.1, and given parametrically by (6.1), the vector $\overrightarrow{AB}$ is the velocity vector of P, and $|\overrightarrow{AB}|$ is the speed of P. As with lines in the plane, we can view a parametric representation of a line in space in two ways "physics" and "geometry":

   (i) The formulas describe a point that moves on a line at constant speed as time t elapses.

   (ii) The formulas describe a line in space as geometric object. In that case, the parameter t is a device to label the points of the line.

2. A line may also be given by a point A and a vector $\vec{b}$ parallel to the line. Then $\vec{r} = \overrightarrow{OA} + t\vec{b}$ is a parametric representation. Here, the vector $\vec{b}$ defines the direction of the line in the same way the vector $\overrightarrow{AB}$ defines the direction of a line given by two points A and B.

Problems

6.2. The line m passes through the point $A(4, 2, -5)$ and is parallel to $\vec{b} = (1, -3, 2)$.

   a) Write down a parametric representation of m.

   b) P and Q are the points on m which are at distance 5 from the point A. Find the coordinates of P and Q. *Hint:* Use a unit vector in the direction of $\vec{b}$.

6.3. A point P moves on a line m at constant speed. At time $t = 5$ sec, its position is $(2, 9, 1)$, and at time $t = 8$ sec, its position is $(6, 0, -1)$. Find the position at time $t = 2$ sec.

Additional Problems

6.4. The line k has parametric representation

$$x = 1 + t, \quad y = 2 - t/2, \quad z = 5 - 2t.$$

   a) Sketch the line k in a standard xyz-system. Then mark on the sketch the points corresponding to $t = -1, 0, 1, 2$.

   b) Calculate the values of t for which the line intersects the xy-plane, the yz-plane, and the xz-plane.

   c) Find the points of intersection of the line k with each of the coordinate planes.

6.5. Find a parametric representation for the line parallel to the vector $\vec{v} = (2, 1, 3)$ that passes through the point $A(5, 0, -1)$ when $t = 1$.

6.6. A particle travels at constant velocity along a straight line. It passes through $A(1, 2, 0)$ at time $t = 0$ sec, and $B(0, 0, 4)$ at time $t = 1$ sec.

   a) Find a formula for the coordinates $(x(t), y(t), z(t))$ of the particle position at time t.

   b) At what speed is the particle traveling, in distance units per second?

   c) Does the particle pass through the point $(-1, -2, 9)$? If so, when?

   d) How close does the particle come to the origin? *Hint:* Find a formula for its distance from the origin as a function of time, and minimize it.

6.7. This problem deals with two lines u and v in space, as shown in Figure 6.2 on the next page. The sketch is not to scale. Also, the shadows of the two lines are not shown. The point P moves on the line u at constant speed. At time $t = 0$ it is at $A(3, 0, 4)$. The components of the velocity vector of P are $\vec{v} = (0.5, 2, 1)$. At the same time, the point Q is moving on a different line v at a different constant speed. At time $t = 0$ the point Q is at $B(7, 2, 0)$. The two lines u and v intersect at $K(4, 4, 6)$. *Both points are at K at the same time. That is, they collide at K.* Find the velocity vector of Q.

6.8. Two lines in space may be positioned relative to each other in three different ways: (1) They are parallel; (2) they intersect at a point; or (3) they are not

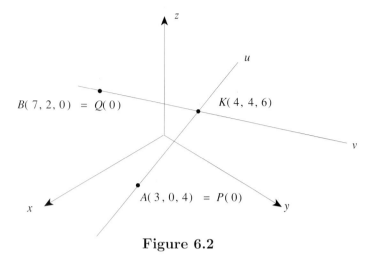

Figure 6.2

parallel, but they do not intersect. Lines such as (3) are called *skew*. This is illustrated in Figure 6.3. The dashed lines are the shadows. We are given the equations of two lines p and q:

$$
\begin{aligned}
x &= 2+t & x &= 4 \\
y &= 3t & y &= 3+3u \\
z &= 4+t & z &= 4+3u
\end{aligned}
$$

Are we in case (1) or (2) or (3)? Work as follows: (i) Check whether the two lines parallel. (ii) If they are not, find out whether they intersect. For this, consider the two parametric representations as as system of *six* equations in the *five* unknowns x, y, z, t, u. The two lines intersect if this system has a solution (x, y, z, t, u). If the system has no solution, the lines are skew (i.e., they do not intersect).

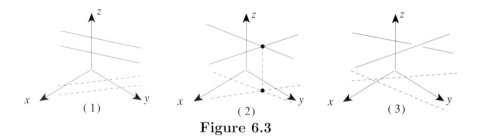

Figure 6.3

7 The Dot Product of Space Vectors

The definition and description of the dot product is practically the same for vectors in the plane and vectors in space. There is a difference, though, when we use the dot product for projections of space vectors. Then we run into questions of space geometry, and space geometry requires more work than geometry in the plane.

☞ How is the dot product of space vectors defined?

Two space vectors $\vec{a}$ and $\vec{b}$ are given. We move them so that their initial points are the same. We write θ for the smaller of the two angles between $\vec{a}$ and $\vec{b}$.

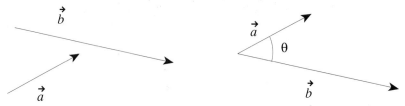

Definition. *The dot product of $\vec{a}$ and $\vec{b}$ is*

- defined *as the* scalar $|\vec{a}||\vec{b}|\cos\theta$,
- written $\vec{a}\cdot\vec{b}$, *and*
- *is* read *"$\vec{a}$ dot $\vec{b}$."*

The preceding is a straightforward copy of the text at the beginning of Section 4, with minimal changes. The same is true for the rules below.

The rules for the dot product of space vectors. For k a scalar, we have:

1. $\vec{a}\cdot\vec{b} = \vec{b}\cdot\vec{a}$
2. $(k\vec{a})\cdot\vec{b} = \vec{a}\cdot(k\vec{b}) = k(\vec{a}\cdot\vec{b})$
3. $\vec{a}\cdot(\vec{b}+\vec{c}) = \vec{a}\cdot\vec{b} + \vec{a}\cdot\vec{c}$

✍ How do we compute the dot product of space vectors?

The vector $\vec{a}$ has components (a_1, a_2, a_3), and $\vec{b}$ has components (b_1, b_2, b_3). We use the rules above to work out $\vec{a}\cdot\vec{b}$, and we obtain what we expect:

$$\vec{a}\cdot\vec{b} = a_1 b_1 + a_2 b_2 + a_3 b_3.$$

Problems

7.1. We consider the three points $A(1, -1, 3)$, $B(2, 1, -1)$ and $C(-1, 1, 2)$.

a) To find the angle between two space vectors $\vec{a}$ and $\vec{b}$ we operate as for vectors in the plane: We solve the equation $\vec{a}\cdot\vec{b} = |\vec{a}||\vec{b}|\cos\theta$ for $\cos\theta$ and obtain

$$\cos\theta = \frac{\vec{a}\cdot\vec{b}}{|\vec{a}||\vec{b}|}.$$

Use this method to find the angle between the vectors $\overrightarrow{AB}$ and $\overrightarrow{AC}$.

b) Find two nonparallel vectors $\vec{p}$, $\vec{q}$ which form a right angle with $\overrightarrow{AB}$. *Hint:* Let $\vec{p} = (u, v, w)$ form a right angle with $\overrightarrow{AB}$. We must have $\overrightarrow{AB}\cdot\vec{p} = 0$. Write out this equation, and use it to find two such vectors $\vec{p}$.

7.2. Given the vector $\vec{w} = (4, 2, -3)$, we consider the collection N of all points $P(x, y, z)$ such that $\overrightarrow{OP}$ is perpendicular to the given vector $\vec{w}$.

a) What does N look like? Describe N in words.

b) Find the equation of N. That is, find the equation that the coordinates (x, y, z) of all points of N must satisfy.

Projections of space vectors

The scalar projection of $\vec{a}$ onto $\vec{w}$. We are given two vectors $\vec{a}$ and $\vec{w}$. We move them so that their initial points coincide. Now we can describe the *scalar projection* of $\vec{a}$ onto $\vec{w}$. It is a scalar. It is defined as $\pm d$ where d is the projection of $\vec{a}$ onto the line of $\vec{w}$, as shown below. We take the plus sign if the angle θ is less than $\frac{\pi}{2}$, and the minus sign if θ is greater than $\frac{\pi}{2}$.

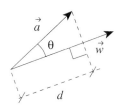
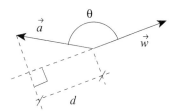

Scalar projection of $\vec{a}$ onto $\vec{w}$: d ⠀⠀⠀ Scalar projection of $\vec{a}$ onto $\vec{w}$: $-d$

We compute the scalar projection of space vectors as in the plane:

The scalar projection of $\vec{a}$ onto $\vec{w}$ is computed as $\vec{a} \cdot \vec{w}_0$ where $\vec{w}_0$ is the unit vector in direction of $\vec{w}$.

Problems

7.3. a) Find the scalar projection of $\vec{m} = (3, 1, 4)$ onto $\vec{q} = (-2, 1, 5)$.

b) Find the scalar projection of $\vec{q} = (-2, 1, 5)$ onto $\vec{m} = (3, 1, 4)$.

c) Your neighbor says: "I am confused. Did the professor not say in class that $\vec{m} \cdot \vec{q} = \vec{q} \cdot \vec{m}$? If that is true, why do we get different answers to a) and b)?" Write out a response to clear up the confusion.

7.4. Questions a) and b) below ask for a vector with given properties. If there is such a vector, find one. If there is no such vector, explain why. In both questions, $\vec{p} = (5, 0, -1)$.

a) Find a nonzero vector $\vec{r}$ such that the scalar projection of $\vec{r}$ onto $\vec{p}$ is zero.

b) Find a vector $\vec{t}$ so that the scalar projection of $\vec{p}$ onto $\vec{t}$ equals 7.

Vector projection of $\vec{a}$ onto $\vec{w}$. The *vector projection* of $\vec{a}$ onto $\vec{w}$ is the vector $\vec{p}$ defined in words by

"(scalar projection of $\vec{a}$ onto $\vec{w}$) times (unit vector in the direction of $\vec{w}$)"

This is illustrated in Figure 7.1 on the following page. As for vectors in the plane, we find the formula for computing the vector projection by translating the verbal definition into formulas.

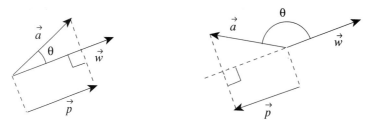

$\vec{p}$: Vector projection of $\vec{a}$ on $\vec{w}$

Figure 7.1

We obtain: *The vector projection of $\vec{a}$ onto $\vec{w}$ is given by*

$$\vec{p} = \left(\vec{a} \cdot \frac{\vec{w}}{|\vec{w}|} \right) \frac{\vec{w}}{|\vec{w}|}.$$

Note that the vector projection of $\vec{a}$ onto $\vec{w}$ is a scalar multiple of $\vec{w}$.

Problems _____

    7.5. Let $\vec{b} = (7, 2, 3)$.

        a) Find the vector projection of $\vec{b}$ onto $\vec{j}$.

        b) Find the vector projection of $\vec{b}$ onto $\vec{k}$.

        c) Write $\vec{b}$ as a sum $\vec{u} + \vec{v}$ so that $\vec{u}$ is parallel to $\vec{j}$ and $\vec{v}$ is perpendicular to $\vec{j}$.

    7.6. We are given two vectors $\vec{m} = (2, 1, -3)$ and $\vec{a} = (3, -1, 0)$. We want to write $\vec{m}$ as a sum $\vec{u} + \vec{v}$ so that $\vec{u}$ is parallel to $\vec{a}$ and $\vec{v}$ is perpendicular to $\vec{a}$. Find $\vec{u}$ and $\vec{v}$. *Hint:* See Problem 4.7 of Section 4.

Additional Problems

Dot product

    7.7. We are given the vectors $\vec{a} = (3, m, -2), \vec{b} = (1, 4, t), \vec{c} = (6, 1, 2)$. Consider the expressions (i) - (iv):

        (i) $(\vec{a} + \vec{c}) \cdot \vec{b}$ (ii) $(\vec{c} \cdot \vec{b}) \vec{b}$ (iii) $\vec{b} \cdot \vec{c} + \vec{a} \cdot \vec{b}$ (iv) $\left((\vec{a} \cdot \vec{c}) \vec{a} \right) \cdot \vec{c}$

        a) Say whether each expression represents a scalar or a vector.

        b) Evaluate each expression. Your answers will contain letters.

    7.8. Consider the three points $A(1, 5, 3)$, $B(3, 5, 5)$, $C(1, 9, 4)$.

        a) Evaluate all dot products of the vectors $\overrightarrow{AB}$, $\overrightarrow{AC}$, $\overrightarrow{BC}$.

        b) Are all angles of the triangle ABC less than $\frac{\pi}{2}$ or is one of the angles greater than $\frac{\pi}{2}$? Write down reasons for your answer. *Hint:* Go back to Problem 4.8 b) and c).

        c) Find the cosine of the angle formed by $\overrightarrow{AB}$ and $\overrightarrow{BC}$.

d) $\vec{u}$ is a unit vector in the direction of $\overrightarrow{AB}$, and $\vec{v}$ is a unit vector in the direction of $\overrightarrow{BC}$. Find the cosine of the angle formed by $\vec{u}$ and $\vec{v}$.

7.9. $\vec{p}$ is the unit vector with components $\frac{1}{9}(4,1,8)$. Write α (read as "alpha") for the angle between $\vec{p}$ and $\vec{\imath}$, β ("beta") for the angle between $\vec{p}$ and $\vec{\jmath}$, and γ ("gamma") for the angle between $\vec{p}$ and $\vec{k}$.

a) Find $\cos\alpha$, $\cos\beta$, and $\cos\gamma$.

b) Find $\sqrt{\cos^2\alpha + \cos^2\beta + \cos^2\gamma}$.

c) Now let $\vec{v} = (a, b, c)$ be any vector, not necessarily a unit vector. Again, we write α, β, γ for the angles which $\vec{v}$ forms with $\vec{\imath}$, $\vec{\jmath}$, $\vec{k}$. Examine your work in a) to write down the cosines of the three angles.

d) The cosines that you found in c) are often called "the direction cosines of $\vec{v}$." They are nothing but the components of a unit vector in the direction of $\vec{v}$. We have the following information on a vector $\vec{m}$: It forms an angle of $\frac{\pi}{3}$ with both the x-axis and the z-axis, and its length is 27. Find the components of $\vec{m}$. *Hint:* Use the direction cosines of $\vec{m}$; in particular, use b) to find $\cos\beta$.

7.10. Given the vector $\vec{p} = (3, 6, 2)$, find

a) Three different vectors which are perpendicular to $\vec{p}$, and such that no two of them are parallel.

b) Two vectors which form an angle greater than $\frac{\pi}{2}$ with $\vec{p}$.

c) A *unit* vector $\vec{a} = (x, y, z)$ (a vector of length one) which forms an angle of $\frac{\pi}{3}$ with $\vec{p}$, in two steps:

(i) Write down the system of equations for (x, y, z).

(ii) In (i) you obtained a system of *two* equations for *three* unknowns x, y, and z. We can choose one unknown, say z, and set it equal to 0. Now you have a system of *two* equations for *two* unknowns. There are two ways to find $\vec{a} = (x, y, 0)$: Either you work with formulas, or you interpret the two equations as a line and a circle in the xy-plane and do ruler-and-pencil work with Figure 7.2.

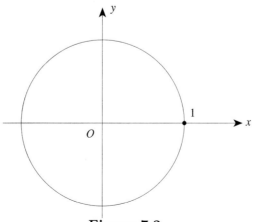

Figure 7.2

Projections

In 7.11–7.13 try to "see" the answer before working out the numbers.

7.11. $P(a, b, c)$ is a point in space and $P_0(a, b, 0)$ its shadow in the xy-plane, as shown in Figure 7.3.

   a) Find the scalar projection of $\overrightarrow{OP}$ onto $\overrightarrow{OP_0}$. Your answer will be a formula in terms of the components of $\overrightarrow{OP}$.

   b) The same for the vector projection. Your answer is a vector whose components will be formulas in terms of the components of $\overrightarrow{OP}$.

   c) Draw a point Q and its shadow Q_0 so that the scalar and vector projections of $\overrightarrow{OQ}$ onto $\overrightarrow{OP_0}$ are the same as those you found in a) and b).

   d) Give the coordinates of a point such as Q of c).

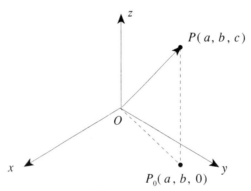

Figure 7.3

7.12. a) Find the coordinates of a point P in space that satisfies the following two conditions at the same time:

   (i) The scalar projection of $\overrightarrow{OP}$ onto $\vec{\imath}$ equals 2;

   (ii) $|\overrightarrow{OP}| = 3$.

   Hint: Look for such a point P in the xz-plane or in the xy-plane.

   b) Now we consider not only *one* such point as in a), but the collection K of *all* such P. What is the shape of K?

   c) Q is one of the points of the collection K. Find the cosine of the angle which $\overrightarrow{OQ}$ forms with the positive x-axis.

7.13. P is a point in space whose three coordinates (a, b, c) are all positive (see Figure 7.4 on the next page).

By the angle POP_0, we mean the angle at O formed by $\overrightarrow{OP}$ and $\overrightarrow{OP_0}$, and similarly for the other angles.

   a) Which of the following angles are right angles:

$$(1)\ POP_0 \quad (2)\ ROP_0 \quad (3)\ SPP_0 \quad (4)\ SP_0O \quad (5)\ QP_0P \quad (6)\ RPQ$$
$$(7)\ POQ \quad (8)\ PQO \quad (9)\ PSP_0$$

   Hint: Construct a makeshift xyz-system. Use your fingertips to mark the points S and P. You will see which angles are right, and which are not.

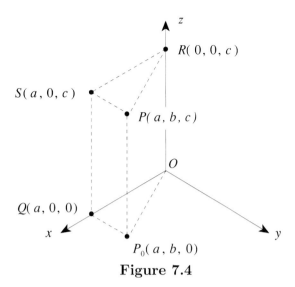

Figure 7.4

b) We want to write $\overrightarrow{OP}$ as $\overrightarrow{OP} = \vec{u} + \vec{v}$, where $\vec{u}$ is parallel to $\vec{\imath}$ and $\vec{v}$ is perpendicular to $\vec{\imath}$. Find the components of $\vec{u}$ and $\vec{v}$. *Hint:* Look at Figure 7.4.

7.14. We consider a line m in space and a point P that does not lie on m. *How do we find the distance of P from m?* The distance d is the length of the segment PQ shown in Figure 7.5. PQ is perpendicular to the line m, but in the picture the right angle at Q may not look like one. To find the distance we reason as follows:

1. Pick a point A on m.
2. $\overrightarrow{AQ}$ is the vector projection of $\overrightarrow{AP}$ onto a vector which is parallel to m.
3. $\overrightarrow{QP} = \overrightarrow{AP} - \overrightarrow{AQ}$, and $|\overrightarrow{QP}|$ is the wanted distance.

Apply this method to find the distance of $P(4, 1, 7)$ from the line m given by $(x, y, z) = (1, 0, 5) + t(6, 3, 2)$.

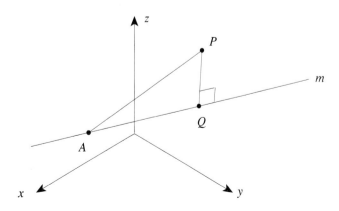

Figure 7.5

8   The Cross Product

There is another product of vectors, called the *cross product*. The cross product
of two vectors is a vector. Like the dot product, it is used in many different areas
of physics and engineering. To define the cross product we need what is called the
Right-Hand Rule.

The Right-Hand Rule.   We are given two nonzero vectors $\vec{a}$ and $\vec{b}$ that are not
parallel (their line segments are not parallel). We move the vectors so they have the
same initial point. Then they define a plane that we call α (read "alpha"), as shown
below right. Like any plane, α has many vectors that are normal to it. Among
these, we take two vectors $\vec{n}_1$ and $\vec{n}_2$, that point to different sides of α. The vectors
$\vec{n}_1$, $\vec{n}_2$ are called *normal vectors* of α or simply *normals* of α.

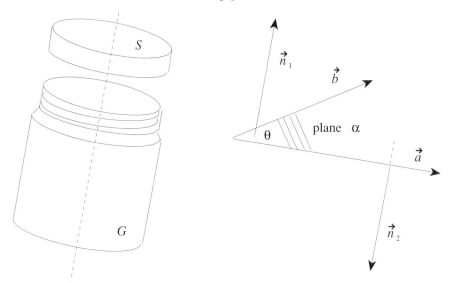

The Right-Hand Rule is a rule that selects one of the two normals, $\vec{n}_1$ or $\vec{n}_2$, for
the given pair of vectors $(\vec{a}, \vec{b})$. In the pair $(\vec{a}, \vec{b})$, the order "first $\vec{a}$, second $\vec{b}$"
matters. There are two steps. Step 2 comes in two versions. Pick the one you like
better.

Step 1: We imagine turning the vector $\vec{a}$ into $\vec{b}$ in the plane α. Note that we only
imagine turning $\vec{a}$, but we actually leave it where it is. There are two ways
of turning $\vec{a}$ into $\vec{b}$, namely, the "long" way of turning $\vec{a}$ by an angle greater
than π, and the "short" way with an angle less than π. *We write θ for the
angle of the shorter way.*

Step 2: Screw top version. We hold a glass jar G and its screw top S such that
the jar rests on the plane α, as shown above left. We put the cap onto
the jar and turn the cap in the same direction by which $\vec{a}$ turns into $\vec{b}$ the
short way. As we turn the cap, it will advance in the direction of either
$\vec{n}_1$ or $\vec{n}_2$. *The Right-Hand Rule assigns to our vectors $\vec{a}$ and $\vec{b}$ the normal
vector which points in the direction of the cap's movement.* If in the figure
above $\vec{a}$ pointed toward us and $\vec{b}$ away from us, we would unscrew the cap.
That is, the cap would move in the direction of $\vec{n}_1$.

Step 2: Fist and thumb version. Make a fist with your *right* hand as follows:

> *Your fingers other than the thumb are parallel to the plane α and point in the direction of the shorter way of turning $\vec{a}$ into $\vec{b}$. Your thumb sticks up.*

This is shown below left.

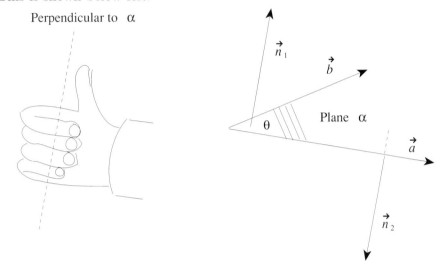

The Right-Hand Rule assigns to our vectors $\vec{a}$ and $\vec{b}$ the normal vector, which points in the direction of your thumb.

Problem _____

8.1. a) The vectors $\vec{j}$ and $\vec{k}$ form the yz-plane. The two normals of this plane are $\vec{\imath}$ and $-\vec{\imath}$. Apply the Right-Hand Rule to $\vec{j}$ and $\vec{k}$. Which normal of the yz-plane do you get? Use the figure below left to visualize the situation.

b) Repeat part a for the vectors $\vec{\imath}$ and $\vec{k}$, in that order.

c) Apply the Right-Hand Rule to the vectors $\vec{a} = (2, 1, 0)$ and $\vec{k}$, in that order (see the following figure, right). You obtain a normal vector $\vec{n}$. Is the x-component of $\vec{n}$ positive, zero, or negative? What about the other two components?

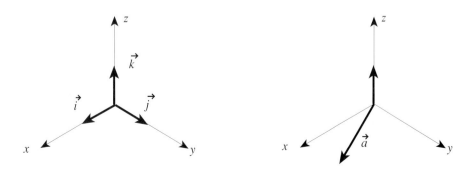

☞ Definition of the cross product

The *cross product* of $\vec{a}$ and $\vec{b}$

- is *defined* as the *vector* $\left(|\vec{a}||\vec{b}| \sin\theta\right) \vec{n}_0$,
- is *written* $\vec{a} \times \vec{b}$, and
- is *read* "$\vec{a}$ cross $\vec{b}$."

The meaning of θ and $\vec{n}_0$ in this definition is as follows.

- θ is the *smaller* angle of $\vec{a}$ and $\vec{b}$ in the plane defined by $\vec{a}$ and $\vec{b}$.
- $\vec{n}_0$ is the normal, of magnitude 1, obtained from $\vec{a}$ and $\vec{b}$ by the Right-Hand Rule.

Problem

8.2.   a) Find $\vec{\imath} \times \vec{\jmath}$ and $\vec{\jmath} \times \vec{\imath}$

       b) $\vec{a} = (672.55, 423.39, 0)$, $\vec{b} = (0, 0, -911.04)$. Is the x-component of $\vec{a} \times \vec{b}$ positive, zero, or negative? What about the other two components? Make a sketch of $\vec{a}$ and $\vec{b}$ that is not to scale but shows vectors with components (*positive, positive, zero*) and (*zero, zero, negative*).

Facts about the cross product

(i) Problem 8.2 part a shows that $\vec{\imath} \times \vec{\jmath} \neq \vec{\jmath} \times \vec{\imath}$. It means: For the cross product, the order of the factors matters.

(ii) Obviously, $\vec{a} \times \vec{0} = \vec{0}$.

(iii) If $\vec{a}$ is parallel to $\vec{b}$ (i.e., if the line segments of the two vectors are parallel), then $\vec{a} \times \vec{b} = \vec{0}$ because $\theta = 0$. It means: The cross product of two nonzero vectors may be zero.

(iv) The magnitude of $\vec{a} \times \vec{b}$ is given by $|\vec{a} \times \vec{b}| = |\vec{a}||\vec{b}| \sin\theta$. This is the area of the parallelogram spanned by $\vec{a}$ and $\vec{b}$. Why? $|\vec{a}| \sin\theta$ is the height h of the parallelogram, as shown below.

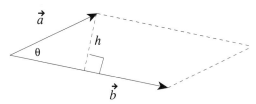

Rules for the cross product.   There are three basic rules, as follows (k stands for a scalar):

1. $\vec{a} \times \vec{b} = -\vec{b} \times \vec{a}$
2. $(h\vec{a}) \times (k\vec{b}) = (hk)\vec{a} \times \vec{b}$
3. $\vec{a} \times (\vec{b} + \vec{c}) = \vec{a} \times \vec{b} + \vec{a} \times \vec{c}$

We do not prove the rules. Rules (1) and (2) are straightforward, for (3) see, for example, R. Courant, *Differential and Integral Calculus, Volume II*, pages 14–17.

✍ Computation of the cross product

We consider the vectors $\vec{a}$ and $\vec{b}$:

$$\vec{a} = (a_1, a_2, a_3) = a_1\vec{i} + a_2\vec{j} + a_3\vec{k}, \quad \vec{b} = (b_1, b_2, b_3) = b_1\vec{i} + b_2\vec{j} + b_3\vec{k}$$

and work out their cross product with the rules 1 to 3 above. There will be nine terms of which we show only the first three:

$$\begin{aligned} \vec{a} \times \vec{b} &= (a_1\vec{i} + a_2\vec{j} + a_3\vec{k}) \times (b_1\vec{i} + b_2\vec{j} + b_3\vec{k}) \\ &= a_1b_1\vec{i} \times \vec{i} + a_1b_2\vec{i} \times \vec{j} + a_1b_3\vec{i} \times \vec{k} + \ldots \\ &= \vec{0} + a_1b_2\vec{k} - a_1b_3\vec{j} \ldots . \end{aligned}$$

If we check the remaining six terms and collect, we obtain

$$(a_1, a_2, a_3) \times (b_1, b_2, b_3) = (a_2b_3 - b_2a_3, a_3b_1 - b_3a_1, a_1b_2 - b_1a_2).$$

How to recall the components of $\vec{a} \times \vec{b}$. There are two methods which we call the *Determinant* and *Cross-Out Methods*.

The Determinant Method. Write down

$$\begin{vmatrix} \vec{i} & \vec{j} & \vec{k} \\ a_1 & a_2 & a_3 \\ b_1 & b_2 & b_3 \end{vmatrix}$$

and work it out using the rules for calculating determinants:

$$\vec{a} \times \vec{b} = \vec{i}(a_2b_3 - b_2a_3) - \vec{j}(a_1b_3 - b_1a_3) + \vec{k}(a_1b_2 - b_1a_2).$$

The Cross-Out Method. Write the components of each vector twice:

$$\begin{matrix} a_1 & a_2 & a_3 & a_1 & a_2 & a_3 \\ b_1 & b_2 & b_3 & b_1 & b_2 & b_3 \end{matrix}$$

Then cross out the first and last column:

$$\begin{matrix} a_2 & a_3 & a_1 & a_2 \\ b_2 & b_3 & b_1 & b_2 \end{matrix}$$

The first component of $\vec{a} \times \vec{b}$ is obtained by multiplying the two terms connected with the arrow pointing down and subtracting from it the product of the two terms with the upwards arrow:

$$a_2b_3 - b_2a_3$$

For the second component we use the second and third column, and for the third component we use the last two columns:

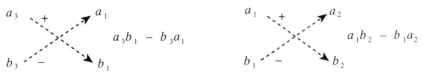

We end up with the same formula as for the Determinant Method:

$$\vec{a} \times \vec{b} = (a_2b_3 - b_2a_3, \ a_3b_1 - b_3a_1, \ a_1b_2 - b_1a_2)$$

Problems

  8.3.   a) Work out the cross products $(2,0,-3)\times(1,-4,5)$ and $(1,1,1)\times(u,v,w)$.

         b) The points $P(1,-1,0)$, $Q(2,1,-1)$ and $R(-1,1,2)$ determine the tri-angle PQR shown below. (Note that the units are *not* 1 cm.) Find the area of PQR. *Hint:* Go back to the definition of the cross product – what is the magnitude of $\overrightarrow{PQ} \times \overrightarrow{PR}$?

         c) Find a unit vector $\vec{n}_0$ normal (i.e., perpendicular) to the plane defined by the triangle PQR.

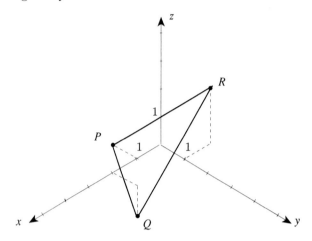

8.4. In Problem 7.14, you found the distance of a point P from a line m. Here we solve the same problem in a a different way. Figure 8.1 shows the line m, the point P, and the point Q closest to m. In addition, $\vec{u}$ is a vector parallel to m. To find the distance we reason as follows:

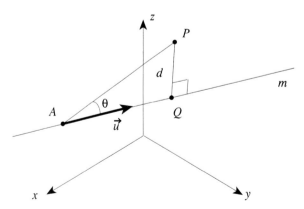

Figure 8.1

1. The distance d is $|\overrightarrow{AP}| \sin\theta$, where θ is the angle between $\vec{u}$ and $\overrightarrow{AP}$.
2. $|\vec{u} \times \overrightarrow{AP}| = |\vec{u}||\overrightarrow{AP}| \sin\theta = |\vec{u}|\, d$.
3. Solve the preceding equation for d.

Apply this method to find the distance of $P(4,1,7)$ from the line m given by $(x,y,z) = (1,0,5) + t(6,3,2)$.

Additional Problems

Definition of the cross product

8.5. The three points A, B, C are on the coordinate axes, as shown below. We consider the vector $\vec{u} = \vec{AB} \times \vec{BC}$. Are the components of $\vec{u}$ positive, or zero, or negative? Write (pos, neg, neg) if you think that the x-component is positive, the y-component negative, etc.

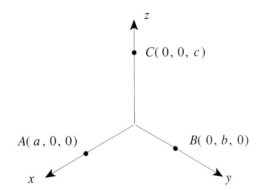

8.6. The vector $\vec{a}$ is defined by $\vec{a} = 2\,\vec{\imath}$. We turn $\vec{a}$ counterclockwise in the xy-plane while keeping the initial point all the time at the origin. We write $\vec{a}(\theta)$ for $\vec{a}$ turned by the angle θ. For example, $\vec{a}(180°) = -\vec{a}$. We will consider the cross product

$$\vec{a} \times \vec{a}(\theta)$$

for various values of θ. *Hint:* In an xyz-system, draw the vector $\vec{a} \times \vec{a}(\theta)$, with initial point at the origin, for several values of θ.

  a) For which θ is $\vec{a} \times \vec{a}(\theta) = \vec{0}$?

  b) We let θ increase from zero to 2π. Describe the direction of $\vec{a} \times \vec{a}(\theta)$. Your answer will be like this: "Between $\theta = \ldots$ and $\theta = \ldots$, the direction of $\vec{a} \times \vec{a}(\theta)$ will be $\ldots$, etc."

  c) As θ increases from 0 to 2π, what can you say about $|\vec{a} \times \vec{a}(\theta)|$? Does the magnitude increase, stay constant, decrease, increase from $\theta = \ldots$ to $\theta = \ldots$, then decrease from $\theta = \ldots$ to $\theta = \ldots$, etc.?

8.7. $\vec{a}$, $\vec{b}$ and $\vec{c}$ are three nonzero vectors. We are told that

$$|\vec{a} \times \vec{b}| = 3, \quad |\vec{a} \times \vec{c}| = 0.$$

  a) Is $\vec{a}$ perpendicular to $\vec{b}$? Is it not? Is it possible to decide on the basis of the given information?

  b) Is $\vec{a}$ parallel to $\vec{b}$? Is it not? Is it possible to decide?

  c) Is $\vec{a}$ perpendicular to $\vec{c}$? Is it possible to decide?

  d) Is $\vec{a}$ parallel to $\vec{c}$? Is it possible to decide?

  e) Is $\vec{a} \cdot \vec{b}$ equal to zero? Is it possible to decide?

  f) Is $\vec{a} \cdot \vec{c}$ equal to zero? Is it possible to decide?

8.8. Throughout this problem, a, b, and c are positive scalars.

Compute the cross products a) to d) below "by inspection." By this we mean: Figure out the direction and the magnitude by looking at the vectors and applying the definition of the direction and magnitude of a cross product. As an example, we show how to work a).

a) $(a, 0, 0) \times (0, b, 0)$

Look at the xyz-system below. Use the Right-Hand Rule to decide whether $(a, 0, 0) \times (0, b, 0)$ points in the direction of $\vec{k}$ or $-\vec{k}$. Into the system draw the parallelogram spanned by $a\vec{\imath}$ and $b\vec{\jmath}$. Its area is the magnitude of $(a, 0, 0) \times (0, b, 0)$.

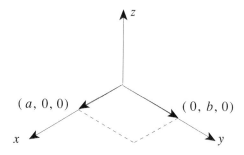

b) $(0, 0, 4c) \times (0, b, 0)$

c) $(0, 3b, 0) \times (0, 2b, 0)$

d) $(0, 0, 2c) \times (7a, 0, 0)$

Evaluation of cross products

8.9. Let

$$\vec{a} = (2, 1, 0), \quad \text{and} \quad \vec{b} = (3, 3, 3).$$

Evaluate the following expressions.

a) (i) $(\vec{a} \times \vec{b}) \times \vec{\imath}$ (ii) $\vec{a} \times (\vec{b} \times \vec{\imath})$

b) (i) $(\vec{a} \times \vec{b}) \cdot (2\vec{\imath} + \vec{k})$ (ii) $\vec{a} \cdot (\vec{b} \times (2\vec{\imath} + \vec{k}))$

c) $(\vec{a} \times \vec{\imath}) \cdot (\vec{b} \times \vec{\jmath})$

8.10. The two lines p and q are given by the equations

$$\begin{aligned} x &= 3 - t, & y &= 2, & z &= 1 + 2t; \\ x &= 3, & y &= 2 + 4u, & z &= 1 - 5u. \end{aligned}$$

They intersect at $D(3, 2, 1)$. Therefore, they define a plane α.

a) Find the equation of a line n which passes through D and which is normal (i.e., perpendicular) to α.

b) Find the equation of the line m which has the following two properties at the same time:

(i) It lies in the plane α and passes through D.

(ii) It is perpendicular to $\vec{k}$.

Hint: Write $\vec{n}$ for a vector normal to α. Then any line in α is perpendicular to $\vec{n}$.

8.11. We define a function $F(t)$ of one variable t by $F(t) = |(3, 1, 4) \times (3, 1, t)|$.

a) Find $F(0)$.

b) Find $F(4)$.

c) Find all zeroes of the function $F(t)$. *Hint:* You can answer this question without any computation if you realize that $F(t)$ is the the magnitude of a cross product. What is the geometric meaning of the cross product?

9   Planes

Normal vectors and normal lines of a plane

A plane α has many vectors $\vec{n}$ that are normal (perpendicular) to it, as shown below.

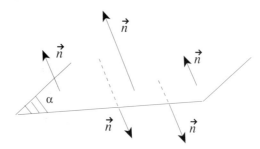

Any of these normal vectors is called a *normal* of α or a *normal vector* of α. A normal vector is perpendicular to any straight line lying in α.

By a *normal line* of α, or *a line normal to* α, is meant a line that is perpendicular to α. A normal line of α is perpendicular to any straight line which lies in α.

Different ways of constructing a plane

A plane α is determined

(i) by three points A, B, and C which do not lie on a straight line, or

(ii) by a straight line g and a point A which does not lie on g, or

(iii) by a point A and a normal vector $\vec{n}$.

This is shown below.

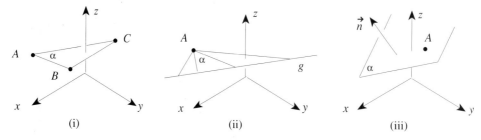

The main question we consider in this subsection is: *Given a plane in any of these ways, how do we find its equation?* We will consider points whose location in a plane α is variable or unspecified. Such a variable point of α is called a *general* point of α. We are going to answer the main question first for (iii), then for (i), and finally for (ii).

Equation of a plane given by a point and a normal vector (case iii). The given point A has coordinates (a_1, a_2, a_3), and the given normal vector $\vec{n}$ has components (n_1, n_2, n_3). We write $P(x, y, z)$ is the general point of the plane α.

The points A and P lie in α. Therefore, $\overrightarrow{AP} = (x - a_1, y - a_2, z - a_3)$ must be perpendicular to $\vec{n}$, as shown in Figure 9.1 on the facing page. Therefore, $\vec{n} \cdot \overrightarrow{AP}$ must be zero:

$$\vec{n} \cdot \overrightarrow{AP} = 0 = n_1(x - a_1) + n_2(y - a_2) + n_3(z - a_3).$$

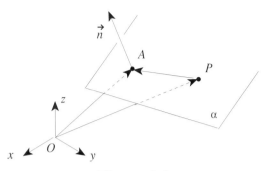

Figure 9.1

We work out the expression and find

$$n_1 x + n_2 y + n_3 z = n_1 a_1 + n_2 a_2 + n_3 a_3.$$

The right side $n_1 a_1 + n_2 a_2 + n_3 a_3$ is a constant determined by the given point and the given normal vector. We write K for this constant. For clarity, we rewrite the equation with (D, E, F) as the components of $\vec{n}$.

Result: *The equation of a plane α is of the form*

$$Dx + Ey + Fz = K,$$

where (D, E, F) are the components of a normal vector of α and K is a constant.

Problem

9.1. a) The plane α contains the point $A(1, 3, 4)$, and the vector $\vec{n} = (5, 3, -2)$ is normal to α. Find the equation of α.

b) Find the coordinates of four different points $Q_1, \ldots, Q_4$ that lie in the plane α.

c) Find the coordinates of four different points $R_1, \ldots, R_4$ that do *not* lie in α.

Equation of a plane given by three points (case i). In this case, the plane is determined by three points A, B, and C that do not lie on a straight line. The vector

$$\vec{n} = \overrightarrow{AB} \times \overrightarrow{AC}$$

is normal to α. We are back to "a plane given by one point and a normal" (case iii), which we worked out in the preceding paragraph.

Problem

9.2. Find the equation of the plane defined by the points

$$A(2, 1, 2), \qquad B(3, 0, 5), \qquad C(0, 1, 4).$$

Equation of a plane given by a point and a line (case ii). In this case, we have a line g in the plane, and a point A in the plane which is not on the line g. Pick two points B and C on the line g, and you are back to "a plane given by three points" (case iii).

Problem _____

9.3.   a) The plane α is defined by the point $A(2,3,0)$ and the line g given by $(x,y,z) = (-4t, 3+3t, 3+t)$. Find the equation of α.

   b) k is a line in α different from the line g. It passes through $A(2,3,0)$. Write down a parametric representation of g. There are many answers.

How to draw planes.   We draw a plane α by drawing the lines in which α intersects the three coordinates planes $x = 0$, $y = 0$, $z = 0$. To draw these lines, we draw the points in which the coordinate axes intersect α, and then connect the points.

For example, we draw the plane α given by $4x + 6y + 3z = 12$. It intersects the x-axis at $A(3,0,0)$, the y-axis at $B(0,2,0)$, and the z-axis at $C(0,0,4)$. The lines connecting A, B and C are the lines in which α intersects the coordinate planes, as shown below.

Note: The triangle ABC represents only a small part of α. The rest of α extends in all directions *behind* and *below* the coordinate planes which we face.

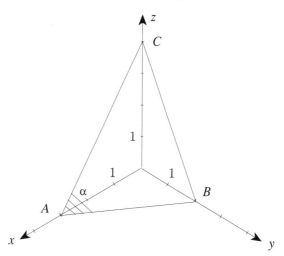

Problem _____

9.4. The plane β is given by $4x + 2y + 8z = 8$. Draw β into the system above.

Topics from the geometry of planes

Planes normal to one of the coordinate planes.   All points (x,y,z) that satisfy the equation

$$3x + 2y = 6$$

form a plane α with normal vector $\vec{n} = (3,2,0)$. *What can we say about our plane α?* The z-component of $\vec{n}$ is zero. This means that $\vec{n}$ is parallel to the xy-plane. *Because $\vec{n}$ is parallel to the xy-plane, our plane α must be normal to the xy-plane.* We draw α by drawing first the intersection $3x + 2y = 6$, $z = 0$ with the xy-plane. The intersections with the other coordinate planes are parallel to the z-axis, as shown in Figure 9.2 on the next page.

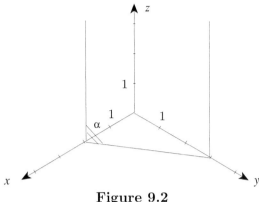

Figure 9.2

Problem _____

9.5. Use a colored pen to draw the plane $x+2z = 4$ into the system in Figure 9.2.
Hint: The plane is normal to the xz-plane (why?).

The distance of a point from a plane. We are given a plane α and a point Q.
How do we find the distance of Q from α? We can choose a viewpoint so that the
plane α is seen "on edge." Then α shows as a straight line (see Figure 9.3).

Pick an arbitrary point A in α. Then the distance d is the absolute value of
scalar projection of $\overrightarrow{AQ}$ onto the normal $\vec{n}$. We have to take the absolute value
because a distance is never negative.

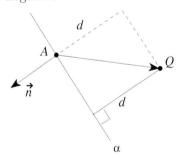

Figure 9.3

Problem _____

9.6. Let α be the plane given by $5x - y + 3z = 10$ and let Q be the point
$Q(2, 5, -3)$. Find the distance of Q from α.

Lines intersecting a plane and lines parallel to a plane. A line m and a
plane α may be mutually positioned in three different ways:

(i) The line m intersects the plane α in a point D.

(ii) The line m is parallel to the plane α, but m does not lie in α.

(iii) The line m lies in the plane α.

For example, in the left portion of Figure 9.4, the line m intersects the plane α in
the point D (case i). In the right portion of Figure 9.4, the line m is parallel to the

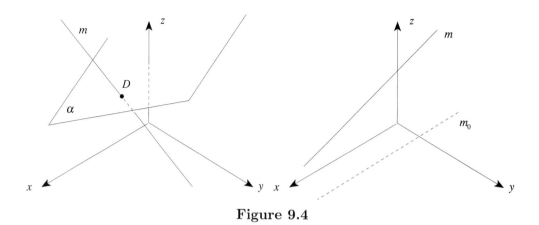

Figure 9.4

xz-plane, but does not lie in the xz-plane (case ii). *Remember:* To say that a vector is parallel to a plane means that the vector is perpendicular to the normal of the plane.

Given a line and a plane, how do we decide in which case we are? The line m is given by

$$(x, y, z) = (a, b, c) + t(p, q, r),$$

and the plane α by

$$Ex + Fy + Gz = K.$$

The line m has velocity vector $\vec{v} = (p, q, r)$. The plane has normal vector $\vec{n} = (E, F, G)$. In both cases ii and iii, the line m is parallel to the plane α, and therefore perpendicular to its normal vector $\vec{n}$. We can test for this using the dot product. If $\vec{v} \cdot \vec{n} \neq 0$, then $\vec{v}$ and $\vec{n}$ are *not* perpendicular, so we are in case i. If $\vec{v} \cdot \vec{n} = 0$, we are in either case ii or case iii. To see which one, just check whether the point (a, b, c) is on the plane. If yes, then we are in case iii. If no, then we are in case ii.

If we are in case i (where the line intersects the plane in a single point), you can *find* the point of intersection as follows. You make the substitution $x = a + tp$, $y = b + tq$, $z = c + tr$ in the equation $Ex + Fy + Gz = K$. You obtain an equation in which all letters other than t are given constants. Solve for the parameter t for which the line meets the plane.

Problem _____

9.7. The plane α is given by $x + 2y + 5z = 8$. We consider three lines

$$
\begin{aligned}
m_1 &: \quad (x, y, z) = (0, 4, 0) + t(1, 2, -1) \\
m_2 &: \quad (x, y, z) = (2, 3, 4) + t(1, 2, 1) \\
m_3 &: \quad (x, y, z) = (1, 1, -1) + t(1, 2, -1).
\end{aligned}
$$

Does m_1 intersect α? If yes, find the point of intersection D. If no, is m_1 parallel to α or does m_1 lie in α? Answer the same questions for the other two lines.

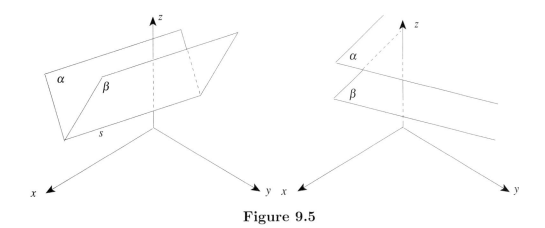

Figure 9.5

Parallel planes and planes which intersect. Consider two planes given by the equations

$$
\begin{aligned}
\alpha &: \quad Ax + By + Cz = D \\
\beta &: \quad Ex + Fy + Gz = K.
\end{aligned}
$$

There are three cases:

(i) α and β intersect in a line s.

(ii) α and β are parallel, but different.

(iii) The equations of α and β represent the same plane.

For example, at left in Figure 9.5, α and β intersect in the line s (case i). At right in Figure 9.5, α and β are parallel and different (case ii). *Given two planes, how do we decide in which case we are?* We look at their normals:

Step 1: Are the normals of the two planes are parallel? Check whether (A, B, C) is parallel to (E, F, G) or whether $(A, B, C) \times (E, F, G) = \vec{0}$. If yes, go to step 2; if no, go to step 3.

Step 2: Pick a point $Q = (q_1, q_2, q_3)$ of α. If Q lies in β, then the two equations represent the same planes. If Q does not lie in β, then the two equations represent different planes.

Step 3: The two planes have nonparallel normals. They intersect in a line s. The line s is normal to (A, B, C) because s lies in α. Similarly, s must be normal to (E, F, G). Therefore, s has the direction of $(A, B, C) \times (E, F, G)$. To find one point of s, you find one solution (x, y, z) of the *two* linear equations in the *three* unknowns (x, y, z):

$$
\begin{aligned}
Ax + By + Cz &= D \\
Ex + Fy + Gz &= K
\end{aligned}
$$

You have more unknowns than equations. Therefore you pick a value for one of the unknown and determine the other two by solving the system above.

Problem

9.8. Are the planes in a) and b) parallel? If they intersect find the line s of intersection. Or do the equations represent the same plane?

   a) $x + y - 2z = 5$ and $3x - z = 12$.

   b) $10x - 5y + 15z = 6$ and $-4x + 2y - 6z = 4$.

Additional Problems

9.9. The plane α contains the point $(1, 1, 2)$ and is parallel to the plane $4x - y + 2z = 2$. Find the equation of α.

9.10. Find the equations of the planes α and β shown in Figure 9.6. Note that β is normal to the yz-plane.

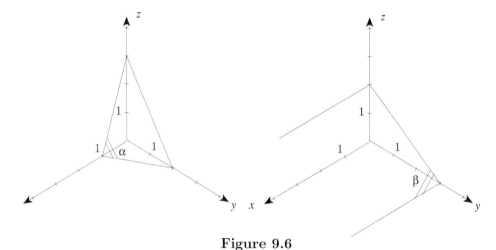

Figure 9.6

9.11. We have the following two pieces of information on a plane α: (i) α contains the point $T(2, 1, 3)$; (ii) the lines p: $(x, y, z) = (2, 1, 9) + t(3, 0, 4)$ and q: $(x, y, z) = (0, 3, 1) + u(4, 1, 1)$ are parallel to α. Find the equation of α. *Hint:* What does it mean to say that a line (or a vector) is parallel to a plane?

9.12. The plane α is given by $3x + 6y + 2z = 6$.

   a) Find all the points on the x-axis that are at distance 3 from α. *Hint:* The points you seek are of the form $Q(t, 0, 0)$, and you have to determine t. Compute the distance of Q from α and set it equal to 3. There will be two such points Q.

   b) Find the equations of the two planes β_1 and β_2 that are parallel to α, and are at distance 3 from α.

9.13. Draw the following into a standard xyz-system: The plane α: $x + 2y = 4$, the plane β: $x = 3$, and their line of intersection s. *Hint:* Find a point in the xy-plane that lies in both planes. The line of intersection s must pass through this point. What is the direction of s?

9.14. In a), b), c) we consider three pairs of equations of planes. Decide for each pair: Do the two planes intersect? Are they parallel? Do the two equations represent the same plane? If they intersect, find a parametric representation of the line of intersection s.

   a) $4x + y = 0, \quad -y + 3z = 8$

   b) $4x + 6y - 2z = -8, \quad -6x - 9y + 3z = 11$

   c) $z = 3, \quad y = 5$

9.15. After you have done Problem 9.5, Figure 9.2 on page 69 contains the two planes $3x + 2y = 6$ and $x + 2z = 4$. Figure 9.7 shows the two planes.

[b]

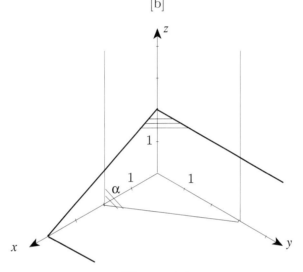

Figure 9.7

   a) Draw the line of intersection s and its shadow s_0 of the two planes. *Hint:* Plane No.1 intersects the xz-plane in a line m_1, and plane No. 2 intersects the xz-plane in a line m_2. The lines m_1 and m_2 are both in the xz-plane and intersect in a point K. This K lies in both planes and is therefore one point of the line of intersection. To find a second point L you repeat with the yz-plane (or the xy-plane) what you just did with the xz-plane.

   b) Write down a parametric representation of the line s of intersection. *Hint:* You can work in two ways: (i) In your drawing, read off the coordinates of two points of s, and use these points for a parametric representation. (ii) Use the general formula procedure for the line of intersection (s is parallel to the cross product of the two normal, find one point of s from the equations of the planes).

10   Curves in Space

Curves in space work much the same way as curves in the plane in Section 2, except that now we have three coordinates and components instead of two.

A point P moves on a curve K in space. As in Section 2, we write

- $P(t)$ for the position of P at time t sec, and
- $\vec{r}(t) = (x, y, z) = (a(t), b(t), c(t))$ for the position vector $\overrightarrow{OP}(t)$ of P,

as shown in Figure 10.1. As in Section 2, we define the velocity vector $\vec{r}\,'(t)$. It is tangent to the curve K at $P(t)$, and its length is the speed of P at time t.

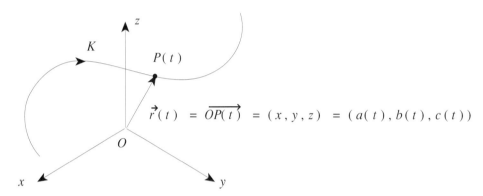

Figure 10.1

As time t passes, the point $P(t)$ outlines the curve C in space. The formulas $x = a(t)$, $y = b(t)$, $z = c(t)$ describe K and are called a *parametric representation* of the curve K or *parametric equations* for K and t is called the *parameter*. We write such a parametric representation in the form

$$(x, y, z) = (a(t), b(t), c(t)).$$

Mathematically, the two concepts "formulas giving the position of a point P moving on a curve K" and "parametric representation of a curve K" are the same. The first is a physical interpretation, and the second is a geometrical interpretation.

If K: $(x, y, z) = ((a(t), b(t), c(t))$ is a parametrized curve, then its derivative $\vec{r}\,'(t) = (a\,'(t), b\,'(t), c\,'(t))$ is a tangent vector at the point $P(t))$ of K. Obviously, $(37)\vec{r}\,'(t)$ and $-(0.076)\vec{r}\,'(t)$ are also tangent vectors.

Note:

1. If we consider $(x, y, z) = (a(t), b(t), c(t))$ as formulas that describe the movement of a point P on the curve K, then $\vec{r}\,'(t)$ tells us something about the way the point moves on the curve.

2. If we consider $(x, y, z) = (a(t), b(t), c(t))$ as parametric representation of a curve, then $\vec{r}\,'(t)$ has no physical significance. If we need a tangent vector, any of $\vec{r}\,'(t)$, $(37)\vec{r}\,'(t)$, $-(0.076)\vec{r}\,'(t)$, … will do.

Problem _____

10.1. A point $P(x, y, z)$ moves on a space curve K as follows: Its shadow $P_0(x, y, 0)$ in the xy-plane moves at constant speed on a circle of center $(x, y, 0) = (0, 0, 0)$ and radius 1; P_0 makes $\frac{1}{2\pi}$ revolutions per second (one revolution takes 2π seconds). The point P itself rises vertically at the constant rate of $\frac{1}{4}$ units per second. At time $t = 0$ the position of P is $(1, 0, 0)$. Below K is shown from $t = 0$ to $t = 4\pi$. It looks like a spiral spring and is called a *helix*.

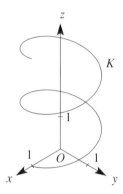

a) Write down a parametric representation of K. *Hint:* First, describe the movement of $P_0(x, y, 0)$ on the circle in the xy-plane. Then look at the z-coordinate.

b) Write down the components of the velocity vector of P.

c) Find a time t at which the velocity vector is parallel to the xz-plane.

d) Write down a parametric representation of the tangent line to K at the point which corresponds to $t = \frac{\pi}{6}$. *Hint:* Do not confuse the parameter t of the curve K with the parameter of the tangent line. We already use t as name for the curve parameter. Therefore, use u or any other letter for the tangent line parameter.

_____

The linear approximation for curves. It works for space curves the way the tangent line approximation works for a function $y = f(x)$ of elementary calculus.

A point P moves on the space curve $K : \vec{r}(t) = (x, y, z) = (a(t), b(t), c(t))$. At time t_0 sec, the point is at $P(t_0)$. We write P_0 for $P(t_0)$, as shown in Figure 10.2 on the following page. Until further notice we keep t_0 fixed. If t_0 is fixed, so is P_0. We let time increase from t_0 to $t_0 + h$. Then the point moves along K from P_0 to the endpoint of $\vec{r}(t_0 + h)$. However:

If during these h seconds the point moved along the tangent line at P_0 with constant velocity vector $\vec{r}'(t_0)$ it would be at Q, the endpoint of $\vec{r}(t_0) + h\vec{r}'(t_0)$.

This is shown in Figure 10.2 on the next page. For small values of h, the movement along the tangent line is so close to the actual movement that we can use the movement along the tangent line as approximation to the actual movement. In other words:

The vector $\vec{r}(t_0) + h\vec{r}'(t_0)$ is an approximation of the vector $\vec{r}(t_0 + h)$.

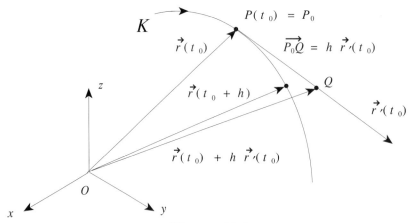

Figure 10.2

We use for this the symbol $\approx$ (read "approximately equal to") to express the approximation:

$$\vec{r}(t_0 + h) \;\approx\; \vec{r}(t_0) + h\vec{r}\,'(t_0). \tag{10.1}$$

The vector on the right side of this approximate equality is called the *linear approximation* of $\vec{r}(t)$ at t_0. Note that in the formula for the linear approximation h is a variable because we can use $\vec{r}(t_0) + h\vec{r}\,'(t_0)$ for different values of h.

Now we write out the formula (10.1) in components. We do it in two steps.

Step 1: We look at each side of the approximate equality:

$$\vec{r}(t_0 + h) \;=\; (a(t_0 + h), b(t_0 + h), c(t_0 + h))$$
$$\vec{r}(t_0) + h\vec{r}\,'(t_0) \;=\; (a(t_0), b(t_0), c(t_0)) + h(a\,'(t_0), b\,'(t_0), c\,'(t_0))$$

Step 2: We compare the components:

$$
\begin{aligned}
a(t_0 + h) &\approx a(t_0) + h\,a\,'(t_0)\\
b(t_0 + h) &\approx b(t_0) + h\,b\,'(t_0)\\
c(t_0 + h) &\approx c(t_0) + h\,c\,'(t_0)
\end{aligned}
$$

For each component we have what is called the linear approximation or tangent line approximation of elementary calculus.

Problem _____

10.2. The curve K is given by $(x, y, z) = (\sin t, t^2 + 2, \cos t)$.

    a) Write down the components of the linear approximation of K at $t_0 = \pi$, that is, at the point P_0 with coordinates $(0, \pi^2 + 2, -1)$.

    b) Use the linear approximation to estimate the coordinates of the point of K which corresponds to $t = \pi + 0.09$.

    c) Use a calculator to evaluate the components for $t = \pi + 0.09$. Compared with the calculator value, by how many percent are the components off if you use the linear approximation?

A different notation for the linear approximation. We go back to the general situation and consider a curve $K : (x, y, z) = (a(t), b(t), c(t))$ and a fixed value t_0. The point $P(t_0)$ is called the *base point*. We will rewrite the linear approximation formula (10.1) by writing $t - t_0$ for h. The result is

$$\vec{r}(t) \approx \vec{r}(t_0) + (t - t_0)\vec{r}'(t). \tag{10.2}$$

This is illustrated below, where "Notation 1" and "Notation 2" refer to formulas (10.1) and (10.2), respectively. You should get used to both notations.

Notation 1
Base point: $P(t_0)$
Variable point: $P(t_0 + h)$
Change in t: $\Delta t = h$

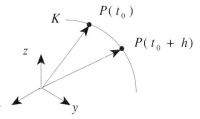

Notation 2
Base point: $P(t_0)$
Variable point: $P(t)$
Change in t: $\Delta t = t - t_0$

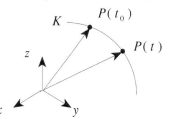

Additional Problems

Space curves and their tangents

10.3. The plot below gives the x, y, and z coordinates of a moving particle as functions of time t in seconds.

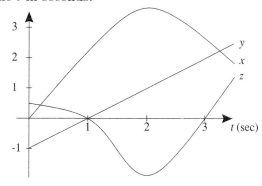

    a) Does the particle ever pass through the origin? If so, when?

    b) Is the particle ever below the xy-plane? If so, when?

    c) Does the particle ever come to rest (velocity zero)? If so, when?

    d) Is the velocity ever parallel to the y-axis? If so, when?

    e) Estimate the velocity and speed when $t = 1$ sec.

10.4. The circle C has center $A(2, 1, 0)$ and radius 3, and is parallel to the xz-plane.

    a) Find a parametric representation for C.

    b) Which parameter values represent the part of the circle above the xy-plane?

10.5. The curve C has parametric representation

$$x = 2t, \quad y = e^{-t}\cos t, \quad z = e^{-t}\sin t.$$

a) Find the point P where the curve intersects the yz-plane.

b) Find the a parametric representation for the tangent line to C at the point P.

10.6. C is the curve given by $(x, y, z) = (\cos(\pi t), t^2, 3t)$. We write $P(t)$ for the general point of C (the point that corresponds to the general value t).

a) At $P(t)$, we consider the tangent line b of C. Write down a parametric representation of b. *Hint:* Read the hint in Problem 10.1.d) concerning the letter name of parameters.

b) Find the coordinates of a point K that has the following properties:

K lies on the tangent to C at $P(1)$.

K lies at the distance $2|\vec{r}'(1)|$ from $P(1)$, in the direction of $\vec{r}'(1)$.

The figure below illustrates the situation, *but it is not to scale. Hint:* Work with the formulas you found in part a.

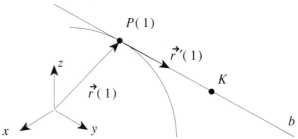

Linear approximation

10.7. The curve C is given by $(x, y, z) = (e^t, t^3 + 4t, \tan t)$.

a) Go back to the discussion of notation for the linear approximation on 77.

(i) Write out the linear approximation of C at $t_0 = 1$ in Notation 1. Your answer will contain the letter h.

(ii) Write out the linear approximation of C at $t_0 = 1$ in Notation 2. Your answer will contain the letter t.

b) Write out the linear approximation of C at $t_0 = 0$ in the notation of your choice.

c) Use b) to estimate the coordinates of the point of C which corresponds to $t = -0.14$.

10.8. A point P is moving on a curve $K: (x, y, z) = (a(t), b(t), c(t))$. We do not know the formulas for the functions $a(t)$, $b(t)$, and $c(t)$, but we have the following information: At time $t = 5$, the point P has coordinates $(3, 7, 4)$, and the velocity vector has components $(-1, 5, 2)$. The diagram in Figure 10.3 on the facing page is not to scale but shows schematically what is going on.

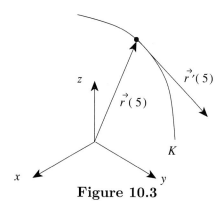

Figure 10.3

 a) Determine as well as you can the coordinates of P at time $t = 5.2$. Is your answer an estimate or is it exact?

 b) Determine as well as you can a parameter representation of the tangent line of the curve K at $t = 5$. Is your answer an estimate or is it exact?

10.9. A point P moves on a curve K in space. As always, we write $\vec{r}(t)$ for the position vector of the point P at time t.

 We do not know the formula for $\vec{r}(t)$, but we are told that $\Delta\vec{r} = \vec{r}(8.4) - \vec{r}(8)$ has components $(0.5, -0.3, 0.2)$. The diagram below is not to scale.

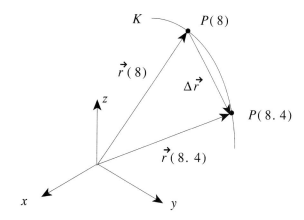

 a) Determine as well as you can the components of the velocity vector at time $t = 8$ *sec*. Is your answer an estimate or is it exact?

 b) Determine as well as you can the speed of P at the same time. Is your answer an estimate or is it exact?

10.10. You are given that the position and velocity of a particle at time $t = t_0$ sec are

$$\begin{aligned} \vec{r}(t_0) &= 5\vec{\imath} - 2\vec{\jmath} + 3\vec{k} \\ \vec{r}'(t_0) &= 2\vec{\imath} + \vec{\jmath} - \vec{k}. \end{aligned}$$

Give your best estimate of the position of the particle 0.2 seconds later.

11   Normals of Curves

Curves in the plane

Below are shown a curve C in the plane and a point P on C. We construct the following objects:

(i) The tangent line b of C at P;

(ii) The line n perpendicular to b at P.

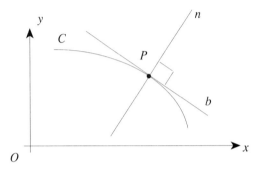

The tangent line represents the direction of C at P, and the line n represents the direction perpendicular to C at P. Instead of saying "n is normal to the direction of C," we just say "n is normal to C at P." The line n is called the *normal line* of C at P or simply the *normal* of C at P.

Similarly, we can consider vectors instead of lines. A vector perpendicular to the tangent vector at P is called a *normal vector* of C at P.

Problem _____

11.1. The curve C in the plane is given by

$$x = (0.1)t^2 + t + 1, \qquad y = \sin\left(\frac{\pi t}{2}\right),$$

and P is the point corresponding to $t = 1$.

a) (i) Write down a parametric representation of the tangent line b at P. Use u for the parameter to distinguish it from the parameter of C.

(ii) Write down the equation of the tangent line at P in the form $Ax + By + C = 0$.

b) (i) Write down a parametric representation of the normal line n at P. Use v as parameter to distinguish it from the parameters of C and b.

(ii) Write down the equation of the normal line at P in the form $Ax + By + C = 0$.

c) Is there a point Q on C where the normal vector is parallel to $\vec{\imath}$? If yes, find the coordinates; if no, explain.

Curves in space.

We ask: *What is the normal of a curve in space?* To begin with, we consider a line b in space and a point P of b. We construct the plane α, that passes through P

and for which the line b is a normal line. The plane α consists of all lines that pass through P and that are perpendicular to b. This is suggested in Figure 11.1. The plane α is called the *normal plane* at P of the line b. Thus, a line in space does not have a normal line but a normal plane.

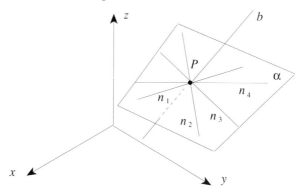

$n_1, n_2. \ldots$ perpendicular to b

Figure 11.1

Now we look at a curve K in space and at a point P of K. We construct the tangent line b to the curve K at the point P, as shown in Figure 11.2. The normal plane at P of the tangent line b consists of all lines which are normal to K at P. This plane is called the *normal plane* at P of the space curve K. In Figure 11.2, the normal plane is suggested by dotted lines normal to the tangent line b. Note that the normal vector of the normal plane of K is the tangent vector to the curve K at the point P.

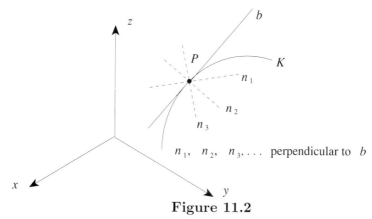

$n_1, \quad n_2, \quad n_3, \ldots$ perpendicular to b

Figure 11.2

Problem _____

11.2. The curve C given by $(x, y, z) = (\cos t, \sin t, \frac{1}{4}t)$ is the helix of Problem 10.1.

  a) Write down the equation of the normal plane at the point Q which corresponds to $t = \frac{\pi}{6}$.

  b) Is there a point R on C at which the normal plane is perpendicular to the xz-plane? If yes, find the coordinates of R; if no, explain.

  c) Repeat part b for the xy-plane.

How to find tangents and normal lines/planes of curves—summary

Curves in the Plane

(i) *Curve given as graph $y = f(x)$ of a function $f(x)$:*
   Slope of tangent line: $f'(x)$;   tangent vector: $(1, f'(x))$
   slope of normal line: $-\frac{1}{f'(x)}$;   normal vector: $(f'(x), -1)$

(ii) *Curve given by a parametric representation $(x, y) = (x(t), y(t))$:*
   tangent vector: $(x'(t), y'(t))$
   normal vector: $(-y'(t), x'(t))$

(iii) *Curve given by an equation $F(x, y) = 0$ (curve given implicitly[1]):*
   For the slope dy/dx of the tangent line, use implicit differentiation: Consider y as function of x, differentiate with respect to x, and solve for y'. Then use the formulas in item 1 for tangents and normals.

Curves in Space

   Curve given by parametric representation $(x, y, z) = (x(t), y(t), z(t))$:
   Tangent vector: $(x'(t), y'(t), z'(t))$
   normal plane: The tangent vector is the normal of the normal plane.

Problem

11.3. The line m passes through $A(1, 3, 2)$ and $B(5, 4, 6)$. We consider m as a space curve.

   a) Write out the equation of the tangent line of m at A.

   b) Write out the equation of the normal plane of m at A.

Additional Problems

11.4. Returning to Problem 10.5 on page 78, find an equation for the normal plane to the curve C at the point P in the form $ax + by + cz = d$.

11.5. There are six questions, each of the following kind:

> Given a curve C in the plane or in space, can you find a point on C at which the tangent line or normal line/plane has certain properties?

If there is such a point, find it; if not, explain. Remember what "parallel" for vectors means: The nonzero vectors $\vec{a}$ and $\vec{b}$ are parallel if they are scalar multiples of each other, that is, if $\vec{a} = c\vec{b}$ for a scalar $c \neq 0$.

   a) *Elementary calculus.* We consider the graph of $f(x) = \sin x$.

   (i) Can you find x so that the normal line of the graph at the point $(x, f(x))$ has slope $\frac{2}{3}$?

   (ii) Can you find x so that the tangent to the graph at the point $(x, f(x))$ has slope -0.5?

---

[1] Note to instructor. These curves will appear as level curves in Section 12, and their normals will be discussed later on via the gradient.

b) *Parametrized curves in the plane.* We consider the curve C given by $(x, y) = (3t^2 + 5t, t^2 + 4t)$.

    (i) m is the line given by $(x, y) = (1, -5) + u(-2, 3)$. Is there a point on C where the normal line is parallel to m?

    (ii) p is the line given by $(x, y) = (0, 4) + u(3, 1)$. Is there a point on C where the tangent line is parallel to p?

c) *Curves in space.* The curve C is given by $(x, y, z) = (t^3, 5t, 2t^2)$.

    (i) Can you find a point on C where the normal plane is parallel to the plane $12x + 5y + 8z = 24$?

    (ii) Can you find a point on C where the tangent line is parallel to the plane $5x + 5y + 5z = 15$?

11.6. Here you are asked to find the formulas for the tangent line and normal line/plane at the general point of a curve. When dealing with a parametrized curve, we will write $P(t)$ for the point corresponding to the general value t of the parameter.

a) *Elementary calculus*

    (i) We write C for the graph $y = \cos x$. Find the equation of the tangent line to C at the general point $(x_0, \cos(x_0))$. Write the equation in the form $y = Ax + B$.

    (ii) The coefficients A and B of the equation you found in (i) contain x_0. Explain in one or more complete sentences why this is the case.

    (iii) Find a parametric representation of the normal line of C at the point $(x_0, \cos(x_0))$.

b) *Parametrized curves in the plane.* C is the curve $(x, y) = (t^3 + t, t^2)$. Write down a parametric representation of the normal line at the point corresponding to the general value t. Remember to use a letter other than t for the parameter of the normal line.

c) *Curves in space.* The curve C is given by $(x, y, z) = (5t, e^t, -t^2)$.

    (i) Write down a parametric representation of the tangent line to C at $P(t)$.

    (ii) Find the equation of the normal plane at $P(t)$. Write it in the form $Ax + By + Cz = D$. Why are the coefficients A, B, C, D functions of t?

Chapter 2

Functions and Differentiation

12 Functions of Two Variables

Example 1. You take out a loan of \$10,000 at an annual interest rate of r. The amount of your monthly payment P depends on the interest rate r and on the number N of monthly payments you have to make. We say that P is a *function of the two variables* r and N, and we write for that $P = f(r, N)$. The monthly payment depends on *two* quantities N and r. Of these, one, or both, may change. How do such changes influence P? This is discussed in Problem 12.1.

Problem _____

12.1. In each of the following scenarios, say how P changes when you vary r and/or N as indicated. Possible answers are: "P increases," "P decreases," "P does not change," or "cannot tell based on the information given."

    a) r increases and N does not change.

    b) r increases and N increases.

    c) r does not change and N increases.

    d) r decreases and N increases.

_____

The monthly payment P is given by

$$P = \frac{10000\,\frac{r}{12}}{1 - \left(1 + \frac{r}{12}\right)^{-N}}.$$

In this formula, the interest rate is not expressed in percent, but in decimal form (e.g., 6% is written as 0.06). For a detailed discussion of this formula, see, for example, *Business Mathematics*, Schaum's Outlines, p. 182–183. ∎

Example 2. The mean, or average, of two numbers a, b is given by $\frac{a+b}{2}$. This expression is a function of the two variables a and b. We may call this function $m(a, b)$: $m(a, b) = \frac{a+b}{2} = \frac{1}{2}a + \frac{1}{2}b$. The function is of the form *"scalar times the first variable plus scalar times the second variable."* Such a function of two variables is called a *linear function*. An example of a linear function $f(x, y)$ of the variables x and y is $f(x, y) = 2x - y + 4$. The general form of such a linear function is $f(x, y) = Ax + By + C$. ∎

Example 3. The function $d(x,y) = \sqrt{(x-4)^2 + (y+1)^2}$ is the distance of the point with coordinates (x,y) from the point with coordinates $(4,-1)$. The formula $g(u,v) = \sqrt{(u-4)^2 + (v+1)^2}$ defines the same function because we have only changed notation.   ■

Example 4. The total surface area S of a solid straight circular cylinder of radius r and height h is given by $S = 2\pi r^2 + 2\pi r h$. That is, S is the function of the two variables r and h, which is defined by $f(r,h) = 2\pi r^2 + 2\pi r h$. Again, $p(x,y) = 2\pi x^2 + 2\pi x y$ defines the same function because we have only changed notation.   ■

Terminology. The variables r, N of Example 1 are called the *independent variables* of the function $f(r,N)$. Similarly, the variables a, b and x, y of Example 2, u and v of Example 3, and r and h of Example 4 are the independent variables of the functions in question.

When we write $P = f(r,N)$ in Example 1, or $z = \sqrt{(x-4)^2 + (y+1)^2}$ in Example 3, then P, or z, is called the *dependent variable*.

Given a function $z = f(x,y)$ of the two independent variables, we ask

(i) *For which pairs (x,y) is the function $f(x,y)$ defined?*

(ii) *For which values of z can the equation $z = f(x,y)$ be solved for (x,y)?*

The collection of all pairs (x,y) for which $f(x,y)$ is defined is called the *domain* of the function $f(x,y)$ (question i). The collection of all z for which $z = f(x,y)$ can be solved for (x,y) is called the *range* of the function $f(x,y)$ (question ii).

Problem _____

    12.2.   a) Use a complete sentence to describe the domain of $f(x,y) = -2x - y + 4$ (Example 2a).

                b) Use a complete sentence to describe the range of the function $g(x,y) = \sqrt{4 - x^2 - y^2}$.

_____

How do we represent functions of two variables?

In elementary calculus, we represent a function of one variable by a graph in the xy-plane. These graphs are helpful. They allow us to "look" at the function, and to "grasp" or "understand" it in one swoop. *Can we similarly represent functions of two variables?* We have two methods, called *graphs* and *level curves*.

The Graph Method

This method called is patterned after the way we construct the graph of a function $y = f(x)$ of one variable. So we ask: How do we get the graph of $y = f(x)$? There are three steps:

Step 1: We represent the values of x as points on a line called the x-axis.

Step 2: We represent the values of $y = f(x)$ as points on the y-axis.

Step 3: In the xy-plane, we draw the y-axis perpendicular to the x-axis and record the value of $f(x)$ perpendicularly above (or below) the value of x. If we do this for all x, we obtain a curve C in the plane in the familiar way, as shown below.

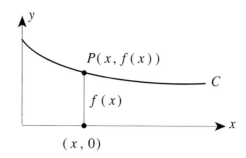

The corresponding three steps for a function $z = f(x, y)$ of *two* variables are as follows:

Step 1: We represent the pairs (x, y) of independent variables as points of the xy-plane.

Step 2: We represent the values of $z = f(x, y)$ as points on the z-axis.

Step 3: In xyz-space, we draw the z-axis perpendicular to the xy-plane and record the value of $f(x, y)$ perpendicularly above (or below) the point (x, y) of the xy-plane. If we do this for all (x, y), we obtain a *surface S*, as shown below left. It is called the *graph* of $z = f(x, y)$.

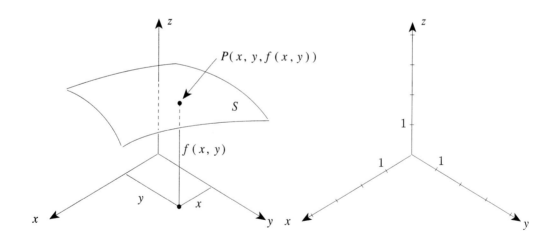

Problems

12.3. Into the system above right draw the graph of $z = f(x, y) = -2x - y + 4$.
Hint: Write $z = -2x - y + 4$ in the form $-2x - y - z + 4 = 0$. What is the shape of the graph?

12.4. a) Describe in words the shape of the graph of $z = f(x, y) = \sqrt{4 - x^2 - y^2}$.
Hint: Square the equation $z = \ldots$ and move all terms which contain x and y to the left side.

b) Write down a parametric representation of the curve C in which the graph of part a intersects the xy-plane.

The Method of Level Curves

Pick a scalar c. In the xy-plane, draw all points (x, y) for which the function takes the value c. That is, draw all points (x, y) such that $f(x, y) = c$. These points usually make up a curve which is called the *level curve* of *level c* of the function $f(x, y)$.

Problems _____

12.5. $f(x, y) = -2x - y + 4$ is the function of Problem 12.3.

    a) Write down the equation of the level curves of $f(x, y)$ of the following levels c: $c = 0$; $c = 4$; $c = -1$.

    b) Draw these level curves into an xy-system. Label each curve with its level (write "$c = 0$" next to the curve $c = 0$, etc.).

12.6. $f(x, y) = \sqrt{4 - x^2 - y^2}$ is the function of Problem 12.4.

    a) Write down the equation of the level curves of $f(x, y)$ of the following levels c: $c = 0$; $c = -1$; $c = 1$.

    b) Draw these level curves into an xy-system. Label the curves.

12.7. We return to Example 1. *Remember:* The function $P = f(r, N)$ gives us the monthly payment P to amortize \$10,000 at an annual interest rate of $r \cdot 100\ \%$ in N payments. Below are shown some level curves. The levels are in dollars.

    a) What is the monthly payment if the interest rate is 9% and you have 10 years to pay?

    b) The interest rate is increased from 9% to 11%, and you increase the time from 10 to 13 years. Will the monthly payment go up or down? By how much?

    c) The interest rate is 8%, and you want to pay \$100 a month. How many payments does it take to pay off the \$10,000?

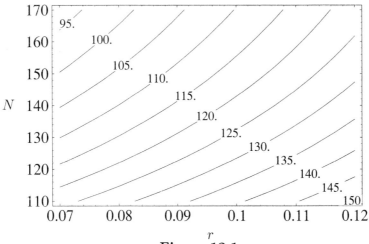

Figure 12.1

How are graphs and level curves related to each other?

First, we observe that the methods of graphing and level curves produce different outputs:

The graph $z = f(x, y)$ is a *surface in xyz-space*.

The level curves of $f(x, y)$ are a *family of curves in the xy-plane*.

The level curve $f(x, y) = c$ in the xy-plane is related to the graph $z = f(x, y)$ in space as follows:

The plane $z = c$ is parallel to the xy-plane and lies at height c above the xy-plane. It intersects the graph $z = f(x, y)$ in a curve K. The shadow of K in the xy-plane is the level curve $f(x, y) = c$.

This is illustrated in Figure 12.2 below for the function $f(x, y) = (1/2)(x^2 + y^2 + 4)$.

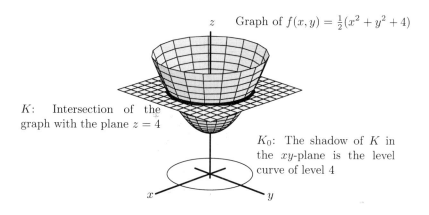

z Graph of $f(x, y) = \frac{1}{2}(x^2 + y^2 + 4)$

K: Intersection of the graph with the plane $z = 4$

K_0: The shadow of K in the xy-plane is the level curve of level 4

x y

Figure 12.2

Terminology. The words *contour line* or simply *contour* are sometimes used instead of "level curve." They mean the same thing. A picture showing a family of several level curves for the same function is called a *contour plot*. For example, Figure 12.1 is a contour plot for the monthly payment function $f(r, N)$. A contour plot for the function $f(x, y) = \frac{1}{2}(x^2 + y^2 + 4)$ is shown below. The contour illustrated in Figure 12.2 is drawn with a heavy line.

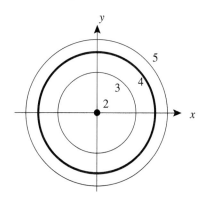

Problem ——————————————————————————

12.8. Figure 12.3 shows the graph $z = -2x - y + 4$. Draw the intersection of the graph with the plane $z = 1.5$ and the corresponding level curve $f(x, y) = 1.5$ in the xy-plane.

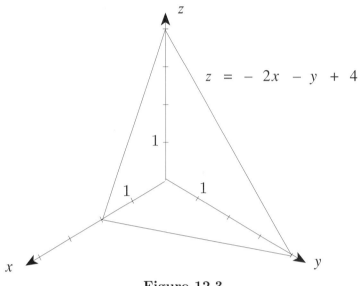

Figure 12.3

Now we discuss two kinds of functions that are often used. They are linear functions and functions defined by composition.

Linear functions

What is a linear function of two variables? It is a function of the form $f(x, y) = Ax + By + C$, where A, B, C are constants.

Problems ——————————————————————————

12.9. Let $f(x, y) = -3x - y + 6$

    a) Find $f(1, 3)$, $f(3a + 6, 22)$.

    b) Find points $P_1(x_1, y_1)$ and $P_2(x_2, y_2)$ so that $f(x_1, y_1) = f(x_2, y_2) = 17$.

12.10. The function $g(x, y)$ defined by $g(x, y) = -x + 4$.

    a) Is $g(x, y)$ a linear function or not? If yes, find the constants A, B, C; if no, explain.

    b) Find $g(1, 5)$, $g(1, 36)$, $g(36, b + c)$, $g(b + c, 36)$.

What is the graph of a linear function of two variables? The graph of $f(x, y) = Ax + By + C$ is the surface in xyz-space defined by $z = Ax + By + C$. If you rewrite $z = Ax + By + C$ in the form $Ax + By - z = -C$ you realize that the graph is a plane.

Problem ——————————————————————————

12.11. We consider the linear functions $f(x, y) = -3x - y + 6$ and $g(x, y) = -x + 4$. Draw the graphs of $f(x, y)$ and $g(x, y)$.

What are the level curves of a linear function of two variables? The level curve of level k of $f(x,y) = Ax + By + C$ is the curve $Ax + By + C = k$ in the xy-plane. It is a line (*remember:* "line" = "straight line").

Problem ──

12.12. Take the two functions $f(x,y) = -3x - y + 6$ and $g(x,y) = -x + 4$ of Problem 12.11. Through the point $P(3,8)$ passes a level curve of $f(x,y)$. What is its level? What is its equation? Answer the same questions for $g(x,y)$.

Functions of two variables defined by composition

We are given a function $u(t)$ of *one* variable and a function $q(x,y)$ of *two* variables. We can substitute $q(x,y)$ for t in $u(t)$, and we obtain a new function $w(x,y) = u(q(x,y))$ of *two* variables. It means: Wherever the t is in $u(t)$, we write the formula for $q(x,y)$. The operation of substituting one function in another one is called *composition of functions.*

Example 5. Take $u(t) = e^t$ and $q(x,y) = 2x - 7y + 1$. Now we define a new function $f(x,y)$ by

$$f(x,y) = u(q(x,y)) = u(2x - 7y + 1) = e^{2x - 7y + 1}.$$

What is $f(3,2)$? **Answer:** $f(3,2) = e^{6 - 14 + 1} = e^{-7}$. ∎

Example 6. With $u(t) = \sqrt{t}$, $q(x,y) = x^2 + y^2$, we define a new function $g(x,y)$ by

$$g(x,y) = u(q(x,y)) = u(x^2 + y^2) = \sqrt{x^2 + y^2}.$$

What is $g(4,7)$? **Answer:** $g(4,7) = u(4^2 + 7^2) = \sqrt{4^2 + 7^2} = \sqrt{65}$. ∎

Example 7. The function $w(t)$ is defined by the graph shown below in Figure 12.4, and $q(x,y)$ is defined by $q(x,y) = x^2 + x + y + 2$. We define a new function $h(x,y)$ by

$$h(x,y) = w(q(x,y)) = w(x^2 + x + y + 2).$$

What is $h(1,2)$? **Answer:** $h(1,2) = w(1^2 + 1 + 2 + 2) = w(6) = 2.6$. ∎

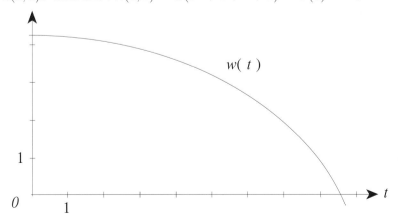

Figure 12.4

Problem ――――――――――――――――――――――――――――

12.13. $f(x,y)$, $g(x,y)$, $h(x,y)$ are the three functions of Example 5, 6, 7.

     a) Find a point $P(x,y)$ on which $f(x,y)$ takes the value 1.

     b) Find the equation of the level curve of level 1 of $f(x,y)$.

     c) Describe in words the level curve of level 64 of $g(x,y)$.

     d) Write down the equation of the level curve of level 1 of $h(x,y)$.

Additional Problems

Linear functions

12.14. Consider the function $z = f(x,y) = 4 - 2x - y$.

     a) Sketch the level curves $z = 2, 3, 4, 5, 6$.

     b) Sketch the graph of f by drawing its intersection with each of the three coordinate planes.

12.15. Two level curves for the linear function $f(x,y)$ are shown below.

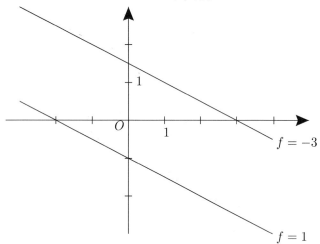

     a) Find a formula for the function $f(x,y)$ of the form $f(x,y) = Ax + By + C$.

     b) With the values of A and B you found in part a, draw the vector $\vec{v} = A\vec{i} + B\vec{j}$ into the figure above.

     c) Draw into the figure above a vector $\vec{w}$ that points in the direction of "steepest ascent" for the function f. By this we mean the direction that will take you from one level line to the next with the least displacement. Comment on any relation you observe between the vector $\vec{v}$ from part b and the vector $\vec{w}$.

     d) Use a ruler to estimate the growth rate of f along the direction of steepest ascent, by dividing the change in f by the distance between the contour lines. Compare the result with the magnitude of the vector $\vec{v}$ from part b, and comment on what you observe.

e) Based on your observations in this problem, explain as clearly as you can what you think is the geometric meaning of the vector $A\vec{\imath} + B\vec{\jmath}$ for a general linear function $f(x, y) = Ax + By + C$.

12.16. The plane α shown below is normal to the xy-plane, and the plane β is normal to the yz-plane.

    a) Is α the graph of a function $z = f(x, y)$? If yes, find the function; if no, explain.

    b) Repeat part a, but for the plane β.

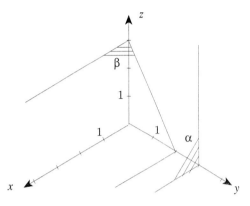

12.17. Some values of a function $f(x, y)$ are tabulated below. Decide whether the function could be linear, and if so, find a formula for $f(x, y)$ that is consistent with the tabulated values.

$x \setminus y$	2.0	2.5	3.0	3.5	4.0
1.0	5.0	6.5	8.0	9.5	11.0
1.2	4.0	5.5	7.0	8.5	10.0
1.4	3.0	4.5	6.0	7.5	9.0

12.18. Contour plots for two functions are shown in Figure 12.5. For each plot, decide whether the function could be linear, and if so, find a formula for the function.

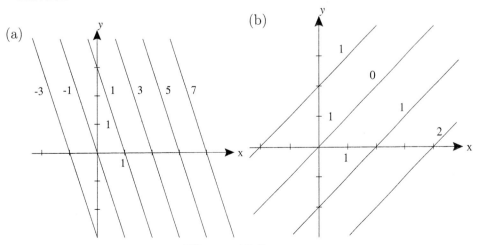

Figure 12.5

Nonlinear functions

12.19. Figure 12.6 shows some level curves of a function $z = f(x, y)$. To answer the questions, use a ruler and a *sharpened* pencil, and work as well as you can.

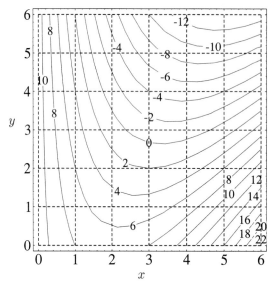

Figure 12.6

a) Find $f(5, 2)$.

b) Find $f(2, 4)$.

c) Find two solutions (x, y) of the equation $f(x, y) = -4$.

12.20. The graph of $z = f(x, y)$ is shown below. It sits over the rectangle $0 \le x \le 5$ and $0 \le y \le 4$ in the x-y plane.

a) Sketch the level curve through each of the points P, Q, and R. A high degree of precision is not required. Do the best you can working from the picture. *Hint:* Imagine slicing the surface by horizontal planes through each of the marked points.

b) Sketch the intersections of the graph with the planes through the the point P that are parallel to the x-z plane and the y-z plane.

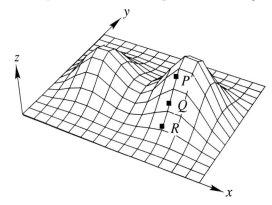

12.21. A function $f(x, y)$ is called *radial* if it can be expressed in the form $f(x, y) = g(\sqrt{x^2 + y^2})$, where g is a function of one variable.

 a) You're given the following information about $f(x, y)$:

 - f is radial;
 - $f(3, 0) = 1$, $f(0, 4) = -1$, and $f(3, 4) = 2$.

 Find $f(0, -3)$, $f(4, 0)$, and $f(5, 0)$.

 b) With $f(x, y)$ as in part a, sketch and label as much as you can of the level curves $f = 1$, $f = 4$, and $f = 5$. (You do not have enough information for a complete sketch.)

 c) Complete the following sentence, which gives a geometric description of radial functions: $f(x, y)$ is radial if and only if the value of f at the point (x, y) is determined by the distance from _____ to _____.

12.22. A long rod initially (at time $t = 0$) has a hot spot at the origin, but as time marches on, the heat initially concentrated at the origin spreads out along the entire length of the rod. The temperature at time t and position x is given by

$$u(t, x) = \frac{1}{\sqrt{t}} e^{-x^2/t}.$$

 a) Make a sketch of temperature as a function of time at the origin $x = 0$. Repeat for positions $x = 1$ and $x = 2$.

 b) What is the initial temperature (at time $t = 0$) at positions $x = 1$ and $x = 2$? Would your answer be different for any other fixed value x?

 c) At the positions $x = 0$, $x = 1$, and $x = 2$, what happens to the temperature as $t \to \infty$? Would your answer be different for any other position x?

 d) What is the peak temperature at each of the positions $x = 0$, $x = 1$, and $x = 2$?

 e) Make a sketch of temperature as a function position at time $t = 1$. Repeat for times $t = 2$ and $t = 3$.

 f) Find the maximum temperature at time t.

12.23. The function $u(t, x)$ is given by $u(t, x) = f(x - ct)$, where c is a positive constant, and $f(x)$ is the function of one variable graphed below.

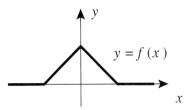

 a) Sketch the graphs of the one variable functions $f_0(x) = u(0, x)$, $f_1(x) = u(1, x)$, and $f_2(x) = u(2, x)$ on the same xy-system.

 b) Now suppose t represents time, and imagine a movie that shows the graph of $f_t(x) = u(t, x)$ at time t. Describe in one or more sentences what the movie shows.

12.24. Below are shown some level curves of a function $z = f(x, y)$.

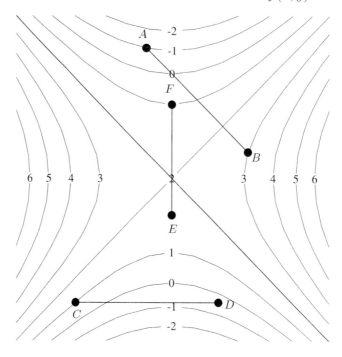

We consider three planes α, β, and γ, as follows. Each is normal to the xy-plane; α passes through the line segment AB, β through CD, and γ through EF. α intersects the graph $z = f(x, y)$ in a curve. Into the system below left draw the part of this curve from A to B as seen in the plane α. Then use the other two systems to do the same for the intersections of β and γ with the surface $z = f(x, y)$.

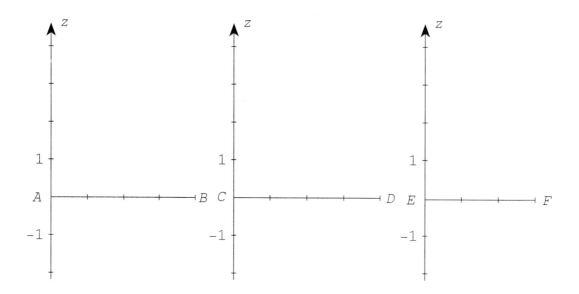

13   Functions of Three Variables

In many ways, functions of three variables work like functions of two variables.

Example 1. The triangle ABC shown below has area S. S depends on the length a, b of the sides and the angle θ, with θ ranging from 0 to π.

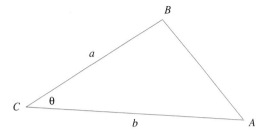

S is a function of the three independent variables a, b, and θ. We write $S = t(a, b, \theta)$ for this function.   ■

Problem _____

13.1. In Example 1, what happens to S when we increase or decrease a, b, θ? Will S increase or decrease? Will it stay the same? Is it impossible to say anything? In each of the scenarios below, decide what happens. The possible answers are "increases," "decreases," "does not change," and "cannot tell."

   a) a decreases, with b and θ held fixed.

   b) θ increases, with a and b held fixed.

   c) a and θ increase, and b decreases.

_____

Example 2. The function $f(x, y, z) = 4x + y + z - 4$ is a *linear function* of the three variables x, y, z. Another example of a linear function is $m(a, b, c) = \frac{a+b+c}{3} = \frac{1}{3}a + \frac{1}{3}b + \frac{1}{3}c$. It gives the average (mean) of the three numbers a, b, c.   ■

Example 3. Let $h(p, q, r) = \sqrt{p^2 + q^2 + r^2}$. Here the independent variables are p, q, r.   ■

Example 4. Problem 12.1 deals with amortizing a debt of \$10,000. Now we want to amortize a debt of A dollars. Then the monthly payment P is a function $P = f(A, r, N)$ of the three variables A (debt), r (interest rate expressed as a decimal), and N (number of payments). The formula for $P = f(A, r, N)$ is given by

$$P = \frac{A\frac{r}{12}}{1 - (1 + \frac{r}{12})^{-N}} \ .$$

■

Representations of a function of three variables

We can represent a function $z = g(x, y)$ of two variables as a surface in xyz-space. *Can we do the same with a function $w = f(x, y, z)$ of three variables?* We cannot because we would have to operate in $xyzw$-space of four dimensions. In four dimensions, we can work theoretically but not visually. If we try we run into trouble.

However, the method of level curves can be adapted to functions of three variables, as follows.

The Method of Level Surfaces

Pick a scalar c. In xyz-space, draw all points (x, y, z) for which the function takes the value c. That is, draw all points (x, y, z) such that $f(x, y, z) = c$. These points usually make up a surface in space which is called the *level surface* of level c of the function $f(x, y, z)$.

Problems

13.2. $f(x, y, z) = 4x + y + z - 4$ is the function of Example 2 above.

  a) Write down the equation of the level surfaces of $f(x, y, z)$ of levels $c = 0$ and $c = 2$. What is the shape of these level surfaces?

  b) Draw the level surfaces of a) into and xyz-system. Mark them $c = 0$ and $c = 2$.

13.3. $f(x, y, z) = \sqrt{x^2 + y^2 + z^2}$ is the function of Example 3. Describe in words the level surfaces of $f(x, y, z)$ of level $c = 0$, $c = -4$, and $c = 4$.

Linear functions of three variables

Consult the corresponding part of Section 12 when you work Problem 13.4 below.

Problem

13.4. Answer the following questions in complete sentences:

  a) What is a linear function of three variables?

  b) What are the level surfaces of a linear function of three variables?

  c) What is the graph of a linear function of three variables? Describe its shape in words.

Functions of three variables defined by composition

If we plug a function $q(x, y, z)$ of *three* variables into a function $u(t)$ of *one* variable, we get a new function $u(q(x, y, z))$ of *three* variables.

Example 5. With $u(t) = e^t$ and $q(x, y, z) = 2x - 7y - z + 1$, we define a new function $f(x, y, z)$ by

$$f(x, y, z) = u(q(x, y, z)) = u(2x - 7y - z + 1) = e^{2x - 7y - z + 1}.$$

Example 6. With $u(t) = \sqrt{t}$, $q(x, y, z) = x^2 + y^2 + z^2$, we define a new function $g(x, y, z)$ by

$$g(x, y, z) = u(q(x, y, z)) = u(x^2 + y^2 + z^2) = \sqrt{x^2 + y^2 + z^2}.$$

Example 7. With $w(t)$ the function of one variable given by the graph in Figure 12.4 on page 91, and $q(x,y,z) = x^2 + xz + yz^3$, we define a new function $h(x,y,z)$ by

$$h(x,y,z) = w(q(x,y,z)) = w(x^2 + xz + yz^3).$$

∎

Problem _____

13.5. $f(x,y,z)$, $g(x,y,z)$, $h(x,y,z)$ are the functions of Examples 5, 6, and 7.

  a) Find a point $P(x,y,z)$ in space on which $f(x,y,z)$ takes the value 2.

  b) Describe in words and formulas the level surface of level 2 of $f(x,y,z)$.

  c) Find a point $Q(x,y,z)$ which lies on the level surface of level 3 of $g(x,y,z)$.

  d) Write down the equation of the level surface of level 3 of $h(x,y,z)$.

Graphs and level objects

A function $g(x,y)$ of two variables has level curves in the xy-plane. A function $h(x,y,z)$ of three variables has level surfaces in xyz-space. Thus, functions of two and three variables come with built-in "level objects." The table below displays the graphs and level objects of functions of two and three variables.

Problem _____

13.6. Complete the table by filling in the graphs and level objects of a function $f(x)$ of one variable.

	Graph	Level Objects
Function $f(x)$ of one variable		
Function $g(x,y)$ of two variables	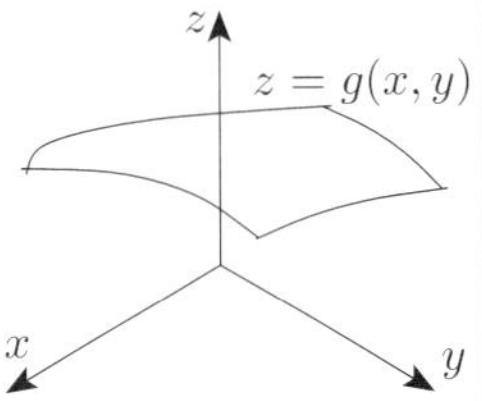 Surface in xyz-space	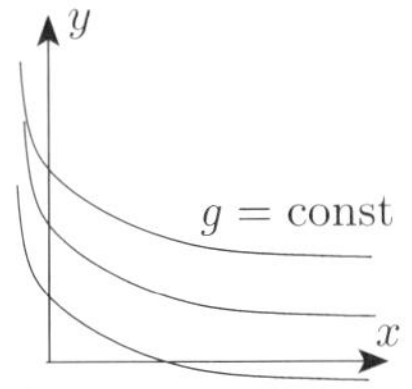 Curves in xy-plane
Function $h(x,y,z)$ of three variables	Not available	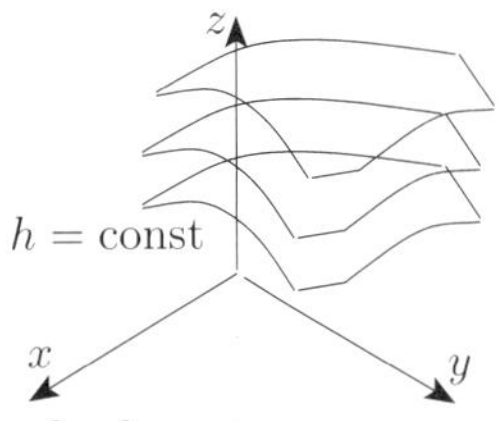 Surfaces in xyz-space

Additional Problems

Linear functions

13.7. Sketch and label the level surfaces $f = 0$, $f = 1$, and $f = 3$ for the linear function $f(x, y, z) = 2x + y + z - 1$ by drawing their intersections with the coordinate planes.

13.8. A linear function $f(x, y, z)$ satisfies the following conditions:

   (i) $f(1, 2, 3) = 4$.

   (ii) With y and z held fixed, f increases by 2 units for each 1 unit increase in x.

   (iii) With x and z held fixed, f decreases by 1 unit for each 1 unit increase in y.

   (iv) With x and y held fixed, f increases by 3 units for each one unit increase in z.

   a) Find a formula for $f(x, y, z)$.

   b) Find a normal vector for the level surfaces of the function $f(x, y, z)$. (*Note:* The level surfaces of $f(x, y, z)$ form a family of parallel planes.)

   c) The level surface $f(x, y, z) = 0$ is also the graph of a function $g(x, y)$ of two variables. Find a formula for the function $g(x, y)$.

Nonlinear functions

13.9. The function

$$\varphi(x, y, z) = \frac{1}{\sqrt{x^2 + y^2 + z^2}}$$

could represent the electrostatic potential associated with a point charge situated at the origin (or the gravitational potential of a point mass).

   a) Describe the level surfaces $\varphi = 1/2$, $\varphi = 1$, and $\varphi = 2$.

   b) Let $\vec{r} = x\vec{i} + y\vec{j} + z\vec{k}$ be the position vector of the point $P(x, y, z)$. Express $\varphi(x, y, z)$ in terms of $\vec{r}$, without explicitly mentioning x, y, or z.

   c) Give a geometric description of $\varphi(x, y, z)$ using words only. No formulas allowed!

   d) Find a formula for the electrostatic potential of the same charge situated at the point $(1, 2, 3)$. *Hint:* Use the physical description you gave in part c.

13.10. A *radial* function $f(x, y, z)$ is one that has the form

$$f(x, y, z) = g\left(\sqrt{x^2 + y^2 + z^2}\right)$$

where g is a function of *one* variable.

   a) Verify that the function $\varphi(x, y, z)$ of Problem 13.9 is radial. What is the function g in this case?

b) If $g(t)$ is a decreasing function of t, what can you say about the level surfaces $f(x, y, z) = c$? What about if $g(t)$ is increasing?

c) If g is the function shown below, describe the level surfaces $f = 0$, $f = 1$, and $f = 2$.

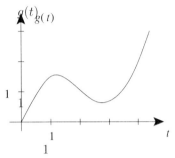

13.11. a) The function $t(x, y, z)$ of three variables x, y, z is defined by $t(x, y, z) = \sqrt{x^2 + z^2}$. Describe the level surface of level 2. *Hint:* $\sqrt{x^2 + z^2} = 2$ means $x^2 + z^2 = 4$. Question 1: Where are all points in the xz-plane that satisfy $x^2 + z^2 = 4$? Question 2: Are there points (x, y, z) outside the xz-plane that satisfy $x^2 + z^2 = 4$? This is similar to the story of a plane such as $x + z = 5$. Work Questions 1 and 2 for this plane.

b) The function $v(x, y, z)$ of three variables x, y, z is defined by $v(x, y, z) = z^2$. Draw the level surface of level 9 into and xyz-system.

13.12. Charge is spread out uniformly along a long straight wire. The resulting electrostatic potential $\varphi(P)$ at a point P in space is the natural logarithm of the distance from P to the wire, which we think of as infinitely long.

a) Describe the level surfaces of the function φ. *Note:* Since we haven't told you how the wire is situated with respect to any xyz-coordinate system, your answer must use geometric language, without reference to coordinates.

b) If the wire runs along the x-axis, find $\varphi(0, 1, 0)$, $\varphi(0, 0, 1)$, $\varphi(1, 1, 0)$, $\varphi(0, 1, 1)$, and $\varphi(1, 1, 1)$.

c) Find a formula for $\varphi(x, y, z)$ if the wire runs along the x-axis.

d) Find a formula for $\varphi(x, y, z)$ if the wire runs along the line defined by the equations $y = 1$ and $z = 2$.

13.13. This problem is about the domain and range of functions of three variables.

a) $w = f(x, y, z)$ is a function of three variables. Write down a definition for the domain and the range of this function.

b) The range of a function $w = f(x, y, z)$ is a collection of (choose one):
points in xyz-space   *points in the xy-plane*   *scalars*

c) The domain of a function $w = f(x, y, z)$ is a collection of (choose one):
points in xyz-space   *points in the xy-plane*   *scalars*

d) Describe the domain and range of the function

$$f(x, y, z) = e^{\sqrt{x^2 + y^2 + z^2 - 4}}.$$

14   Limits and Continuity

This section is about a property of functions which some functions have and others do not have. The property is called *continuity*. Speaking very roughly, a function is continuous if small changes of the independent variable result in only small changes of the function. The question *"Is this function continuous or not?"* arises often in the physical sciences and engineering, as well as in mathematics.

Many functions used in the sciences and engineering are continuous but by no means all of them. For an example of a function that is not continuous, consider an electrical device that we switch on. At this moment, the current starts flowing instantaneously. That is, there is a sudden jump of current vs. time. We describe this by saying "at this moment, the function current vs. time is not continuous." For another example, consider a pot of boiling water. Because the liquid water has a higher density than the water vapor in the bubbles, the density changes abruptly along the surfaces of the bubbles. For this reason, the density, as a function of position, is not continuous along the bubble surfaces.

To deal with continuity we need the concept of *limit*. It is one of the cornerstones of calculus.

There are three subsections: **Examples**, **Limits**, and **Continuity**. In each, we first consider functions of one variable, and then we work through functions of two (or more) variables.

Examples

Example 1.   Switching power off and on.   An electrical device such as a light bulb requires a certain amount of electric current to operate. The current is measured in units called Ampères, abbreviated as amps.

We are working with a flashlight. It powers a 6 watt bulb with a 12 volt battery. When the battery is fully charged, a current of 0.5 amp runs through the bulb. Experience tells us that if we leave the light on for some time, the current provided by the battery decreases so that the light will dim. After the light is on for 6 hours, the battery is exhausted. Now we ask:

> *How does the current that flows through the bulb vary over time? In particular, what happens when we switch the flashlight off and on?*

We study the flow of current by means of a graph in the tI-coordinate system shown below. On the vertical axis we record the current in amps.

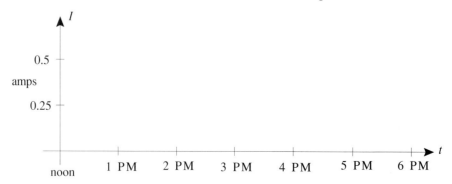

Problem

14.1. a) At noon the light is already on and we leave it on. Write $a(t)$ for the current flowing through the bulb. We know that $a(t)$ decreases and that $a(6) = 0$. But we do not know exactly *how* $a(t)$ decreases. In the Iw-system draw what you think is a reasonable guess of the graph $I = a(t)$.

b) We switch the flashlight off and on as follows: At noon it is already on. Exactly at 2 PM, we switch it off. Exactly at 3 PM we switch it on again. We write $s(t)$ for the current flowing through the bulb. Now use a pen of a different color to draw the "switched" graph $I = s(t)$.

Comparison between the "always-on" function $a(t)$ and the "switched" function $s(t)$

Always on: If you measure the current in the always-on situation at two times t and t', you will get different currents $a(t)$ and $a(t')$. If the times t and t' are close together, then the currents will be close together as well; if the times t and t' are *very* close together, so will be the currents. So a small change in time results in only a small change of current. Expressed in mathematical terms: Small changes in the independent variable t result in only small changes of the dependent variable $I = a(t)$. This pattern "small changes of input result in only small changes of output" is the hallmark of the property of functions called *continuity*. Our function $a(t)$ follows this pattern at all times t. Therefore, one says: The function $a(t)$ is continuous for all t. A function which is continuous for all values of the independent variable is called a continuous function. Thus, our always-on function $a(t)$ is a continuous function.

Switched: Now we look at the switched situation $I = s(t)$. At 2 PM, the flashlight is switched off. At this moment, the current shoots instantly from a nonzero value down to zero. Thus an extremely small change in time results in a comparatively large change in current. We say that the current $s(t)$ is *not continuous* or is *discontinuous* at $t = 2$, or that the function $s(t)$ has a *discontinuity* at $t = 2$. Another discontinuity occurs at $t = 3$ when we switch the flashlight on again. A function that has a discontinuity is called a *discontinuous* function or is said to be *not continuous*. Thus $s(t)$ is a discontinuous function. Note that in the above discussion we have not defined exactly what is meant by "continuity." We have only given an intuitive description. In Problem 14.2, we ask you to work in the same intuitive way.

Problem

14.2. We use the 12 volt battery of Example 1 to switch off and on two bulbs called A and B, connected in parallel. Each draws a current of 0.5 amp when the battery is fully charged. If both A and B are on they draw together 1 amp, and after both are left on for three hours, the battery is exhausted. We write $I = c(t)$ for the total amount of current flowing from the battery. Now we switch A and B off and on as follows: From noon to 1 PM, no bulb is on. At 1 PM, we switch on both A and B. At 2:30 PM, we switch off A but leave

B on. At 4 PM we switch on A again. At 5 PM, we switch off both bulbs and do not turn them on any more.

a) Draw a possible graph $I = c(t)$ of total power flow from noon to 6 PM.

b) For which t would you say that $c(t)$ is discontinuous? Write down a justification of your answer.

∎

Example 2. Polar coordinates and the principal polar angle. A point (x, y) in the xy-plane has polar coordinates (r, θ), and the (x, y)-coordinates are related to the polar coordinates by the formulas $x = r \cos \theta$, $y = r \sin \theta$. The angle θ is called the *polar angle*. Any point in the plane other than the origin has many different polar angles, and any two of them differ by a multiple of 2π. If we want to work with polar coordinates we have to choose one of the many polar angles. We choose the polar angle θ which satisfies $-\pi < \theta \leq \pi$. It is often called the *principal value* of the polar angle. Thus, for any point (x, y) other than the origin, we define $\theta(x, y)$ to be the principal polar angle of the point (x, y). By assigning to a point (x, y) the principal value $\theta(x, y)$ we define a function of two variables (x, y). We now ask: Is $\theta(x, y)$ a continuous function of (x, y)? That is: Is it true for all points that a small change of (x, y) will result in only a small change of $\theta(x, y)$? There are two cases to consider.

Case 1: *The point (x, y) is on the negative part of the x-axis, as illustrated below.* In this case, $\theta(x, y) = \pi$. We consider a nearby point (x', y') that is just slightly below the x-axis. Then $\theta(x', y')$ is very close to $-\pi$. Thus, $\theta(x', y')$ is *not* close to $\theta(x, y)$ even though (x', y') *is* close to (x, y). It means that $\theta(x, y)$ is discontinuous at points along the negative part of the x-axis. The principal polar angle of (x', y') equals $-\pi + \varepsilon$ where ε is a small positive number.

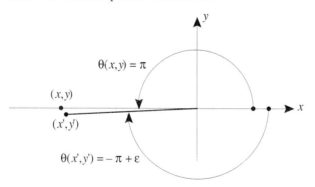

Case 2: *The point (x, y) is not on the negative part of the x-axis.* In this case, a small change in (x, y) will produce only a small change in $\theta(x, y)$. It means that $\theta(x, y)$ is continuous at all points which are *not* on the negative half of the x-axis.

∎

Problem _____

14.3. Let K be the region in the xy-plane defined by $x^2 + y^2 \leq 1$. Define a function $f(x, y)$ by letting $f(x, y) = 1$ if (x, y) is in the region K, and $f(x, y) = 0$ if (x, y) is *not* in the region K.

a) What is $f(0,0)$? What is the value of $f(x,y)$ when (x,y) is near $(0,0)$? Is $f(x,y)$ continuous at the point $(0,0)$?

b) What is $f(1,0)$? What are the values of $f(x,y)$ for points (x,y) which are close to $(1,0)$? Is $f(x,y)$ continuous at the point $(1,0)$?

c) Describe the set of points (x,y) in the plane at which $f(x,y)$ fails to be continuous.

Limits

Limits of functions of one variable

There are two parts to this subsection. In the first part we work through in detail an example of a limit which occurs in elementary calculus in connection with differentiation of trigonometric functions. In the second part we lay out the precise definition limit.

The expression $\frac{\sin h}{h}$ for values of h close to zero. In elementary calculus, the derivative of a function $f(x)$ of one variable at a point $x = x_0$ is obtained by examining the ratios

$$\frac{f(x_0 + h) - f(x_0)}{h}$$

for small, but nonzero values of h. We will examine this ratio for the function $f(x) = \sin x$ at the point $x_0 = 0$. Since $\sin 0 = 0$, this leads to the ratio

$$S(h) = \frac{\sin h}{h}. \tag{14.1}$$

The derivative $f'(0)$ can be approximated as closely as we wish by calculating the above ratio for sufficiently small, but nonzero, values of h. We now pose the following question:

How can we make sense of the exact *value of $f'(0)$?*

Notice that we cannot answer this question by simply putting $h = 0$ in (14.1), since this would lead to the indeterminate expression $\frac{0}{0}$. In the process of finding the answer, we will develop the precise definition of a limit of a function of one variable. There will be several steps. The punch line is in step 3.

Step 1: Here we use elementary trigonometry to get estimates for $\frac{\sin h}{h}$. Below you find part of a unit circle in an xy-system.

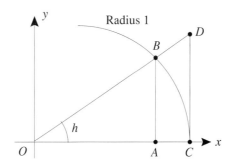

The area of the triangle OAB is $\frac{1}{2}\cos h \sin h$, that of the pie slice OCB is $\frac{h}{2}$, and that of the triangle OCD is $\frac{1}{2}\tan h$. The area of the pie slice lies between the areas of the triangles OAC and OD:

$$\frac{1}{2}\cos h \sin h < \frac{1}{2}h < \frac{1}{2}\tan h.$$

Multiply through by 2, take the reciprocals so that the inequality signs are switched, and multiply through by $\sin h$. You obtain

$$\cos h < \frac{\sin h}{h} < \frac{1}{\cos h}. \tag{14.2}$$

Notice that we assumed that $0 < h < \frac{\pi}{2}$ in obtaining this inequality, but since all three terms are even functions of h, (14.2) also holds for $-\frac{\pi}{2} < h < 0$. The idea now is to use the inequality (14.2) to obtain information about $\frac{\sin h}{h}$. We do this by looking at the left and right terms $\cos h$ and $\frac{1}{\cos h}$ in steps 2 and 3.

Step 2: Now look at the left and right sides of (14.2), and ask what happens as h gets close to 0. We expect that the left side will be close to $\cos 0$, which is 1, and that the right side will be close to $\frac{1}{\cos 0}$, which is also 1. Since $\frac{\sin h}{h}$ is squeezed between these two values, which are both close to 1, we expect that $\frac{\sin h}{h}$ will also be close to 1 when h is near 0. The question now is: *How close to 1 will $\frac{\sin h}{h}$ be when h is near 0?* For example, can we be certain that $\frac{\sin h}{h}$ differs from 1 by less than 0.01 when h is near 0? Let us check.

Since $\cos h$ decreases from 1 to 0 as h increases from 0 to $\frac{\pi}{2}$, there is exactly one solution to the equation $\cos h = .99$ in the interval $0 < h < \frac{\pi}{2}$. Call it δ_1. It then follows that $\cos h > 0.99$ for $0 \le h < \delta_1$, and since $\cos h$ is an even function of h, the same inequality holds for $-\delta_1 < h < 1$. Thus, we get

$$\cos h > 0.99 \qquad \text{for} \qquad -\delta_1 < h < \delta_1. \tag{14.3}$$

Now we look at the term on the right in (14.2). Since $\frac{1}{\cos h}$ increases from 1 to infinity as h increases from 0 to $\frac{\pi}{2}$, there is exactly one solution to the equation $\frac{1}{\cos h} = 1.01$ in the interval $1 < h < \frac{\pi}{2}$. Call it δ_2. It follows that

$$\cos h < 1.01 \qquad \text{for} \qquad -\delta_2 < h < \delta_2. \tag{14.4}$$

We now let δ be the minimum of the two positive numbers δ_1 and δ_2. Combining (14.2), (14.3), and (14.4), we get

$$0.99 < \cos h < \frac{\sin h}{h} < \frac{1}{\cos h} < 1.01$$

provided that h is in the interval $-\delta < h < \delta$, and $h \ne 0$. (We must exclude the value $h = 0$, since the middle term doesn't make sense when $h = 0$.) In other words, we are guaranteed that $\frac{\sin h}{h}$ will be between 0.99 and 1.01 as long as h is between $-\delta$ and δ.

Step 3: In step 2, we found a positive number δ with the property that $\frac{\sin h}{h}$ is guaranteed to be within 0.01 of 1, provided that h is in the interval $-\delta < h < \delta$. But we can find such a δ for *any* positive number ε and not just 0.01. Why is that so? In case ε is greater than 0.01, we take the δ we just found and it will work. For if $\frac{\sin h}{h}$ is within 0.01 of 1, it will be also within ε of 1. In case ε is less than 0.01 we do the equation solving with 0.99 and 1.01 replaced by $1 - \varepsilon$ and $1 + \varepsilon$, respectively. The result is: Given an ε we can find a δ such that for all h in the interval from $-\delta < h < \delta$ the value of $\frac{\sin h}{h}$ lies between $1 - \varepsilon$ and $1 + \varepsilon$. This is illustrated in the figure below. Given the ε, the graph of $w = \frac{\sin h}{h}$ is guaranteed to lie in the strip $ABCD$ for h between $-\delta$ and δ. Note that for $h = 0$ the graph is not defined because $S(0)$ is not defined.

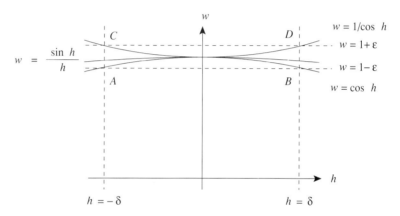

The important point is that we are able to find such a δ no matter what the given ε is. *That is, we have established not one fact, but infinitely many facts, viz. one for each choice of ε.* The property of $\frac{\sin h}{h}$ that for any given positive ε we can find such a δ is expressed in mathematical language as "the limit of $\frac{\sin h}{h}$ as h tends to zero equals 1." Here it is spelled out again:

> Given any positive number ε, no matter how small or how big, we can find a positive number δ so that $\frac{\sin h}{h}$ lies between $1 + \varepsilon$ and $1 - \varepsilon$ provided that h lies between $-\delta$ and $+\delta$, and $h =\neq 0$.

An equivalent formulation in terms of approximations reads as follows:

> For any desired accuracy ε, we can find a small interval I from $h = -\delta$ to $h = \delta$ so that a sampled value of $\frac{\sin h}{h}$ is *guaranteed* to approximate 1 with the specified accuracy ε provided that $h \neq 0$ is sampled from the interval I.

These limit statements are abbreviated as:

$$\lim_{h \to 0} \frac{\sin h}{h} = 1.$$

Already at this point we can see what the concept limit is all about. The key point is that the limiting value of 1 can be approximated to *any* degree of accuracy we please (as measured by ε) by choosing h to be sufficiently near 0 (as measured by δ).

14.4. Go back to step 2 above. Find a numerical value of δ. That is, find a positive number δ with the following property: A sampled value of $\frac{\sin h}{h}$ is guaranteed to approximate 1 with an accuracy $\varepsilon = 10^{-2}$ provided that $h \neq 0$ is sampled from the interval $-\delta < h < \delta$.

The limit of a function of one variable. We already formulated the limit statement for $S(h) = \frac{\sin h}{h}$. Now we do that for a generic function $f(x)$:

Definition. *The statement "as x tends to x_0, the limit of $f(x)$ equals L" is written as $\lim_{x \to x_0} f(x) = L$ and means the following:*

> *Given any positive number ε, no matter how small or how big, we can find a positive number δ so that $f(x)$ lies between $L+\varepsilon$ and $L-\varepsilon$ provided that x lies between $x_0-\delta$ and $x_0+\delta$ and $x \neq x_0$. That is: If $x_0-\delta < x < x_0+\delta$ and $x \neq x_0$, then we must have $L - \varepsilon < f(x) < L + \varepsilon$.*

An equivalent formulation in terms of approximations reads as follows:

> *For any desired accuracy ε, we can find a small interval I centered at $x = x_0$ so that a sampled value of $f(x)$ is guaranteed to approximate L with the specified accuracy ε provided that x is sampled from the interval I except possibly at x_0.*

Note: Given a function $f(x)$ and a value x_0 we ask: What is the limit of $f(x)$ as x tends to x_0? The possible answers are "the limit equals $L = \ldots$" or "the limit does not exist." In detail:

- To show that the limit L exists we have to do this: For any given number $\varepsilon > 0$ we have to *find* another number $\delta > 0$ so that $f(x)$ lies between $L - \varepsilon$ and $L + \varepsilon$ for all $x \neq x_0$ between $x_0 - \delta$ and $x_0 + \delta$.

- To show that the limit at x_0 does *not* exist, we would have to show the following. Given any real number L there is a positive ε such that for any positive δ the interval $x_0 - \delta < x_0 + \delta$ contains at least one x other than x_0 itself such that $y = f(x)$ lies outside the interval $L - \varepsilon < y < L + \varepsilon$.

14.5. The function $f(x)$ is given by $f(x) = 6x+5$. We want to approximate $f(\sqrt{7})$ with accuracy $d = 4 \cdot 10^{-6}$. Find a value of δ so that $f(x)$ is guaranteed to approximate $f(\sqrt{7})$ provided x lies in the interval of length 2δ centered at $x_0 = \sqrt{7}$.

14.6. Does $\lim_{x \to 0} \frac{1}{x}$ exist? If yes, find the limit. If no, explain. *Hint:* Go back to the note just before Problem 14.5.

Limits of functions of two variables

The idea of the limit of a function of two variables is exactly the same as for a function of one variable. Again, we want a precise way to say that a function value becomes close to some limiting value as the independent variables get close to some specified numbers. In the case of functions of two variables, "closeness" for the independent variables x and y is measured by distance in the xy-plane. Thus, the distance between (x, y) and (x_0, y_0) is given by the usual distance formula:

$$\text{distance} = ((x - x_0)^2 + (y - y_0)^2)^{1/2}.$$

The set of points within a specified distance, say δ, of a given point (x_0, y_0) is a disk with center (x_0, y_0) and radius δ. It is defined by the inequality

$$(x - x_0)^2 + (y - y_0)^2 < \delta^2.$$

Definition. *We say "as (x, y) tends to (x_0, y_0), the limit of $f(x, y)$ equals L," and write $\lim_{(x,y) \to (x_0,y_0)} f(x, y) = L$ to mean the following:*

> *Given any positive number ε, no matter how small or how big, we can find a positive number δ so that $f(x, y)$ lies between $L + \varepsilon$ and $L - \varepsilon$ provided that (x, y) lies in the disk D_δ defined by $(x - x_0)^2 + (y - y_0)^2 \leq \delta^2$ and $(x, y) \neq (x_0, y_0)$. That is: If (x, y) lies in D_δ and $(x, y) \neq (x_0, y_0)$, then we must have $L - \varepsilon < f(x, y) < L + \varepsilon$.*

An equivalent formulation in terms of approximations reads as follows:

> *For any desired accuracy ε, we can find a small disk D centered at (x_0, y_0) so that a sampled value of $f(x, y)$ is* guaranteed *to approximate L with the specified accuracy ε provided that (x, y) is sampled from the disk D except possibly at (x_0, y_0).*

Note: Given a function $f(x, y)$ and a point (x_0, y_0), the limit of $f(x, y)$ as (x, y) tends to (x_0, y_0) may or may not exist.

Problem _____

14.7. Here we use the function $f(x, y)$ of Problem 14.3. It is defined as follows: Let K be the region in the xy-plane defined by $x^2 + y^2 \leq 1$. Define a function $f(x, y)$ by letting $f(x, y) = 1$ if (x, y) is in the region K, and $f(x, y) = 0$ if (x, y) is *not* in the region K. Evaluate each of the limits below. In each case your answer should be either a number or the assertion "The limit does not exist."

    a) $\lim_{(x,y) \to (0,0)} f(x, y)$

    b) $\lim_{(x,y) \to (1,1)} f(x, y)$

    c) $\lim_{(x,y) \to (1,0)} f(x, y)$

Continuity

Continuity of functions of one variable

We first go back to the current of the flashlight discussed at the beginning. In Figure 14.1 on the following page you find an enlarged copy of part of the graph $I = s(t)$ (light on at noon and then switched off at 2 PM exactly). We have picked a time b between noon and 2 PM. *What is the limit of $s(t)$ as t tends toward b?* As t tends toward b, there seems to be nothing else for $s(t)$ to tend to but $s(b)$. So

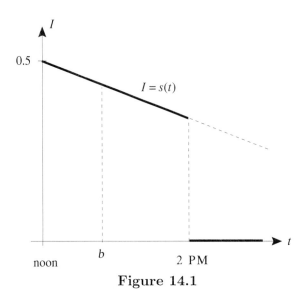

Figure 14.1

we guess that the limit is $s(b)$. To justify this we have to do the following: For any $\varepsilon > 0$, we have to find a $\delta > 0$ such that for all t between $b - \delta$ and $b + \delta$ the values $s(t)$ lie between $s(b) - \varepsilon$ and $s(b) + \varepsilon$. You find $\delta = 12\varepsilon$.[1] Thus, we have

$$\lim_{t \to b} s(t) = s(b).$$

How does this situation relate to the discussion of limits in the earlier subsection? There we talked about the limit L as $f(x)$ as x tends to x_0. Here the L is $f(x_0)$, the value of $f(x)$ at the x_0 to which x tends. We continue with $s(t)$ and write out what $\lim_{t \to b} s(t) = s(b)$ means:

> For any desired accuracy ε, we can find a small interval J centered at $t = b$ so that a sampled function value $s(t)$ is *guaranteed* to approximate $s(b)$ – the value of $s(t)$ at b – with the desired accuracy ε provided that x is sampled from the interval J. We can condense this statement as follows: $s(b)$, *the value of $s(t)$ at $t = b$, can be approximated by values $s(t)$ as closely as we want to by making sure that t is sufficiently near b.*

A function $f(x)$ that at $x = x_0$ has the property that $s(t)$ has at $t = b$ is said to be *continuous* at $x = x_0$. The definition of continuity for functions of one variable reads as follows.

Definition of continuity of $f(x)$ at a point $x = x_0$. *A function $f(x)$ is said to be continuous at $x = x_0$ if it satisfies these two conditions at the same time:*

1. *$f(x_0)$ is defined.*

2. *$\lim_{x \to x_0} f(x) = f(x_0)$. In words: For any desired accuracy ε, we can find an interval I centered at $x = x_0$ for which the following is true: A function value $f(x)$ is* guaranteed *to approximate $f(x_0)$ with the desired accuracy ε*

---

[1] $s(t) = -\frac{t}{12} + \frac{1}{2}$; as in Problem 14.5, $\frac{\varepsilon}{\delta}$ is the slope of the graph. Note that we always take $\delta > 0$.

provided that x is sampled from the interval I. Or more briefly: $f(x_0)$ can be approximated by values $f(x)$ as closely as we want to by taking x sufficiently near x_0.

Note:

1. At the beginning of this section, we described continuity roughly by saying that small changes in input result in small changes of output. This rough description is too vague to work with precisely, because it does not specify the meaning of "small." The precise definition given above addresses this deficiency by saying that the change in output is guaranteed to be *as small as you please* provided that the change in input is sufficiently small.

2. If one of the two conditions 1, 2 of continuity fails to hold for $f(x)$ at x_0, then $f(x)$ is *discontinuous* (or *not continuous*) at x_0.

Problems

14.8. We go back to the flashlight function $I = s(t)$ and see what happens at 2 PM when we switch off the flashlight. To orient yourself use the graph of $s(t)$ as shown above. We have to decide what $s(2)$ is. We say "there is current up to and including $t = 2$." Therefore, we set $s(2) = 4$. For $t > 2$ we have $s(t) = 0$. Is $s(t)$ continuous at $t = 2$? If yes, justify your answer with a limit argument. If no, explain why that is the case. We outline the reasoning in parts a to d, and you fill in the details.

    a) What do you think the answer will be? What makes you think so?

    b) Look up the conditions 1 and 2 in the definition of continuity. Is 1 satisfied?

    c) Now we come to condition 2 and check the limit. *Hint:* Take $\varepsilon = 0.1$. Can you find a δ such that $s(t)$ is within 0.1 of $s(2)$ for all t between $2 - \delta$ and $2 + \delta$? If yes, find a δ. If no, explain.

    d) Look at your work in part c: Did you settle the question whether $s(t)$ is continuous at $t = 2$?

14.9. Your friend says: "The reasoning you just gave that $s(t)$ is not continuous at $t = 2$ does not convince me. Can we not take any accuracy ε we want? Sure we can. So I take $\varepsilon = 7$. As δ I take $\frac{1}{2}$. Then for all t between $2 - \frac{1}{2}$ and $2 + \frac{1}{2}$, the value of $s(t)$ lies between $2 - \varepsilon = -5$ and $2 + \varepsilon = 9$ for sure. Does that not tell us that $s(t)$ *is* continuous at $t = 2$?" What do you say? Write down a response.

Continuity of functions of two variables

Continuity for functions $f(x)$ is defined by means of limits of functions $f(x)$. In exactly the same way, continuity for functions $f(x, y)$ is defined by limits of functions $f(x, y)$. The definition reads as follows.

Definition of continuity at a point. *The function $f(x, y)$ is said to be continuous at $(x, y) = (x_0, y_0)$ if it satisfies these two conditions at the same time:*

    1. $f(x_0, y_0)$ is defined.

2. $\lim_{(x,y)\to(x_0,y_0)} f(x,y) = f(x_0,y_0)$. *In words: For any desired accuracy ε, we can find a disk of radius δ centered at (x_0,y_0) for which the following is true: A sampled function value $f(x,y)$ is* guaranteed *to approximate $f(x_0,y_0)$ with the desired accuracy ε provided that (x,y) is sampled from the disk. Or more briefly: $f(x_0,y_0)$ can be approximated by values $f(x,y)$ as closely as we want to by choosing (x,y) sufficiently near (x_0,y_0).*

Note:

1. As with functions of one variable, the definition of continuity of a function of two variables gives us a precise, quantitative way of saying that "small" changes in input produce only small changes in output.

2. If one of the two conditions 1, 2 of continuity fails to hold for $f(x,y)$ at (x_0,y_0), then $f(x,y)$ is *discontinuous* (or *not continuous*) at (x_0,y_0).

Problems

14.10. We go back to the function $f(x,y)$ of Problems 14.3 and 14.7. It is defined as follows: K is the region in the xy-plane defined by $x^2 + y^2 \leq 1$. Then $f(x,y) = 1$ if (x,y) is in the region K, and $f(x,y) = 0$ if (x,y) is *not* in the region K. Look at the points $(x,y) = (0,0)$, $(1,1)$, $(1,0)$. Is $f(x,y)$ continuous or discontinuous at these points?

14.11. Does $\varphi(x,y) = \ln(x^2 + y^2)$ have points of discontinuity? If yes, find one. If no, explain.

Limits and continuity for functions of three and more variables

The ideas underlying limits and continuity for functions of more than two variables are the same as for one and two variables, with adjustments similar to those required to pass from one variable to two. For example, as you pass from two to three variables, disks $(x - x_0)^2 + (y - y_0)^2 \leq \delta^2$ in the xy-plane have to replaced by solid balls $(x - x_0)^2 + (y - y_0)^2 + (z - z_0)^2 \leq \delta^2$ in xyz-space. For four variables, the solid balls have to be replaced by "solids" $(x - x_0)^2 + (y - y_0)^2 + (z - z_0)^2 + (u - u_0)^2 \leq \delta^2$ in four-dimensional $xyzu$-space.[2]

How to recognize a continuous function

Some functions are given by formulas, others by physical description. In either case, how do we recognize whether a given function is continuous or where it has discontinuities?

Functions defined by formulas. The idea here is to begin with a few basic examples of continuous functions, and use them as building blocks for more complicated ones.

---

[2] For a development of limits and continuity for functions of more than two variables see W. Rudin, *Principles of Mathematical Analysis*, Chapter 4.

Constant functions. The constant function of one variable $f(x) = b$ is continuous at all points $x = x_0$. Why? Given any ε, we have $f(x) = b$ for all x in any interval $\delta - \varepsilon < x_0 < \delta + \varepsilon$. That is, we can take any δ for our the interval $x_0 - \delta < x < x_0 + \delta$. The constant function of two variables $f(x, y) = b$ is also continuous at all points $(x, y) = (x_0, y_0)$. Why? Given any ε, we have $f(x, y) = b$ in any disk $(x - x_0)^2 + (y - y_0)^2 \leq \delta^2$. That is, we can take any δ as radius of our disk.

Identity function, coordinate functions. The function $f(x)$ of one variable defined by $f(x) = x$ is the *identity function*. It is continuous at all points x_0. Why? Given any ε we can use the same ε for the interval: If x lies in the interval $x_0 - \varepsilon < x < x_0 + \varepsilon$ then $f(x) = x$ lies in the same interval.

Moving to functions of *two* variables, the analogues of the identity function of one variable are the two *coordinate functions*

$$c_1(x, y) = x$$
$$c_2(x, y) = y$$

The same idea applies for more than two variables. For example in xyz-space we have three coordinate functions:

$$c_1(x, y, z) = x$$
$$c_2(x, y, z) = y$$
$$c_3(x, y, z) = z$$

We now ask: *Are the coordinate functions continuous?* This is the question of the following Problem 14.12.

Problem
14.12.　a)　Verify that in the xy-plane the coordinate function $c_1(x, y)$ is continuous, by checking the two conditions for continuity, as follows. Clearly $c_1(x, y)$ is defined for all points (x, y). So you have to check only the "approximation" condition at the point (x_0, y_0). It means you have to do this: For any given $\varepsilon > 0$ you have to find a $\delta > 0$ so that $c_1(x, y)$ lies in the interval $x_0 - \varepsilon < x < x_0 + \varepsilon$ for all (x, y) in the disk $(x - x_0)^2 + (y - y_0)^2 \leq \delta^2$. *Hint:* Draw the xy-plane which shows the following: The point (x_0, y_0), the point $x_0 = c_1(x_0, y_0)$ on the x-axis, and the given interval of length 2ε centered at x_0. Look at the drawing and decide how you have to choose δ.

　　　　b)　In the same way, verify that $c_2(x, y)$ is also continuous. *Hint:* Use the drawing you made for a) and add to it.

　　　　c)　This is an optional question about the continuity of functions of three variables. Take the coordinate function $c_1(x, y, z) = x$ defined for all points in xyz-space. To check the approximation condition do the following. First, formulate what you have to do. In doing that, remember that instead of dealing with disks in the xy-plane you have to deal with solid balls in xyz-space (a solid ball of radius b and center (x_0, y_0, z_0) is defined by $(x - x_0)^2 + (y - y_0)^2 + (z - z_0)^2 \leq b^2$. Now go back to a) and see what you did there with $c_1(x, y)$: For a given $\varepsilon > 0$ you found the right radius δ of a *disk* of center (x_0, y_0). Translate this into getting the right radius of a *ball* of center (x_0, y_0, z_0) for the given ε.

Building more complex functions. There are a few basic principles, which can be used to build up more complicated continuous functions from the basic building blocks discussed above. We list them here without justification.

1. Sums and differences of continuous functions are continuous. Thus, for example, the function $f(x, y) = x + y - 3$ is continuous at each point of the xy-plane, because it is the sum of the two coordinate functions and the and the constant function -3.

2. Products of continuous functions are continuous. Thus, for example, the function $f(x, y) = 2xy$ is continuous at each point of the xy-plane, because it is the product of a constant function and two coordinate functions. Similarly, $g(x, y) = 5y^4$ is continuous.

3. The quotient of two continuous functions is continuous at each point where the denominator is not equal to zero. Thus, for example, the function $f(x, y) = x/y$ is continuous except at points on the x-axis, where the function is undefined.

4. The functions x^p, $\ln x$, e^x, the trigonometric, and inverse trigonometric functions are continuous at each point where they are defined.

5. Compositions of continuous functions are continuous. Thus, for example $f(x, y) = \ln(x^2 + y^2)$ is continuous at each point of the xy-plane except $(0, 0)$, where it is undefined.

The upshot of all of this is that any function for which you can write down a formula will be continuous at each point where the formula makes sense.

Problems _____

14.13. The function $f(x)$ is defined by $f(x) = \ln(1 + \sin x)$. At which points x is $f(x)$ discontinuous?

14.14. We abbreviate $\sqrt{x^2 + y^2}$ by r. Then we define functions $f_n(x, y) = r^n$ where n is any negative or positive integer $\ldots, -3, -2, -1, 0, 1, 2, \ldots$ Which of these functions $f_n(x, y)$ are continuous (i.e., continuous at all points of the xy-plane), which have discontinuities and where do these occur?

Functions described physically

Sometimes you will have to deal with functions given by physical descriptions rather than by formulas. For example, $u(t, x, y, z)$ be denote the temperature, t minutes after turning on the furnace, at the point in a room with coordinates (x, y, z). It is unlikely that such a function can be described by a formula.

For functions given by physical description, points of discontinuity may be part of the description itself. Look for:

1. Points where the function "jumps" from one value to another as you cross a point, a line, or a surface;

2. Points where the function "blows up," i.e., approaches infinity.

Problems

14.15. The function $u(x, y, z)$ gives the density of matter in a region of xyz-space which contains a block of ice in water. Describe the points of discontinuity of u. *Hint:* Ice is less dense than water (and that is why it floats, by the way).

14.16. The law of gravitation tells us: A point mass situated at point P_0 in space attracts other mass points P in space with a certain force. The magnitude of this force is inversely proportional to the square of the distance of P from P_0. The magnitude of the force is a function $f(P)$ of points P in space. Does $f(P)$ have discontinuities? If yes, where?

Additional Problems

Note: Problems on functions of one and of two variables are intermixed.

14.17. The function $f(x)$ of one variable is defined by $f(x) = \frac{1}{x}$.
   a) Find all points x_0 where $f(x)$ is discontinuous.
   b) Find $\lim_{x \to 0} f(x)$.
   c) Find $\lim_{x \to 10^{-67}} f(x)$.
   d) Your friend says: "I do not get it. In part c you get a number different from zero, and the number 10^{-67} is pretty close to zero, and I could have taken one which is even closer to zero. Does this not mean that the limit of $f(x)$ for x tending to zero can be approximated as closely as I want to so that $f(x)$ has a limit as x tends to zero?" Write down a response.

14.18. The function $g(x, y)$ of two variables is defined by $g(x, y) = \frac{1}{x} + y$.
   a) Can you find a point (x_0, y_0) so that $\lim_{(x,y) \to (x_0, y_0)} g(x, y) = 3.75$? If not, explain.
   b) Find three points (x, y) where $g(x, y)$ is discontinuous. If three such points cannot be found, explain.

14.19. The function $H(x)$ is defined as follows: $H(x) = 0$ if $x < 0$, $H(x) = 1$ if $x \geq 0$. It is called the Heaviside function. Consider the points $x = 1$, $x = 4.92 \cdot 10^{-4}$, $x = 0$, $x = -4.92 \cdot 10^{-4}$, $x = -1$. At these points, is $H(x)$ continuous? Justify your answer.

14.20. We consider the function $f(x, y) = \dfrac{xy}{\sqrt{x^2 + y^2}}$. The function is not defined at the origin $(0, 0)$. It still *may* have a limit L as $(x, y) \to (0, 0)$. By working out parts a to e, you will find out whether it has a limit and, if yes, what it is.
   a) We look at the values of $f(x, y)$ on the circle C_δ defined by $x^2 + y^2 = k^2$. We will consider different positive values of δ. Introduce polar coordinates (r, θ) and express the values of $f(x, y)$ on the circle C_δ in terms of δ and θ.

b) Find the minimum and maximum values of $f(x, y)$ on the circle C_δ.

c) Now we let the radius δ decrease to $\frac{\delta}{2}$. What happens to the minimum and maximum values of $f(x, y)$ on the circles of decreasing radius?

d) Can you find a value of δ with the following property: The absolute value of $|f(x, y)|$ of all points in the interior of C_δ is less than 10^{-92}? If yes find such a δ. If no, explain.

e) Does $f(x, y)$ have a limit as $(x, y) \to (0, 0)$? If yes, write down the complete limit statement. If no, explain.

14.21. The function $g(x, y)$ is defined as follows: $g(x, y) = 0$ for $0 \le |x| < 1$, $g(x, y) = 1$ for $1 \le |x| < 2$, $g(x, y) = 2$ for $2 \le |x| < 3$, $g(x, y) = 3$ for $3 \le |x| < 4$, etc. Its graph $z = g(x, y)$ looks like a staircase. The treads are of depth 1, and they are parallel to the y-axis.

a) Find $g(24.7, 8)$, $g(-24, 8.5)$, $g(24, 367)$, $g(-23.99999, 0)$

b) Find three points (x, y) where $g(x, y)$ is continuous, and three points where $g(x, y)$ is discontinuous. *Hint:* Make sure you really "see" the graph $z = g(x, y)$. Then you can read off the answers.

14.22. This is a problem on a function of three variables. We consider the function $\varphi(x, y, z)$ defined by $\varphi(x, y, z) = (x^2 + y^2 + z^2)^{-\frac{1}{2}}$.

a) At which points of xyz-space is $\varphi(x, y, z)$ discontinuous?

b) Find $\lim_{(x,y,z) \to (2,6,3)} \varphi(x, y, z)$.

14.23. The function $g(x)$ of one variable is defined as follows: $g(x) = 0$ for $x \le 0$, $g(x) = x^2$ for $0 < x \le 2$, and $g(x) = 5$ for $x > 2$. Your friend says: "This function has two points of discontinuity, viz., $x = 0$ and $x = 2$, right?" What is your response and how would you justify it? *Hint:* Draw the graph $y = g(x)$. It helps you to decide how to reason at $x = 0$ and $x = 2$.

14.24. We consider the interval I of the t-axis defined by $-2 \le t \le 1$. Then we define a function $\chi(t)$ as follows: $\chi(t) = 1$ if x lies in the interval I, and $\chi(t) = 0$ if x does not lie in I.

a) Evaluate $\chi(t)$ for $t = -16$, -2.1, -0.01, 0.99, 1877. *Hint:* Draw a graph $w = \chi(t)$ so as not to get confused.

b) Find a value t_1 with the following property: If you increase or decrease t_1 by 10^{-6} then $\chi(t_1)$ will change by less than $1.27 \cdot 10^{-3}$.

c) Find a value t_2 with the following property: No matter by how little you increase t_2 (for example, by $10^{-3988271}$) the value $\chi(t_2)$ will change by at least 0.83.

d) Which of the following statements are true, which are not?

(i) $\chi(t)$ is a continuous function.

(ii) The function $\chi(t)$ has infinitely many points at which it is continuous.

(iii) The function $\chi(t)$ has infinitely many points where it is discontinuous.

15   Partial Derivatives

The key questions.[1]   Below are shown some level curves of a function $z = g(x, y)$ and the point $P_0(5, 2)$. Until further notice, $g(x, y)$ will be this particular function and P_0 this particular point.   The key questions deal with the values of $g(x, y)$

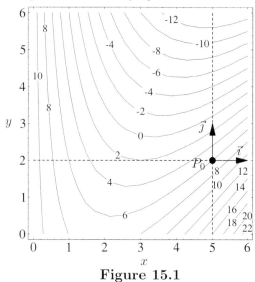

Figure 15.1

as you move (x, y) along horizontal and vertical lines passing through P_0. This is illustrated above and explained below.

> **Key question 1.** *You move a point (x, y) across P_0 on the line $y = 2$ from left to right, and you watch the values $g(x, y)$ all the time. At P_0, is $g(x, y)$ in the course of an increase or a decrease? Or more concisely: At P_0, is $g(x, y)$ increasing or decreasing in the direction of $\vec{\imath}$?*

> **Key question 2.** *You move a point (x, y) across P_0 on the line $x = 5$ from below to above, and you watch the values $g(x, y)$ all the time. At P_0, is $g(x, y)$ in the course of an increase or a decrease? In short: At P_0, is $g(x, y)$ increasing or decreasing in the direction of $\vec{\jmath}$?*

Problems _____

15.1.   a) Use the level curves to answer key question 1.

   b) Use the level curves to answer key question 2.

15.2. The opposite of "increase" is "decrease," and vice versa.

   a) Find a point Q_0 for which the answer to key question 1 is the opposite of the answer to P_0.

   b) Find a point R_0 for which the answer to key question 2 is the opposite of the answer to P_0.

_____

[1]We will use the term *key question* often in introducing new concepts. When we do, you should expect that the formulation of and answer to the key question involve coming to grips with an important and often difficult idea. Do not expect a quick answer. It may take quite a bit of effort just to refine the question, and more still to come up with an answer!

Revised key questions.   Now we want to know not only whether $g(x, y)$ increases or decreases at P_0, but *how fast, or at what rate*, the function increases or decreases.

For the revised key questions, we turn to formulas. The formula for $g(x, y)$ is

$$g(x, y) = (x - 2)^2 - xy + 7.$$

First we do question 1. As (x, y) moves on the line $y = 2$, we have $g(x, y) = g(x, 2)$. It means: On the line $y = 2$ the function $g(x, y)$ of two variables becomes a function $h(x)$ of one variable defined by

$$h(x) = g(x, 2) = (x - 2)^2 - 2x + 7.$$

So the question is "at which rate does $h(x)$ increases or decreases at $x = 5$?" This is a matter of elementary calculus. We take the derivative $h'(x)$ and find $h'(5) = 4$.

Result: *In the direction of $\vec{\imath}$, the function $z = g(x, y)$ is increasing at $P_0(5, 2)$ at the rate of 4 z-units per x-unit.*

We would have arrived at the same result by doing the following:

- In the formula $g(x, y) = (x - 2)^2 - xy + 7$, we treat y as a constant and take the derivative with respect to x.
- Then we set $(x, y) = (5, 2)$.

Problems _____

   15.3.   a) Answer the revised key question 2 by reworking the paragraph "Revised key questions" above. Instead of the line $y = 2$ you have to take the line $x = 5$. Instead of getting a function $h(x)$ and taking $h'(x)$ you will get a function $k(y)$, and you have to take $k'(y)$.

   b) Rewrite the assertions in the bullet list above for key question 2.

   15.4. Go with the same function $g(x, y) = (x - 2)^2 - xy + 7$ as in Figure 15.1, but now K is the point $K(1, 3)$.

   a) At K, is $g(x, y)$ increasing or decreasing in the direction of $\vec{\imath}$? At which rate?

   b) At K, is $g(x, y)$ increasing or decreasing in the direction of $\vec{\jmath}$? At which rate?

Note: Often one talks only about increases and considers a decrease as a negative increase.

Partial derivatives of functions of two variables

Now we look at the general function $f(x, y)$ at the general point (x, y), and we do what we just did with $g(x, y)$ and $P_0(5, 2)$:

In $f(x, y)$ we treat y as a constant and take the derivative with respect to x. Then we obtain what is called the *partial derivative* of $f(x, y)$ with respect to x. It is written as

$$\frac{\partial f}{\partial x} \quad \text{(read as "partial derivative of } f \text{ with respect to } x\text{"), or}$$

$$f_x \quad \text{(read "} f \text{ sub } x\text{"), or}$$

$$z_x \quad \text{(read "} z \text{ sub } x\text{"), or}$$

$$\frac{\partial z}{\partial x} \quad \text{(read "partial of } z \text{ with respect to } x\text{")}$$

The symbol ∂ in ∂f and ∂x tells us that we deal with a *partial* derivative of a function of more than one variable. The partial derivative evaluated at $(x, y) = (a, b)$ is written

$$f_x(a, b) \quad \text{or} \quad \frac{\partial f}{\partial x}\bigg|_{(a,b)} \quad \text{or} \quad z_x(a, b) \quad \text{or} \quad \frac{\partial z}{\partial x}\bigg|_{(a,b)}.$$

What does the partial derivative with respect to x mean?

Answer: The partial derivative f_x evaluated at $(x, y) = (a, b)$ gives us the instantaneous rate of change of $f(x, y)$ at $(x, y) = (a, b)$ in the direction of $\vec{\imath}$. The partial derivative of $f(x, y)$ with respect to y works similarly. It gives the rate of change of $f(x, y)$ in the direction of $\vec{\jmath}$.

Problem

15.5. Let $z = f(x, y) = e^{xy}$

a) Find f_x and $\dfrac{\partial f}{\partial y}$.

b) Find $\dfrac{\partial f}{\partial x}\bigg|_{(1,2)}$ and $z_y(3, 0)$.

Partial derivatives and graphs

A function $f(x, y)$ has partial derivatives, and it has a graph. There *must* be a relation between the partial derivatives and the graph. *What is the relation?*

Imagine you stand on the graph $z = f(x, y)$ at the point $P(a, b, f(a, b))$. You look straight ahead in the direction of the space vector $\vec{\imath} = (1, 0, 0)$ and ask:

Does the surface $z = f(x, y)$ slope up or slope down under my feet?

Actually, you are asking the following question about (a, b) in the xy-plane:

At (a, b), does $f(x, y)$ increase or decrease in the direction of $\vec{\imath} = (1, 0)$?

To answer the latter question you find $f_x(a, b)$, the partial derivative of $f(x, y)$ evaluated at (a, b). Now we can answer the original question about standing on the graph: If $f_x(a, b)$ is positive, then the surface $z = f(x, y)$ will slope up under your feet at $P(a, b, f(a, b))$, and if it is negative, it will slope down. The same reasoning applies when you look in the direction of $\vec{\jmath} = (0, 1, 0)$.

Result: *The partial derivatives of $f(x, y)$ at $P_0(a, b)$ tell us whether at the point $P(a, b, f(a, b))$ the graph $z = f(x, y)$ slopes up or down in the direction of $\vec{\imath}$ and $\vec{\jmath}$.*

It means: The partial derivatives measure the "steepness" of the graph $z = f(x, y)$ in the directions of $\vec{\imath}$ and $\vec{\jmath}$. *What do we mean by "steepness" of the graph?* Look at Figure 15.2 on the next page. It shows part of the graph $z = f(x, y)$ together with the point P_0 in the xy-plane and its corresponding point $P(a, b, f(a, b))$ on the

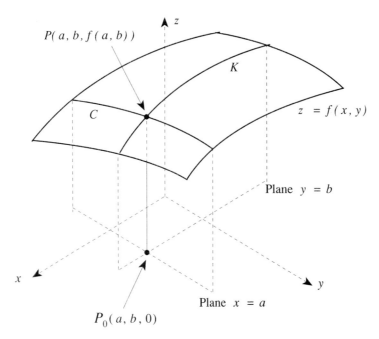

Figure 15.2

surface. The plane $x = a$ intersects the graph in the curve C, and the plane $y = b$ intersects the graph in the curve K. What is the equation of K? It is given by $z = f(x, b), y = b$. What is the slope of K in the plane $y = b$ for $x = a$? It is the derivative of $z = f(x, b)$ with respect to b, and this is the partial derivative $f_x(a, b)$.

Result: *The plane $y = b$ intersects the graph $z = f(x, y)$ in a curve K. As curve in a plane, K has a slope. For $x = a$, this slope is the partial derivative $f_x(a, b)$. Similarly, $f_y(a, b)$ is the slope of the curve C in which the plane $x = a$ intersects the graph $z = f(x, y)$.*

Problems

15.6. We consider the function $g(x, y) = \frac{1}{10}(-x^2 y + 12xy^2 + 5y - 6)$. The plane $y = 1$ intersects the graph $z = g(x, y)$ in the curve K. We look at the plane $y = 1$ in the direction of $\vec{j}$ so that K looks like a curve in an xz-system. Sketch K into an xz-system.

15.7. Go back to Figure 15.2. It explains the geometric meaning of partial derivatives. Use ruler and pencil to find $f_x(a, b)$ and $f_y(a, b)$ as follows. Mark units of 1 cm on all axes. Determine the slopes of the tangents to C and K by measuring their slope in the planes $x = a$ and $y = b$. Note that the broken lines are parallel to a coordinate axis. Therefore, you can measure the coordinates of all points you need with the ruler.

Partial derivatives of functions of three variables

Everything works as for functions of two variables with one exception: A function of three variables does not have a graph. Therefore, we have no way of visualizing a partial derivative as slope of a curve which lies on a surface.

For functions of two variables, there are two key questions 1 and 2, and they are about points (x, y) in the plane. For functions of three variables, there are three key questions, and they are about points (x, y, z) in space.

For example, look at a function $w = m(x, y, z)$ and a point $P_0(a, b, c)$. Then key question 3 reads as follows.

Key question 3. At $P_0(a, b, c)$, is $m(x, y, z)$ increasing or decreasing in the direction of $\vec{k} = (0, 0, 1)$? At which rate?

Problem _____

15.8. In this problem we are working not in xyz-space, but in a space where the coordinates are written a, b, θ. If necessary, think of a as x, b as y, and θ as z. In $ab\theta$-space we have three basis vectors $\vec{i} = (1, 0, 0)$, $\vec{j} = (0, 1, 0)$, $\vec{k} = (0, 0, 1)$ as in xyz-space. Go back to the beginning of Section 13. There, in Example 1, we discuss the function $S = t(a, b, \theta)$ which gives the area of a triangle with sides a, b and angle θ between them. The formula for S is $S = \frac{1}{2} a b \sin \theta$.

We consider the right triangle with $a = 10$ cm, $b = 7$ cm, and $\theta = \frac{\pi}{2}$. $(10, 7, \frac{\pi}{2})$ is a point in $ab\theta$-space. We call it P_0.

a) Write down the three key questions for the function $t(a, b, \theta)$ at the point $P_0(10, 7, \frac{\pi}{2})$.

b) You move through P_0 in the direction of $\vec{j}$. At P_0, is S increasing or decreasing? Explain what your answer means for the triangle.

_____

Now we summarize the facts about partial derivatives of a function $w = f(x, y, z)$ of three variables. We do it only for f_z:

1. **Definition, Notation.** In $f(x, y, z)$ we treat x and y like constants and take the derivative with respect to z. What we obtain is called the partial derivative of $f(x, y, z)$ with respect to z. It is written as

$$\frac{\partial f}{\partial z} \quad \text{(partial of } f \text{ with respect to } z) \quad \text{or} \quad f_z \quad (f \text{ sub } z) \quad \text{or} \quad w_z \quad \text{or} \quad \frac{\partial w}{\partial z}.$$

2. **Evaluation at a point.** The partial derivative evaluated at $(x, y, z) = (a, b, c)$ is written

$$f_z(a, b, c) \quad \text{or} \quad \frac{\partial f}{\partial z}\Big|_{(a,b,c)}, \quad \text{etc.}$$

3. **Meaning.** The partial derivative $f_z(a, b, c)$ is the instantaneous rate of change of $f(x, y, z)$ at $(x, y, z) = (a, b, c)$ in the direction of $\vec{k} = (0, 0, 1)$.

Problems _____

15.9. With $w = x^2 + xyz$, find w_z and $\frac{\partial w}{\partial x}\big|_{(1,2,3)}$.

15.10. The function $p(x, y, z)$ of three variables is defined by $p(x, y, z) = y^2 - 1$. Evaluate all three partial derivatives at $(x, y, z) = (11, -2, 4)$.

15.11. Can you find a function $f(x, y, z)$ which has constant partial derivatives as follows: $f_x = -11$, $f_y = 2$, $f_z = 23$ (that is, $f_x(x, y, z) = -11$ for all (x, y, z), etc.). If yes, find one; if no, explain.

_____

Higher-order partial derivatives

A function $f(x, y)$ of two variables has two partial derivatives: f_x and f_y. Each of them is again a function of two variables, and we can take their partial derivatives again. There are four of them:

$$(f_x)_x \ \text{ written as } f_{xx}, \ \text{ or } \ \frac{\partial}{\partial x}\Big(\frac{\partial f}{\partial x}\Big) \ \text{ written as } \ \frac{\partial^2 f}{\partial x^2};$$

$$(f_x)_y \ \text{ written as } f_{xy}, \ \text{ or } \ \frac{\partial}{\partial y}\Big(\frac{\partial f}{\partial x}\Big) \ \text{ written as } \ \frac{\partial^2 f}{\partial y \partial x};$$

$$(f_y)_x \ \text{ written as } f_{yx}, \ \text{ or } \ \frac{\partial}{\partial x}\Big(\frac{\partial f}{\partial y}\Big) \ \text{ written as } \ \frac{\partial^2 f}{\partial x \partial y};$$

$$(f_y)_y \ \text{ written as } f_{yy}, \ \text{ or } \ \frac{\partial}{\partial y}\Big(\frac{\partial f}{\partial y}\Big) \ \text{ written as } \ \frac{\partial^2 f}{\partial y^2}.$$

These partial derivatives of partial derivatives are called the *second-order partial derivatives* of $f(x, y)$. For example, f_{xy} is read as "second partial of f with respect to x and y." It turns out that the "mixed" derivatives are the same:

$$f_{xy} = f_{yx},$$

provided that all the second-order partial derivatives are continuous functions.[2] Therefore we have only three second-order partial derivatives of $f(x, y)$:

$$f_{xx} \quad f_{xy} \quad f_{yy}.$$

Similarly, a function $g(x, y, z)$ has nine second-order partial derivatives of which only six are possibly different from each other:

$$g_{xx} \quad g_{xy} \quad g_{xz} \quad g_{yy} \quad g_{yz} \quad g_{zz}.$$

Problems

15.12.   a) With $f(x, y) = e^{xy}$, write down the formulas for $f_{xx}, \ f_{xy}, \ f_{yx}, \ f_{yy}$.

        b) With $h(x, y) = x^2 + 3xy + y + 1$ find all (x, y) such that $(h_x, h_y) = (0, 0)$.

15.13. $w = f(m)$ is a function of one variable m. We use $f(m)$ to define a new function $u(x, t)$ of the two variables x and t by $u(x, t) = f(x - t)$. *How do we find the partial derivatives of $u(x, t)$?* For $f(m)$ we use the chain rule of ordinary differentiation, and for $x - t$ we use partial differentiation. The answer will contain the ordinary derivative f'. For example,

$$u_t = f'(x - t)\frac{\partial(x - t)}{\partial t} = -f'(x - t).$$

        a) Evaluate the expression $(u_x - u_t)|_{(x,t)=(3,2)}$ as far as this is possible.

        b) To find higher partial derivatives of $u(x, t)$ we apply the method of a) to u_x and u_y. Write out the formula for $u_{xx} + u_{xt} + u_{tx} + u_{tt}$.

---

[2]For the question why the mixed derivatives of certain functions are the same see, for example, R.C. Buck, *Advanced Calculus*, page 189.

Additional Problems

Evaluation of partial derivatives

15.14. Let $z = f(x, y) = 3x - 0.2y + 1.9$.

  a) Find:

  (i) $f_x(0, 4)$ (ii) $z_x(2, 5)$ (iii) $\dfrac{\partial f}{\partial x}\Big|_{(a+b,a)}$

  (iv) $z_y(0, 0)$ (v) $\dfrac{\partial f}{\partial y}\Big|_{(47,619.5)}$ (vi) $f_y(\sqrt{r + s}, \sqrt{r - s})$

  b) You remember from elementary calculus that the derivative of a linear function $g(x) = Ax + B$ is constant. Formulate the corresponding fact for functions of two variables.

  c) Write down the formula of a linear function $h(w_1, w_2, \ldots w_7)$ of seven variables $w_1, w_2, \ldots w_7$. Then find h_{w_4}.

15.15. The function $z = f(x, y)$ is defined by

$$f(x, y) = x^2 y + 3xy^5 - 6.$$

Find the partial derivatives $\frac{\partial f}{\partial x}\big|_{(1,1)}$ and $z_y(2, 2)$.

15.16. Evaluate $\dfrac{\partial(3\,x\,y\,\cos(\pi y))}{\partial x}$ and $\dfrac{\partial(3\,x\,y\,\cos(\pi y))}{\partial y}\Big|_{(a,1)}$.

15.17. Let K be defined by

$$K = u\,\sqrt{d} + u\,v.$$

Find the following:

  (i) K_u;   (ii) $\dfrac{\partial K}{\partial v}$;   (iii) K_d;   (iv) $\dfrac{\partial K}{\partial x}$

The x in item iv is not a typographical error.

15.18. The function $r = G(p, q)$ is defined by

$$G(p, q) = \frac{p\,e^{p+q}}{2\,q}.$$

  (i) Find G_p.   (ii) Find G_q.

15.19. Let $u(t)$ and $v(t)$ be functions of one variable t. We use these functions to define several new functions $f(x, y), g(x, y), \ldots$ of two variables. The partial derivatives of the new functions $f(x, y), g(x, y), \ldots$ will be formulas that contain the ordinary derivatives u' and v'. Find the partial derivatives with respect to x and y of the new functions below.

  a) $f(x, y) = u(x) \cdot v(y)$

  b) $g(x, y) = u(3x - 5y) + v(xy)$

  c) $h(x, y) = \dfrac{u(x)}{v(y)}$

  d) $i(x, y) = \sin\!\left(u(x^2) \cdot v(y^3)\right)$

15.20. As in Problem 13.10 on page 100, let $f(x, y, z)$ be the radial function defined by $f(x, y, z) = g\left(\sqrt{x^2 + y^2 + z^2}\right)$, where $g(t)$ is the function of one variable graphed in Figure 15.3. Estimate the partial derivatives f_x, f_y, and f_z at the point $P(1, 0, 1)$.

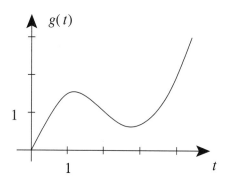

Figure 15.3

15.21. Let $f(t, u, v, w, z) = tu + uv + vw + wz$.

    a) Write out the following partial derivatives: f_u; f_v; f_w; f_t; f_x; f_z.

    b) Find numbers t_0, u_0, v_0, w_0, z_0 so that $f_u(t_0, u_0, v_0, w_0, z_0) = 17$.

15.22. We consider the following function $q = h(a, b, c, u, v, w)$ of the six variables $a, \ldots w$:

$$h(a, b, c, u, v, w) = 2auw + \sqrt{a^2 + b + uvw} + \sin(a + b + c).$$

Write out the following partial derivatives: h_a, h_c, h_v, h_w.

Higher partial derivatives

15.23.   a) For $f(x, y) = 3x - 0.2y + 1.7$, find f_{xx}, f_{xy}, f_{yy}.

    b) The function $p = h(s_1, s_2, \ldots s_{399})$ is defined by

$$p = s_1 + \frac{1}{2}s_2 + \frac{1}{3}s_3 + \frac{1}{4}s_4 + \ldots \frac{1}{399}s_{399}).$$

    Find

$$\frac{\partial^2 h}{\partial s_{187}\partial s_{304}}.$$

    c) Remember from elementary calculus: $\frac{d^2(Ax+B)}{dx^2} = 0$. Formulate the corresponding fact for functions of n variables where n is any integer greater than or equal to 2.

    d) With $h(x, y, z, u) = 2x - 4y + zu$, find $\frac{\partial^2 h}{\partial z \partial u}$. Your colleague asks you: "In part c we found that all second-order partial derivatives of a linear function of any number of variables are zero. Here, we find $\frac{\partial^2 h}{\partial z \partial u} \neq 0$. Isn't there something wrong here? I do not get it." Write down a response.

15.24. As in Problem 15.15, let $f(x, y) = x^2y + 3xy^5 - 6$. Write out the formulas for f_{xx}, f_{xy}, and f_{yy}.

15.25. Let $K = u\sqrt{d} + uv$ as in Problem 15.17. Find the second-order partials $\frac{\partial^2 K}{\partial u \partial v}$ and K_{vv}.

15.26. The function $f(x, y)$ is defined by $f(x, y) = g(2x - 3y)$, where g is a function of one variable. Work out formulas for the partials $f_x(0,0)$, $f_y(0,0)$, $f_{xx}(0,0)$, $f_{yy}(0,0)$, and $f_{xy}(0,0)$ in terms of $g(0)$, $g'(0)$, and $g''(0)$.

Interpretation of partial derivatives

15.27. In Example 4 at the beginning of Section 13, we considered the following function $P = f(A, r, N)$: You borrow A dollars at an annual interest rate of r, and you amortize the debt in N monthly payments. Then the monthly payment P is given by

$$P = f(A, r, N) = \frac{A\frac{r}{12}}{1 - (1 + \frac{r}{12})^{-N}} \ .$$

    a) You can answer the following question by common sense, without using the formula for $f(r, N)$: Is $\frac{\partial P}{\partial r}$ positive, or zero, or negative, or does it depend on the values of r and N at which you evaluate the partial derivative?

    b) Again, use common sense to decide whether $\frac{\partial P}{\partial N}$ is positive, or negative, or whether it will depend on the values of r and N.

    c) Write out the formula for $\frac{\partial P}{\partial N}$. *Remember:* From elementary calculus, you know that $(a^x)' = \ln a \cdot a^x$

    d) Write out the formula for $f_A = \frac{\partial f}{\partial A}$.

15.28. A computer generated plot of a function $f(x, y)$ is shown below, with the point $P(1, 1, f(1, 1))$ marked with a black dot.

    a) Are the partial derivatives $f_x(1, 1)$ and $f_y(1, 1)$ positive or negative? *Hint:* If you start at P and move in the direction of increasing x, does $f(x, y)$ increase or decrease?

    b) Are the second-order partials $f_{yy}(1, 1)$ and $f_{yx}(1, 1)$ positive or negative? *Hint:* If you start at P and move in the direction of increasing x or y, does f_y increase, or decrease?

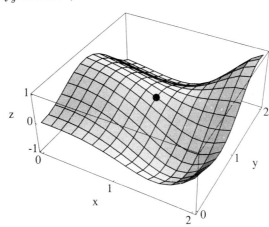

15.29. We consider the function $z = f(x, y) = 4x - y + 2$. We know that $f_x = 4$ and $f_y = -1$ at all points (x, y).

The graph $z = 4x - y + 2$ is a plane in xyz-space. *How can you see from the graph that the partial derivatives of $f(x, y)$ have the same constant value at all points (x, y)?* Write down an answer in complete sentences. *Hint:* Look at the discussion "The partial derivatives of $f(x, y)$ and the graph of $f(x, y)$" which begins on page 119 right after Problem 15.5

15.30. The time T it takes to roast a turkey depends on the weight w of the turkey and the temperature u of the oven. What are the signs of the partial derivatives $\frac{\partial T}{\partial w}$ and $\frac{\partial T}{\partial u}$? Explain.

15.31. The three images below represent three successive frames of a movie of a vibrating string, at times $t = t_0$, $t = t_1$, and $t = t_2$. Let $u(x, t)$ denote the upward displacement of the string at time t and x units from the left end of the string, so that the pictures below are the graphs of the functions $u(x, t_0)$, $u(x, t_1)$, and $u(x, t_2)$, as functions of x with t held fixed.

   a) At the point (x_0, t_1), what are the signs of $\frac{\partial u}{\partial x}$ and $\frac{\partial u}{\partial t}$? Explain.

   b) At the point (x_0, t_1), what are the signs of $\frac{\partial^2 u}{\partial x^2}$, $\frac{\partial^2 u}{\partial t^2}$, and $\frac{\partial^2 u}{\partial t \partial x}$? Explain.

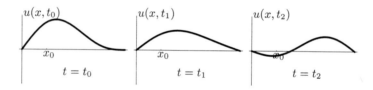

16 Linear Approximation

Outline. In elementary calculus, one studies at length the tangent line of a graph $y = g(x)$ and the tangent line approximation of $y = g(x)$. Here, we discuss the corresponding notions for function of two variables, as shown in the following box.

$y = g(x)$	$z = f(x, y)$
Tangent line of graph $y = g(x)$	Tangent plane of graph $z = f(x, y)$
Tangent line approximation of $g(x)$	Linear approximation of $f(x, y)$

Tangent planes and linearization

What is the linearization of a function of two variables? We work with a generic function $f(x, y)$ of two variables and a fixed point $P_0(a, b)$ in the xy-plane. Our goal is to describe a *linear* function $L(x, y)$ which gives the best possible approximation to $f(x, y)$ when (x, y) is near (a, b). We make two assumptions about the function $f(x, y)$.

1. The surface $z = f(x, y)$ is smooth near the point $(a, b, f(a, b))$. This means that the point $(a, b, f(a, b))$ does not lie on a sharp point or crease in the surface.

2. By item 1, the surface $z = f(x, y)$ has a tangent plane at the point $(a, b, f(a, b))$. We assume that this tangent plane is not vertical, that is, it is not perpendicular to the xy-plane.

By the preceding assumptions, the graph of our function has a nonvertical tangent plane at the point $(a, b, f(a, b))$, which hugs the graph very closely near the point of tangency, as shown in Figure 16.1. Because the tangent plane is not vertical,

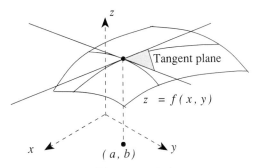

Figure 16.1

it is the graph of a linear function $z = L(x, y)$. The fact that the tangent plane $z = L(x, y)$ hugs the surface $z = f(x, y)$ near the point of tangency $(a, b, f(a, b))$ tells us

$$f(x, y) \approx L(x, y) \qquad \text{for } (x, y) \text{ near } (a, b).$$

The linear function $L(x, y)$ is called the *linearization* of $f(x, y)$ at (a, b), or the *linear approximation* to $f(x, y)$ based at (a, b).

How do I find the linearization? In the previous paragraph, we discussed what the linearization *is*, but we did not tell you how to *find* it. We will work that out now. Since $L(x, y)$ is linear, we can write

$$L(x, y) = A(x - a) + B(y - b) + C \qquad (16.1)$$

where A, B, and C are constants which are yet to be determined. Because the graphs of $z = f(x, y)$ and $z = L(x, y)$ both pass through the point $(a, b, f(a, b))$, we must have

$$f(a, b) = L(a, b).$$

But (16.1) gives $L(a, b) = C$, so we get

$$C = f(a, b).$$

It remains to determine the values of A and B. Because the graphs of $z = f(x, y)$ and $z = L(x, y)$ are tangent at the point $(a, b, f(a, b))$, they have the same slopes in the x and y directions at that point. In other words, both partial derivatives of the functions $f(x, y)$ and $L(x, y)$ agree at the point (a, b):

$$f_x(a, b) = L_x(a, b)$$
$$f_y(a, b) = L_y(a, b).$$

But from (16.1), we have $L_x(x, y) = A$ and $L_y(x, y) = B$, so

$$A = f_x(a, b) \quad \text{and} \quad B = f_y(a, b).$$

Thus we have a formula for the linearization of $f(x, y)$ at the point (a, b):

$$L(x, y) = f(a, b) + f_x(a, b)(x - a) + f_y(a, b)(y - b).$$

Problems

16.1. $g(x, y) = e^{xy} + y$, and $L(x, y)$ is the linear approximation of $g(x, y)$ based at $(x, y) = (2, 1)$.

    a) Find (i) $L(2, 1)$, (ii) $L(1, 0)$.

    b) $L(x, y)$ is a function of two variables. Therefore, it has partial derivatives L_x and L_y. Find L_x and L_y at (i) $(x, y) = (4, 2)$, (ii) $(x, y) = (371, -958)$.

    c) Write out the formula for $L(x, y)$ in the form $Ax + By + C$.

16.2. $g(x, y) = e^{xy} + y$ is the function of Problem 16.1, but we work at a different "basepoint" $Q_0(1, 0)$. $L(x, y)$ is now the linear approximation of $g(x, y)$ based at $Q_0(1, 0)$.

    a) Find (i) $L(1, 0)$, (ii) $L(2, 1)$.

    b) Find L_x and L_y at (i) $(x, y) = (4, 2)$, (ii) $(x, y) = (371, -958)$.

    c) Write out the formula for $L(x, y)$ in the form $Ax + By + C$.

16.3. $L(x, y)$ is the linear approximation of the general function $f(x, y)$ based at the general point (a, b).

    a) Express $L(a, b)$ in terms of $f(x, y)$, a, b.

    b) Express the partial derivatives of L at (a, b) and at $(23, -87)$ in terms of $f(x, y)$, a, b.

Notation for the linear approximation. There is no standard notation for the linear approximation. We use $L(x, y)$. Be forewarned that this notation is ambiguous: It does not tell you what function you are linearizing, or at what point the linearization is based.

How do we find the equation of the tangent plane? We write its equation $z = L(x, y)$ in the form

$$f_x(a, b)(x - a) + f_y(a, b)(y - b) - z = -f(a, b).$$

From this equation we read off that $(f_x, f_y, -1)$ is a normal vector of the tangent plane. Note that the partial derivatives f_x, f_y are evaluated at $(x, y) = (a, b)$.

Problems _____

16.4. With $f(x, y) = xy^2 + x$

    a) Write down the linear approximation of $f(x, y)$ based at $(x, y) = (1, 2)$.

    b) Find a vector perpendicular to the tangent plane of the graph $z = f(x, y)$ at $(1, 2, 5)$.

    c) Find the equation of the tangent plane of the surface $z = f(x, y)$ at the point $(1, 2, 5)$.

16.5. With $g(x, y) = e^{x^2 + y}$

    a) Find the equation of the tangent plane of the surface $z = g(x, y)$ at the point $(1, 1, e^2)$.

    b) Write down the formula for the linear approximation of $g(x, y)$ based at the point $(x, y) = (1, 1)$.

_____

How good is the linear approximation?[1] The approximation formula

$$f(x, y) \approx L(x, y) \tag{16.2}$$

for (x, y) near (a, b) is useful, but is lacking in precision, because it gives no indication how good the approximation (implied by the "$\approx$" sign) really is. The most naive interpretation is simply that $L(x, y)$ can be made as close as we wish to $f(x, y)$ by making sure that (x, y) is sufficiently near (a, b). Although this is true, it is not good enough. There are in fact *many* linear functions that have the above property, but there is only *one* linearization of $f(x, y)$ based at (a, b), and it approximates $f(x, y)$ better than any other. Our goal in this paragraph is to discover the sense in which the linearization of $f(x, y)$ at $P_0(a, b)$ gives the *best* linear approximation to $f(x, y)$ for $Q(x, y)$ near P_0.

Example 1. We will try to gauge the accuracy of the linear approximation to the function $f(x, y) = xy$ based at $P_0(1, 2)$. Since $f_x(1, 2) = 2$ and $f_y(1, 2) = 1$, the linearization of $f(x, y)$ at P_0 is

$$L(x, y) = 2 + 2(x - 1) + (y - 2) = 2 + 2\Delta x + \Delta y \tag{16.3}$$

where

$$\Delta x = x - 1, \quad \text{and} \quad \Delta y = y - 2.$$

Solving for x and y gives

$$x = 1 + \Delta x \quad \text{and} \quad y = 2 + \Delta y,$$

_____

[1]Full justification of the material presented here is beyond the scope of this discussion. For a detailed treatment of these matters, see R.C. Buck, *Advanced Calculus*, pages 180–190.

and substituting in the formula for $f(x, y)$ gives

$$f(x, y) = xy = (1 + \Delta x)(2 + \Delta y) = 2 + 2\Delta x + \Delta y + (\Delta x)(\Delta y). \tag{16.4}$$

Comparing equations (16.3) and (16.4), you obtain

$$f(x, y) - L(x, y) = (\Delta x)(\Delta y).$$

This difference represents the *error* introduced when you approximate $f(x, y)$ by $L(x, y)$. Since $|\Delta x|$ and $|\Delta y|$ are *both* small when (x, y) is near $(1, 2)$, you can conclude that the error is the product of *two* small numbers, and hence *even smaller*.

∎

Problem _____

16.6. a) Linearize the function $f(x, y) = x^2 + 2y^2$ at the point $P_0(-1, 1)$.

b) Express the error $f(x, y) - L(x, y)$ in terms of $\Delta x = x + 1$ and $\Delta y = y - 1$.

c) Calculate the error for each of the following values of (x, y): $(0, 1)$, $(-1, 0)$, $(-1.5, 0.5)$, $(-0.9, 1.1)$, $(-1.01, 1.01)$. Make a table showing (x, y) in column 1, the distance between (x, y) and $(-1, 1)$ in column 2, and the error in the linear approximation in column 3.

_____

You discovered in Problem 16.6 that, as (x, y) gets close to (a, b), the size of the linear approximation error is much smaller than the distance between (x, y) and (a, b). We will now look at this phenomenon more carefully. We use the idea of limit to express precisely the sense in which the linearization of a function is a good approximation to the function. Let $L(x, y)$ be the linearization of $f(x, y)$ based at (a, b). We introduce the symbol $R(x, y)$ to stand for the "error" introduced when we approximate $f(x, y)$ by $L(x, y)$. In other words, $R(x, y)$ is the difference between $f(x, y)$ and $L(x, y)$, so that

$$f(x, y) = L(x, y) + R(x, y). \tag{16.5}$$

The basic fact of life about linear approximation is:

If the function $f(x, y)$ has continuous partial derivatives f_x and f_y at the point (a, b), then the error term $R(x, y)$ is small when compared with $|\overrightarrow{P_0 Q}|$. Here $Q = Q(x, y)$ is the point with coordinates (x, y). More precisely,

$$\lim_{(x,y)\to(a,b)} \frac{|R(x, y)|}{|\overrightarrow{P_0, Q}|} = 0. \tag{16.6}$$

Notice that this is a much stronger assertion than simply saying that the error term $R(x, y)$ approaches 0 as Q approaches P_0. It says that the error term approaches 0 *faster than the distance between Q and P_0*. It is equally important that there is *only one* linear function that approximates $f(x, y)$ this well:

The only linear function $L(x, y)$ such that the error term $R(x, y) = f(x, y) - L(x, y)$ satisfies (16.6) is the linearization of $f(x, y)$ based at P_0.

So what does this mean in practice?

Answer: *If you can find a linear function satisfying (16.6), then you have found the linearization.*

This will often permit you to linearize functions without explicitly calculating partial derivatives. This is illustrated in the following example.

Example 2. Returning to the function $f(x, y) = xy$ from Example 1, we will linearize at $P_0(1, 2)$ *without calculating partial derivatives.* As in Example 1, write $\Delta x = x - 1$ and $\Delta y = y - 2$, and substitute $x = 1 + \Delta x$, $y = 2 + \Delta y$ into the formula $f(x, y) = xy$ to get

$$\begin{aligned} f(x, y) &= (1 + \Delta x)(2 + \Delta y) = 2 + 2\Delta x + \Delta y + (\Delta x)(\Delta y) \\ &= L(x, y) + R(x, y) \end{aligned}$$

with

$$L(x, y) = 2 + 2\Delta x + \Delta y = 2 + 2(x - 1) + (y - 2)$$

and

$$R(x, y) = (\Delta x)(\Delta y).$$

Since $|R(x, y)|$ is small compared with $|\overrightarrow{P_0 Q}| = \sqrt{\Delta x^2 + \Delta y^2}$, it follows that the above linear function $L(x, y)$ *must be the linearization of $f(x, y)$ based at P_0.* Notice also that once you know the linearization, you can read off the partial derivatives $f_x(1, 2)$ and $f_y(1, 2)$ as the coefficients of Δx and Δy respectively: $f_x(1, 2) = 2$ and $f_y(1, 2) = 1$. ∎

Summary: *To linearize a function $f(x, y)$ at the point $P_0(a, b)$ without calculating partial derivatives, substitute $x = a + \Delta x$ and $y = b + \Delta y$, expand $f(x, y)$ in powers of Δx and Δy, and drop all terms of degree two or more in Δx and Δy. The linear function you are left with is the linearization.*

Problem ―――――――――――――――――――――――――――

16.7. Linearize $f(x, y) = x + x^2 y$ at $(a, b) = (-1, 2)$ by expanding in powers of Δx and Δy. Check your work by linearizing using partial derivatives. You should get the same answer.

Summary of Linear Approximation

- The linear approximation of $f(x, y)$ based at $P_0(a, b)$ is a linear function $L(x, y)$ of x and y. It approximates $f(x, y)$ near P_0 better than any other linear function.

- It is defined by $L(x, y) = f(a, b) + f_x(x - a) + f_y(y - b)$. In this formula, the partials f_x and f_y are evaluated at (a, b).

- At the basepoint P_0, the function and its linear approximation have the same value and the same partial derivatives.

- The graph $z = L(x, y)$ is the tangent plane of the surface $z = f(x, y)$ at the point $(x, y, z) = (a, b, f(a, b))$. The normal of that tangent plane has components $(f_x, f_y, -1)$.

- The tangent plane at $(a, b, f(a, b))$ is a "linear version" of the surface $z = f(x, y)$ in the same way the tangent line at $(a, g(a))$ is a "linear version" of the curve $y = g(x)$.

The differential

The differential is nearly the same as the linear approximation. Here we look at the basic approximate equality in the following form:

$$f(x, y) - f(a, b) \approx f_x(a, b)(x - a) + f_y(a, b)(y - b).$$

The right side is a function of the coordinates (x, y) of the point Q. We write df for this function:

$$df = f_x(a, b)(x - a) + f_y(a, b)(y - b).$$

Note that df is a linear function of (x, y). The formula also contains the coordinates (a, b) of P_0 and the partial derivatives of $f(x, y)$ at P_0.

Definition. *The function df is called the differential of $f(x, y)$ based at $P_0(a, b)$.*

How is the differential of $f(x, y)$ based at $P_0(a, b)$ related to the linear approximation of $f(x, y)$ based at $P_0(a, b)$? The differential df approximates how much the function changes between $P_0(a, b)$ and $Q(x, y)$, and the linear approximation approximates the value of $f(x, y)$ for (x, y) near (a, b):

$$
\begin{aligned}
\text{differential} &: & df &\approx f(x, y) - f(a, b) \\
\text{linear approximation} &: & L(x, y) &\approx f(x, y)
\end{aligned}
$$

Thus, the linear approximation and the differential differ only by the constant $f(a, b)$.

What happens if we take the differential of the same function at a different basepoint $T_0(c, d)$? As with the linear approximation, df will change. As a rule, the differentials of the same function at two different basepoints will be different.

Problem _____

> **16.8.** With $g(x, y) = e^{xy} + y$, write out the formula for the differential of $g(x, y)$ based at (i) $P_0(2, 1)$, and (ii) $P_0(1, 0)$.

Notation for the differential. Often $x - a$ is written as dx and $y - b$ as dy so that the formula for the differential reads

$$df = f_x dx + f_y dy$$

The notation df shows the name f of the function. It does not show the basepoint.

Problems _____

> **16.9.** We are given a function $z = f(x, y)$. Its linear approximation based at (a, b) is $L(x, y)$, and its differential based at the same point is df.
>
>   a) $L(x, y)$ and df are related to each other by an equation. Write out that equation.
>
>   b) df is a linear function of x and y. Evaluate df for $(x, y) = (a, b)$.
>
>   c) How are the partial derivatives of the functions $f(x, y)$, $L(x, y)$, and df related to each other?

16.10. We have the following information on the function $z = h(x, y)$: The equation of the tangent plane of the graph $z = h(x, y)$ at the point $Q(3, 2, h(3, 2))$ is given by $4x - y + 2z = 12$. Do we know enough to find the differential dh based at Q? If yes, find the formula; if no, explain why not.

Use of the differential for estimates

We consider a body B in space, for example a satellite or a planet. It has mass M (in kilograms kg), and its distance to the center of the earth is s (in kilometers km). Physicists tell us: The earth's gravitational force attracts B. The magnitude of the force exerted on B is a function $F(M, s)$ of the mass M (in kilograms) and distance s (in kilometers). With properly chosen force units,[2] the formula for $F(M, s)$ is given by

$$F(M, s) = \frac{M}{s^2}.$$

For the rest of this section, we will do the following:

- We ask two questions on the function $F(M, s)$. Question 1 is about "absolute change," and question 2 is about "margins of errors."
- We use the differential to get *estimates*, that is, *approximate* answers.

Question 1 on absolute change. B is a man-made satellite of mass M_0. We change B so that its mass changes by a certain amount $\Delta M = M - M_0$. It is at distance s_0 from the earth. Within the next few seconds, the distance will have changed by $\Delta s = s - s_0$. *What will be the corresponding change*

$$\Delta F = F(M_0 + \Delta M, s_0 + \Delta s) - F(M_0, s_0)$$

of the gravitational force acting on B? The point (M_0, s_0) is our basepoint in the Ms-plane of the independent variables. It stays fixed during the investigation.

Discussion of question 1. We can do two things:

(i) We use the exact formula for ΔF:

$$\Delta F = \frac{M_0 + \Delta M}{(s_0 + \Delta s)^2} - \frac{M_0}{s_0^2} = \frac{(M_0 + \Delta M)s_0^2 - M_0(s_0 + \Delta s)^2}{(s_0 + \Delta s)^2 s_0^2} = \dots$$

(ii) We replace the "exact" ΔF by the differential dF which approximates ΔF:

$$\Delta F \approx dF = \frac{\partial F}{\partial M} \Delta M + \frac{\partial F}{\partial s} \Delta s$$

As function of ΔM and Δs, the differential is easier to handle than the "exact" ΔF.

Problem _____

16.11. We are in the setup of question 1.

    a) Write out the formula for the differential dF as function of ΔM and Δs for $M_0 = 87$ kilograms and $s_0 = 237$ kilometers.

---

[2]By this we mean that, to keep our formulas simple, we use nonstandard force units so that the product of the gravitational constant and the mass of B is 1.

b) Use the formula for dF you found in part a to estimate by how much F will change if the mass is reduced by 1.5 kilograms and the distance is increased by 2 kilometers.

c) Use the differential to estimate by how much F will change if we increase the mass from 87 to 89 kilograms and decrease the distance from 237 to 236 kilometers.

Question 2 on error estimates. If B is a planet, it is difficult to measure M and s. We can measure M and s only up to a margin of error, as follows:

The mass is $M_0 \pm \Delta M$ kg, and the error ΔM is in absolute value less than or equal to a fixed margin of error ΔM_0.

The distance is $s_0 \pm \Delta s$ km, with Δs in absolute value less than or equal to a fixed margin Δs_0.

The fixed margins ΔM_0 and Δs_0 are positive numbers. The way the possible errors ΔM and Δs relate to the fixed margins ΔM_0 and Δs_0 is best seen in an Ms-plane. The true values of M and s represent a point (M, s) somewhere in the rectangle of length $2\,\Delta M_0$ and height $2\,\Delta s_0$, as shown below.

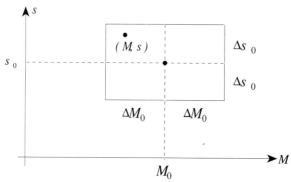

Now we can ask

Question 2 *Given these margins of error, what is the greatest possible error we make when we use the formula for F? In other words: What is the maximum (highest possible value) of*

$$|\Delta F| = |F(M, s) - F(M_0, s_0)|$$

if (M, s) ranges over all points in the rectangle? Or, using ΔM and Δs, what is the maximum of

$$|\Delta F| = |F(M_0 + \Delta M, s_0 + \Delta s) - F(M_0, s_0)|$$

if $|\Delta M| \le \Delta M_0$ and $|\Delta s| \le \Delta s_0$?

Discussion of question 2. Here the values of the independent variables are so uncertain that we do not bother working with the exact ΔF, but only with the differential dF:

$$\Delta F \approx dF = (1/s_0^2)\,\Delta M - (2M_0/s_0^3)\,\Delta s.$$

We temporarily rewrite dF in simpler notation:

$$A = 1/s_0^2 \text{ and } B = 2M_0/s_0^3 \text{ so that } dF = A\,(\Delta M) - B\,(\Delta s).$$

Here A and B are positive. Therefore, $A\,(\Delta M) - B\,(\Delta s)$ is highest if ΔM is highest and Δs is lowest. This is the case for $\Delta M = \Delta M_0$ and $\Delta s = -\Delta s_0$. Similarly, dF is lowest for $\Delta M = -\Delta M_0$ and $\Delta s = \Delta s_0$. The result is

$$|dF| \;\leq\; \frac{1}{s_0^2}\,\Delta M_0 \;+\; \frac{2M_0}{s_0^3}\,\Delta s_0$$

Note that the "+" in front of the second term on the right is not a misprint. It is there because we are interested in the greatest possible error, that is, in the worst possible scenario.

Problem _____

16.12. We have the following measurements for M and s of our planet B:

$$M_0 = 6 \cdot 10^{23} \pm 4 \cdot 10^7 \text{kg}, \quad s_0 = 2 \cdot 10^5 \pm 10^2 \text{km}$$

Estimate the maximal error in the determination of the force F by which the planet is attracted by the earth.

_____

Variation of question 2—the percentage version. We continue with B being a planet for which mass and distance from the earth can be measured only up to certain margins of errors. These margins of error can be given in percentages: We are able to measure M up to a percentage error of at most $m\%$. Similarly, we measure s_0 up to an error of at most $s\%$.

 Given these error margins, what is the greatest possible percentage error in the determination of the force F?

For a percentage change we need a base value. In our case it is $F_0 = F(M_0, s_0)$. A percentage change is (relative change) times 100. In our case, the relative change is

$$\frac{\Delta F}{F_0} = \frac{F(M, s) - F(M_0, s_0)}{F(M_0, s_0)}.$$

Discussion of percentage version of question 2. To work such a question

 (i) You convert the question into one on absolute change (question 2).

 (ii) You solve the converted problem.

 (iii) You reformulate your answer into a statement on relative change by dividing by the base value.

 As an example, we consider a planet whose mass we can determine with a percentage error of at most 2.3%, and whose distance from the earth we can determine with an error of at most 1.5%. We want to estimate the maximal percentage error in the determination of the force F.

 (i) **Conversion to a question on absolute change:** To say "M is measured with a percentage error of at most 2.3% and s with an error of at most 1.5%" means

$$M = M_0 \pm \Delta M, \; |\Delta M| \leq 0.023 M_0\,; \quad s = s_0 \pm \Delta s, \; |\Delta s| \leq 0.015 s_0$$

(ii) **Solution of the converted problem:**

$$|dF| \leq \frac{1}{s_0^2} 0.023 M_0 + \frac{2M_0}{s_0^3} 0.015 s_0$$

(iii) **Reformulation in terms of relative change:** $\dfrac{|dF|}{F_0} \leq \dots$

Problem _____

16.13. a) Finish the computation of the example by dividing the estimate for the absolute change. Use the measured values of M and s from Problem 16.12. Your answer has to be a numerical percentage.

b) Why does one consider percentage changes? If there is an advantage, what is it?

_____

Shortcut for estimating percentage change. You estimate the percentage change in a function $f(x, y)$ from a basepoint $(x, y) = (a, b)$ by estimating the total change and dividing by $f(a, b)$:

$$\text{percent change} = \frac{\text{absolute change}}{f(a, b)} \approx \frac{f_x(a, b)\, dx + f_y(a, b)\, dy}{f(a, b)}$$

$$= \frac{f_x(a, b)}{f(a, b)}\, dx + \frac{f_y(a, b)}{f(a, b)}\, dy.$$

Look closely at the right side. The coefficients of dx and dy are the x and y partial derivatives of $g(x, y) = \ln f(x, y)$ evaluated at $(x, y) = (a, b)$. Therefore, you get

$$\text{percent change} \approx g_x(a, b)\, dx + g_y(a, b)\, dy$$

where $g(x, y) = \ln f(x, y)$. More concisely,

$$\text{percent change in } f \approx d(\ln f)$$

where the differential on the right is based at (a, b). In summary

> *The percentage change in $f(x, y)$ is well approximated by the differential of $\ln f(x, y)$.*

It is often easier to work with $d \ln f$ than to estimate percentage change directly. In the gravitational example $F(M, s) = M/s^2$ worked out above, you have

$$\ln F(M, s) = \ln\left(\frac{M}{s^2}\right) = \ln M - 2 \ln s$$

so

$$d(\ln F) = \frac{1}{M}\, dM - \frac{2}{s}\, ds.$$

Problem _____

16.14. Redo Problem 16.13 part a using the above shortcut.

_____

Additional Problems

16.15. The function $k(r, h)$ is defined by $k(r, h) = r\sqrt{r^2 + h^2}$.

    a) $L(r, h)$ is the linear approximation based at $P_0(2, 5)$. Write out the formula for $L(r, h)$.

    b) Write out an expression which approximates $k(2 + \Delta r, 5 + \Delta h)$. Your answer will be a formula which contains Δr and Δh.

    c) dk is the differential of the function $k(r, h)$ based at the general point (r, h). Write out dk in the form $dk = k_r dr + k_h dh$.

16.16. A quantity of gas is enclosed in a tank with a piston and a heating element, so that the volume v and temperature T can be controlled. The pressure p of the gas is determined by the volume and temperature according to the formula

$$p = \frac{T}{100v}$$

when p is measured in atmospheres, v in liters, and T in degrees Kelvin. Thus, at a temperature of 300 degrees Kelvin and a volume of 3 liters, the pressure p is 1 atmosphere. Use linear approximation or differentials to answer the following question. Starting at a temperature of 300 degrees and a volume of 3 liters, the temperature is increased by 5 degrees and the volume is increased by 0.1 liter. Will the pressure increase, or decrease, and by approximately how much?

16.17. A surface of the shape of an inner tube is called a torus. Below are shown two views of a torus T. The view on the right is from the top. The view on the left is a computer approximation. In reality, there are no corners on T, and everything is round and smooth.

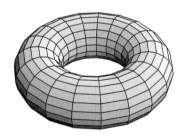 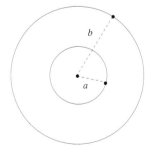

    a) Our torus has an "outer radius" $b = 1$ meter and an "inner radius" $a = 0.4$ meter. We increase b by 5 cm and a by 7 cm. By just looking at the torus, would you think that the surface area stays the same, or increases, or decreases? Or is it impossible to predict?

    b) The surface area A of T is given by $A = \pi^2(b^2 - a^2)$. Use the differential to estimate by how much the surface increases or decreases if we make the changes of a).

    c) We have measured another torus and found $b = 2.5$ meter and $a = 1.6$ meter. The measurement of b is within 2% and the measurement of

a within 1.5%. Estimate the greatest possible percentage error in the determination of the surface area A.

16.18. We have the following information on a function $T(u, v)$:

$$T(2,5) = 4; \qquad T_u(2,5) = 3; \qquad T_v(2,5) = 1.$$

a) Give the best estimate you can for $T(2.2, 4.7)$.

b) We vary u from $u = 2$ by at most 4%, and we vary v from $v = 5$ by at most 2%. Estimate the percentage by which T will vary from $T(2,5) = 4$ at most.

16.19. In elementary calculus, you learned to find approximate solutions to equations of the form $f(x) = a$ by Newton's Method. Starting with a rough estimate for the solution, say $x = x_0$, you linearize $f(x)$ at $x = x_0$, and solve the linear equation $L(x) = a$ to get an approximate solution to the non-linear equation $f(x) = a$. The same idea can be applied to systems of two equations in two unknowns. Consider the following system:

$$
\begin{aligned}
x^2 + y^2 &= 1 \\
e^x \cos y &= 2.
\end{aligned}
$$

An approximate solution, obtained by trial and error, is $x_0 = 0.9$ and $y_0 = 0.5$. Your mission is to obtain a better estimate for the solution by linearizing the equations at $(x_0, y_0) = (0.9, 0.5)$, and solving the linearized equations.

17 The Chain Rule

The Chain Rule for functions of two variables. We are given a function $z = f(x, y)$ of *two* variables and two functions $x = a(t)$, $y = b(t)$ of *one* variable.[1] Then we define a new function $F(t)$ of *one* variable as follows:

$$F(t) = f(a(t), b(t)).$$

It means: Wherever x appears in $f(x, y)$, we write $a(t)$, and wherever y is we write $b(t)$. Letting $z = f(x, y)$, the situation can represented schematically as follows:

$$t \mapsto (x, y) \mapsto z. \tag{17.1}$$

The diagram indicates that z depends on x and y, while x and y depend on t. Thus, the value of z depends ultimately on t. For this reason, we call z the *dependent variable* and t the *independent variable*. The variables x and y are called *intermediate variables*.

Example 1. The function $z = f(x, y)$ is given by $z = f(x, y) = (x - 3)^2 + (y + 5)^2$, and the functions $x = a(t)$, $y = b(t)$ are given by $x = e^t$, $y = t^3$. Then $F(t) = f(a(t), b(t))$ is obtained by substituting $x = e^t$ and $y = t^3$ in the formula for $f(x, y)$:

$$z = f(x, y) = (x - 3)^2 + (y + 5)^2 = (e^t - 3)^2 + (t^3 + 5)^2.$$

∎

Example 2. Let $f(x, y)$ be a function of two variables. By $f(\sin t, \cos t)$ we mean that $f(a(t), b(t))$ where $a(t) = \sin t$ and $b(t) = \cos t$. For example, take $f(x, y) = x + 3y + 1$. Then

$$f(\sin t, \cos t) = \sin t + 3 \cos t + 1.$$

∎

Problems

17.1. A point P moves on the curve $(x, y) = (e^t, t^3)$ (here t stands for time). The function $d(x, y)$ is defined as the distance of the point (x, y) from the point $(3, -5)$. Then the function $D(t) = d(e^t, t^3)$ gives the distance of the curve point P from $(3, -5)$ at time t. Write down the formula for $D(t)$.

17.2. Given $f(x, y) = e^{x + 2y} + 3xy$, write out the formula for $F(t) = f(t^2 + t, \sqrt{t})$.

In the situation diagramed in (17.1), z is ultimately a function of *one* independent variable t. We now ask: *How can we calculate the derivative dz/dt?* In other words, with $F(t) = f(a(t), b(t))$, *how do we find $F'(t)$?* Notice that this is an ordinary one-variable derivative of the kind you studied in elementary calculus. You *could* calculate it by performing the indicated substitutions to get a formula for z as a function of t, and differentiate it by the rules of elementary calculus. This is not always desirable. We will work out a formula for dz/dt in terms of the *partial* derivatives of z with respect to the *intermediate* variables x and y.

---

[1] In this section, we deal only with functions with continuous partial derivatives, so that the linearization makes sense, and provides a good local approximation to the function, as discussed in Section 16.

We begin by asking what happens to the intermediate variables x and y when t is changed by a small amount, say Δt. The resulting changes in x and y are approximately

$$\Delta x \approx \frac{dx}{dt}\,\Delta t \qquad \text{and} \qquad \Delta y \approx \frac{dy}{dt}\,\Delta t. \tag{17.2}$$

These changes in the intermediate variables x and y will, in turn, cause changes in the dependent variable z. The change in z can be estimated using the differential:

$$\Delta z \approx df = \frac{\partial z}{\partial x}\,\Delta x + \frac{\partial z}{\partial y}\,\Delta y.$$

Substituting from (17.2) gives

$$\begin{aligned}
\Delta z \quad &\approx \quad \frac{\partial z}{\partial x}\frac{dx}{dt}\,\Delta t + \frac{\partial z}{\partial y}\frac{dy}{dt}\,\Delta t \\
&= \quad \left(\frac{\partial z}{\partial x}\frac{dx}{dt} + \frac{\partial z}{\partial y}\frac{dy}{dt} \right)\Delta t.
\end{aligned}$$

Dividing through by Δt now gives

$$\frac{\Delta z}{\Delta t} \approx \frac{\partial z}{\partial x}\frac{dx}{dt} + \frac{\partial z}{\partial y}\frac{dy}{dt}$$

and the approximation improves as Δt approaches 0. This gives a version of the *Chain Rule* for the derivative dz/dt.

Chain Rule *If $z = f(x, y)$ with x and y both functions of t, then*

$$\frac{dz}{dt} = \frac{\partial z}{\partial x}\frac{dx}{dt} + \frac{\partial z}{\partial y}\frac{dy}{dt}.$$

Example 3. Returning to the situation of Example 1, we let $f(x, y) = (x - 3)^2 + (y + 5)^2$, and $F(t) = f(e^t, t^3)$. We will calculate $F'(0)$ using two different methods.

For the first calculation, we use the explicit formula for $F(t)$ that was worked out in Example 1:

$$F(t) = (e^t - 3)^2 + (t^3 + 5)^2.$$

From elementary calculus,

$$F'(t) = 2(e^t - 3)e^t + 2(t^3 + 5)(3t^2)$$

so $F'(0) = -4$.

For the second solution, we use the Chain Rule for functions of two variables. Write

$$z = f(x, y) = (x - 3)^2 + (y + 5)^2.$$

Next, write the intermediate variables x and y in terms of the independent variable t:

$$x = e^t, \qquad y = t^3. \tag{17.3}$$

The Chain Rule takes the form

$$\begin{aligned}
\frac{dz}{dt} \quad &= \quad \frac{\partial z}{\partial x}\frac{dx}{dt} + \frac{\partial z}{\partial y}\frac{dy}{dt} \\
&= \quad 2(x - 3)e^t + 2(y + 5)(3t^2).
\end{aligned}$$

We are interested in the value of dz/dt when $t = 0$. To evaluate the preceding expression when $t = 0$, we must determine the values of x and y when $t = 0$. These are easily found from (17.3) to be 1 and 0, respectively. Finally, substituting $t = 0$, $x = 1$, and $y = 0$ into the above formula for dz/dt gives

$$\frac{dz}{dt}\bigg|_{t=0} = 2(1 - 3) \cdot 1 + 2(0 + 5) \cdot 0 = -4.$$

As expected, we get the same answer as before. ∎

Example 4. A fixed quantity of gas is confined in a chamber. The volume V of the chamber and the temperature T of the gas can be controlled. With V measured in liters and T in Kelvin degrees, the pressure P, in atmospheres, is given by

$$P = \frac{T}{V}. \tag{17.4}$$

The volume V is increasing at a rate of 0.2 liters/minute, and the temperature T is increasing at a rate of 20 degrees/minute. We ask: *At the instant when the volume is 2 liters and the temperature is 300 degrees Kelvin, is the pressure increasing, or decreasing, and at what rate?* Here the pressure P depends on V and T, while V and T depend on time t (in minutes). Diagrammatically, we have

$$t \mapsto (V, T) \mapsto P.$$

Here t is the independent variable, V and T are intermediate variables, and P is the dependent variable. The Chain Rule takes the form

$$\frac{dP}{dt} = \frac{\partial P}{\partial V}\frac{dV}{dt} + \frac{\partial P}{\partial T}\frac{dT}{dt}.$$

The values of dV/dt and dT/dt are given to us. Because the volume is increasing at a rate of 0.2 liters/minute, we have $dV/dt = 0.2$. Because the temperature is increasing at a rate of 20 degrees/minute, we have $dT/dt = 20$. Finally, the formulas for $\partial P/\partial V$ and $\partial P/\partial T$ can be worked out from (17.4). Substituting all this into the Chain Rule gives

$$\frac{dP}{dt} = -\frac{T}{V^2}\frac{dV}{dt} + \frac{1}{V}\frac{dT}{dt} = -\frac{T}{V^2} \cdot 0.2 + \frac{1}{V} \cdot 20.$$

At the instant when $V = 2$ and $T = 300$, we get

$$\frac{dP}{dt} = -\frac{300}{2^2} \cdot 0.2 + \frac{1}{2} \cdot 20 = -5.$$

This means that the pressure is decreasing at a rate of 5 atmospheres/minute. ∎

Problems _____

17.3. Given $f(x, y) = xy + x$, we define $F(t)$ by $F(t) = f(t^2 + 1, -t^3)$.

    a) Evaluate $F(1)$.

    b) Evaluate $F'(1)$ by writing out the formula for $F(t)$ and then taking the derivative as in elementary calculus.

    c) Evaluate $F'(1)$ by means of the Chain Rule.

17.4. Let $g(x, y)$ be a function of two variables whose formula we do not know. Push the computation of

$$\frac{d}{dw}\left(g\left(\frac{1}{1+w^2}, w\right)\right)\Bigg|_{w=2}$$

as far as you can. By "as far as you can" we mean: If you have to take a derivative of a function with known formula, do it. If you have to take a derivative of a function of which you only know the name, such as $g(x, y)$, use the symbols g_x, etc.

Notation and terminology: The gradient. Given a function $z = f(x, y)$ of two variables, it is convenient to consider the pair (f_x, f_y) of partial derivatives as components of a vector called the *gradient* of $f(x, y)$. There are two commonly used notations for gradients:

$$\operatorname{grad} f, \quad \text{or} \quad \nabla f \quad \text{(read as "del } f\text{" or "nabla } f\text{")}.$$

With the gradient notation, the Chain Rule can be written as a scalar product:

$$
\begin{aligned}
\frac{df(a(t), b(t))}{dt} &= \frac{dz}{dt} \\
&= \frac{\partial z}{\partial x}\frac{dx}{dt} + \frac{\partial z}{\partial y}\frac{dy}{dt} \\
&= f_x(a(t), b(t))a'(t) + f_y(a(t), b(t))b'(t) \\
&= \operatorname{grad} f \cdot (a'(t), b'(t)).
\end{aligned}
$$

In this formula, remember that $\operatorname{grad} f = (f_x, f_y)$ *has to be taken at the point* $(x, y) = (a(t), b(t))$.

Note:

1. The components of the vector $\operatorname{grad} f$ are *functions* of two variables, and not just numbers.
2. The gradient of a function $z = f(x, y)$ is a vector *in the xy-plane*.

Example 5. We consider polar coordinates (r, θ). Any point P in the xy-plane has Cartesian coordinates (x, y) and polar coordinates (r, θ), as illustrated below. For a review of polar coordinates, see the beginning of Section 26, page 220.

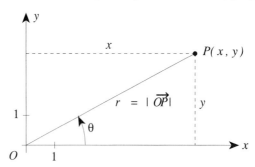

The xy-coordinates and $r\theta$-coordinates of P are related to each other by the equations

$$x = r\cos\theta, \qquad y = r\sin\theta. \qquad (17.5)$$

Given a function $z = f(x, y)$, we can rewrite it in terms of polar coordinates by making the substitutions $x = r\cos\theta$, $y = r\sin\theta$. We obtain a new formula $F(r, \theta)$ for the same function:

$$z = f(x, y) = f(r\cos\theta, r\sin\theta) = F(r, \theta).$$

Here z is the dependent variable. It is determined by the two intermediate variables x and y. These in turn depend on *two* independent variables r and θ. Diagrammatically, we have

$$(r, \theta) \mapsto (x, y) \mapsto z.$$

We now ask: *How can we use the Chain Rule to calculate the* partial *derivatives* $\partial z/\partial r$ *and* $\partial z/\partial\theta$? The answer is straightforward. Suppose, for example, that we want to find $\partial z/\partial\theta$. It is only necessary to recall that this partial derivative is calculated by by treating r as a constant and calculating the ordinary derivative with respect to θ. The Chain Rule gives

$$\frac{\partial z}{\partial\theta} = \frac{\partial z}{\partial x}\frac{\partial x}{\partial\theta} + \frac{\partial z}{\partial y}\frac{\partial y}{\partial\theta}.$$

From (17.5) we get

$$\frac{\partial x}{\partial\theta} = -r\sin\theta, \qquad \frac{\partial y}{\partial\theta} = r\cos\theta.$$

Substituting this in the formula for $\partial z/\partial\theta$ gives

$$\frac{\partial z}{\partial\theta} = -r\sin\theta\frac{\partial z}{\partial x} + r\cos\theta\frac{\partial z}{\partial y}.$$

Since we don't know anything specific about the function $z = f(x, y)$, this is as far as we can go.[2] ∎

Problems

17.5. Given $f(x, y) = (x + y)^2$, find

$$\frac{\partial f(v, u^2 + v^2)}{\partial u}\bigg|_{(u,v)=(1,1)}.$$

17.6. Go back to Example 5 where we convert $f(x, y)$ to polar coordinates:

$$z = f(x, y) = f(r\cos\theta, r\sin\theta) = F(r, \theta).$$

We want to know what the instantaneous rate of change of z with respect to r is. In other words, we want to know the partial derivative

$$\frac{\partial f(r\cos\theta, r\sin\theta)}{\partial r}.$$

Use the Chain Rule to carry out the computation as far as you can.

---

[2]We can, however, using (17.5), express the formula for $\partial z/\partial\theta$ entirely in terms of x and y:

$$\frac{\partial z}{\partial\theta} = -y\frac{\partial z}{\partial x} + x\frac{\partial z}{\partial y}.$$

This may be more convenient for some purposes.

17.7. $z = f(x, y)$ is a function of two variables. We do not know the formula. Push the evaluation of

(i) $\left.\dfrac{\partial f(uv, u+v)}{\partial u}\right|_{(u,v)=(2,3)}$ and (ii) $\left.\dfrac{\partial f(uv, u+v)}{\partial v}\right|_{(u,v)=(2,3)}$

as far as you can.

The Chain Rule for functions of three or more variables. The Chain Rule for functions of more than two variables follows the same pattern as the two variable Chain Rule. The difference is that there will be more than two intermediate variables. Each intermediate variable gives rise to a term in the Chain Rule.

For example, if $w = f(x, y, z)$, where x, y, and z are functions of t, the situation can be diagramed as follows:

$$t \mapsto (x, y, z) \mapsto w.$$

Here t is the independent variable, x, y, z are intermediate variables, and w is the dependent variable. The Chain Rule takes the form

$$\frac{dw}{dt} = \frac{\partial w}{\partial x}\frac{dx}{dt} + \frac{\partial w}{\partial y}\frac{dy}{dt} + \frac{\partial w}{\partial z}\frac{dz}{dt}.$$

Similarly, if $\alpha = f(w, x, y, z)$ with w, x, y, and z depending on t, then we have

$$t \mapsto (w, x, y, z) \mapsto \alpha.$$

There are *four* intermediate variables, so the Chain Rule gives rise to a sum with four terms:

$$\frac{d\alpha}{dt} = \frac{\partial \alpha}{\partial w}\frac{dw}{dt} + \frac{\partial \alpha}{\partial x}\frac{dx}{dt} + \frac{\partial \alpha}{\partial y}\frac{dy}{dt} + \frac{\partial \alpha}{\partial z}\frac{dz}{dt}.$$

Gradient notation. As in the two variable case, the Chain Rule can be expressed concisely in terms of gradients. For a function $f(x, y, z)$ of three variables, the gradient is the vector function

$$\operatorname{grad} f = (f_x, f_y, f_z).$$

Using this notation, the Chain Rule takes the form

$$\frac{d}{dt} f(a(t), b(t), c(t)) = \operatorname{grad} f \cdot (a'(t), b'(t), c'(t)).$$

Problems ───────────────────────────────

17.8. $g(x, y, z)$ is a function whose formula we do not know. Push the computation of $\left.\dfrac{dg(t, t, t)}{dt}\right|_{t=5}$ as far as you can.

17.9. Let $f(x, y, z) = xyz^3$.
   a) Find the partial derivatives f_x, f_y, and f_z at the following points:
      (i) $(0, 1, 4)$ (ii) $(1, 2, 3)$
   b) Evaluate: (i) $f(t^3+3, t^2+t, t)|_{t=1}$ (ii) $\left.\dfrac{d}{dt}\left(\dfrac{f(t^3+3, t^2+t, t)}{dt}\right)\right|_{t=1}$.

Additional Problems

Functions defined by composition

17.10. We are given the following functions:

$$f(x, y) = 3x + y + xy^2 + 7, \quad x(t) = t^2, \quad y(t) = \sqrt{t}.$$

Find $f(x(t), y(t))|_{t=4}$.

17.11. The two functions $w = a(t)$ and $w = b(t)$ are given by their graphs below.

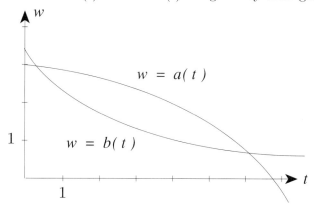

The function $f(x, y)$ is defined by $f(x, y) = 3(x - y)^2$. We define a new function $F(t)$ by $F(t) = f(a(t), b(t))$.

  a) Find $F(2) = f(a(2), b(2))$.

  b) Solve the equation $F(t) = \frac{1}{2}$ for t. If there is no solution, explain why.

17.12. On the left side of Figure 17.1 are shown some level curves of a function $z = f(x, y)$ and the circle C given by $(x, y) = (3 \cos t, 3 \sin t)$. Use the tz-system to the right to sketch the graph $z = f(3 \cos t, 3 \sin t)$ for $0 \le t \le 2\pi$. For the sketch you will have to make many guesses – do as well as you can.

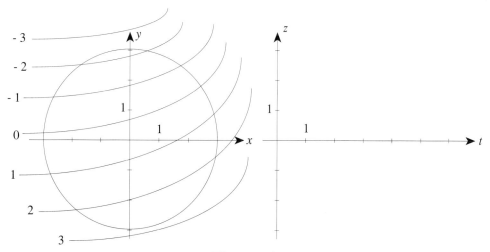

Figure 17.1

The gradient

17.13. Let $z = f(x, y) = 2xy - \frac{1}{3}y^3$.

    a) Find the components of grad f at the point $(x, y) = (2, 2)$.

    b) The vector grad f has different components at different points. This comes out clearly when we *draw* the gradient. By "drawing grad f at the point (x, y)" we mean "drawing the gradient with initial point at (x, y)." In the system below, draw grad f at the following points and mark the points with a heavy dot: (i) $A(2, 2)$; (ii) $B(0, 1)$; (iii) $C(1, 1)$; (iv) $D(-1, 0)$.

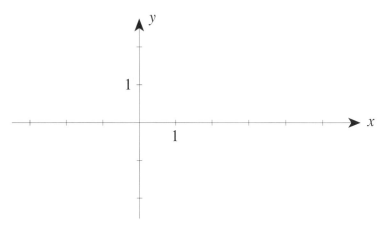

17.14. Let $z = f(x, y) = \sin(x^2 y)$.

    a) Find the components of grad f at the point $(x, y) = (3, 0)$.

    b) Find two points (x_0, y_0) and (x_1, y_1) at which the gradient of $f(x, y)$ is the zero vector. Note that there are many such points.

17.15. Find the gradient of $w = 4x - 2y + 7z - 11$ at the point $(x, y, z) = (p, q, r)$.

17.16. Let $w = f(x, y, z) = \sqrt{x^2 + y^2 + z^2}$.

    a) Write out the components of grad f.

    b) Find the magnitude $|\operatorname{grad} f|$ of the gradient at a generic point (x, y, z).

The Chain Rule

Push the computations as far as you can. By "as far as you can" we mean: If you have to take a derivative of a function with known formula, do it. If you have to take a derivative of a function of which you only know the name, such as $g(x, y)$, use the symbols g_x, etc.

17.17. We return to the functions used in Problem 16.12:

$$z = f(x, y) = 3x + y + xy^2 + 7, \quad \text{with} \quad x(t) = t^2, \quad b(t) = \sqrt{t}.$$

Evaluate $\dfrac{d}{dt}\Big[f(x(t), y(t))\Big]_{t=4}$.

17.18. We go back to Problem 17.11: $f(x, y)$ is the function $f(x, y) = 3(x-y)^2$, $w = a(t)$, and $w = b(t)$ are as shown in Problem 17.11, and $F(t) = f(a(t), b(t))$. Find $F'(1)$. *Hint:* Use ruler and pencil for $a(t)$ and $b(t)$ and their derivatives.

17.19. The function $f(x, y)$ satisfies

$$\frac{\partial f}{\partial x} = y, \qquad \frac{\partial f}{\partial y} = x.$$

Find the point(s) on the circle $x^2 + y^2 = 1$ at which $f(x, y)$ has a larger value than at any other point on the circle. *Hint:* The circle has parametric representation $x = \cos t$, $y = \sin t$. Look at the derivative of the one variable function $f(\cos t, \sin t)$.

17.20. $h(x, y, z) = xyz$. Find $\dfrac{dh(t + 1, t^2 + 2, t^3 + 3)}{dt}$.

17.21. We do not know the formula of $p(x, y)$. Find $\dfrac{\partial p(2u + 3v, u^2)}{\partial u}$.

17.22. The function $a(t)$ is given by its graph shown in the following figure at left. We define a new function $g(x, y)$ by $g(x, y) = a(\frac{1}{2}x^2 + \frac{1}{2}y^2)$. Find $g_y(1, 2)$, that is, the partial derivative g_y evaluated for $(x, y) = (1, 2)$.

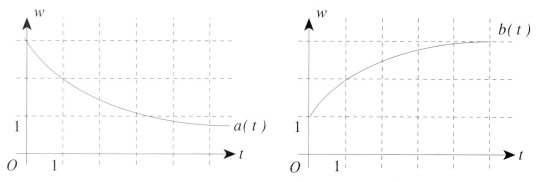

17.23. The function $m(t)$ is given by $m(t) = e^t$. We define a new function $g(x, y)$ by $g(x, y) = m(\sin x \cos y)$. Find g_y.

17.24. $h(x, y) = xy$, and $a(t)$, $b(t)$ are the functions whose graphs are shown above. Evaluate the derivative $\dfrac{dh(a(t), b(t))}{dt}$ for $t = 2$.

17.25. With $g(x, y) = \sin x \sin y$ and $p(u, v)$ a function whose formula we do not know, find $\dfrac{\partial g(p(u, v), p(u, v))}{\partial u}$.

17.26. Imagine an xyz-system fixed rigidly to the earth, with its origin at the earth's center, the z-axis passing through the north pole, and the x-axis passing through the zero longitude line through Greenwich. Then the xyz-coordinates of the point on the earth's surface with latitude u and longitude v are

$$\begin{aligned} x &= R \cos u \cos v \\ y &= R \cos u \sin v \\ z &= R \sin u \end{aligned}$$

where R is the radius of the earth. Let $w = f(x, y, z)$ be a function defined for points (x, y, z) near the surface of the earth.

a) Work out a formula for the latitude/longitude derivatives $\frac{\partial w}{\partial u}$ and $\frac{\partial w}{\partial v}$ in terms of the xyz derivatives $\frac{\partial w}{\partial x}$, $\frac{\partial w}{\partial y}$, and $\frac{\partial w}{\partial z}$.

b) If w depends only on z, and not on x or y, what can you say about the longitudinal derivative $\frac{\partial w}{\partial v}$?

17.27. The radius of a cylinder is increasing at a rate of 2 cm/min while the height is decreasing at a rate of 1 cm/min. At the instant when the radius is 10 cm and the height is 5 cm, is the volume increasing or decreasing? At what rate?

18   Directional Derivatives: Functions of Two Variables

In Section 15, we considered a function $z = f(x, y)$ and a point $P_0(a, b)$ in the xy-plane. There, the key question was (see the following figure, left):

> *At $P_0(a, b)$, does $f(x, y)$ increase or decrease in the direction of $\vec{\imath}$ (or $\vec{\jmath}$)?*
> *At which rate does $f(x, y)$ increase or decrease?*

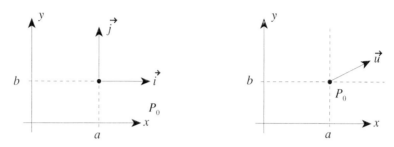

The key question of this section asks the same thing for *any* vector $\vec{u}$, and not just $\vec{\imath}$ and $\vec{\jmath}$, as follows (see above right):

> **The key question.** At $P_0(a, b)$, does $f(x, y)$ increase or decrease in the direction of a given vector $\vec{u}$? At which rate does $f(x, y)$ increase or decrease?

For the key question of Section 15 we used partial derivatives. For the key question of the present section, we are going to use what is called the *directional derivative* of $f(x, y)$ at $P_0(a, b)$ in the direction of $\vec{u}$. A directional derivative measures the rate at which $f(x, y)$ changes as you move in the direction of the vector $\vec{u}$.

Problem _____

18.1. Below are shown some level curves of a linear function $f(x, y)$.

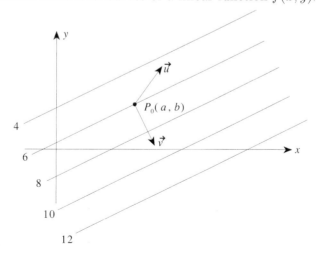

a) Is the partial derivative f_x positive, zero, or negative? What about the partial derivative f_y?

b) At P_0, does $f(x, y)$ increase or decrease in the direction of the vector $\vec{u}$?

c) In part b, you have answered the following question: "At the point P_0, is the directional derivative of $f(x, y)$ in the direction of $\vec{u}$ positive, or zero, or negative?" Now answer the following question for the vector $\vec{v}$, which is different from $\vec{u}$: At the point P_0, is the directional derivative of $f(x, y)$ in the direction of $\vec{v}$ positive, or zero, or negative?

d) Can you find a vector $\vec{w}$ such that the directional derivative of $f(x, y)$ at P_0 is zero?

☞ What is a directional derivative?

To talk about a directional derivative we need the following items:

1. A function $f(x, y)$;
2. The point $P_0(a, b)$ in the xy-plane where we want to take the directional derivative;
3. The vector $\vec{u} = (u_1, u_2)$ in the direction of the required directional derivative.

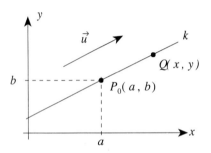

Figure 18.1

Items (2) and (3) are shown in Figure 18.1. We draw the line k that passes through P_0 and that is parallel to $\vec{u}$. Now imagine a point $Q(x, y)$, which moves along the line k with unit speed in the direction determined by the vector $\vec{u}$. We place the point Q at P_0 at time $t = 0$, so that the coordinates of Q at time t are given by

$$x(t) = a + \frac{u_1}{|\vec{u}|} t$$

$$y(t) = b + \frac{u_2}{|\vec{u}|} t.$$

We watch the values of $f(x, y)$ as the point $Q(x, y)$ moves as described earlier. The *directional derivative* of $f(x, y)$ at the point P_0 in the direction $\vec{u}$ is defined to be the rate of change of $f(x(t), y(t))$ as the point Q passes through the point P_0 at time $t = 0$. More concisely, it is

$$\left. \frac{d}{dt} f(x(t), y(t)) \right|_{t=0}. \tag{18.1}$$

Notice that because $\frac{\vec{u}}{|\vec{u}|}$ is a unit vector, the point $(x(t), y(t))$ moves along the line with speed 1. This tells us that we travel 1 distance unit for each time unit. It follows that the directional derivative may be viewed as the rate of change of $f(x, y)$ with respect to distance traveled as you move along a line in the direction of the vector $\vec{u}$.

Note:

1. For $\vec{u} = \vec{i}$ we get the partial derivative f_x, and for $\vec{u} = \vec{j}$ we get the partial derivative f_y.

2. The directional derivative in the direction of $-\vec{u}$ is the negative of the directional derivative in the direction of $\vec{u}$ (why?). In general, if we change $\vec{u}$ to a different vector $\vec{v}$ then we will get a different directional derivative. This means: *At a point, one and the same function has many different directional derivatives.*

✍ How do we compute a directional derivative?

Since (18.1) gives the directional derivative as a derivative of a composition of two functions, we can calculate it by the Chain Rule. We use the gradient form of the Chain Rule to calculate the derivative in (18.1). We get that the directional derivative of $f(x, y)$ in the direction $\vec{u}$ at the point (a, b) is

$$\frac{d}{dt} f(x(t), y(t)) \bigg|_{t=0} = \text{grad}\, f(x(0), y(0)) \cdot (x'(0), y'(0))$$

$$= \text{grad}\, f(a, b) \cdot \frac{\vec{u}}{|\vec{u}|}.$$

Remember: The dot product of a vector $\vec{a}$ with a unit vector $\vec{b}$ is the scalar projection of $\vec{a}$ onto $\vec{b}$.

Result: *The directional derivative of $f(x, y)$ at $P_0(a, b)$ in the direction of $\vec{u}$ is computed as the scalar projection of the gradient of $f(x, y)$ at P_0 onto $\vec{u}$.*

Problems

18.2. With

$$h(x, y) = (x - 2)^2 - xy + 7,$$

find the directional derivative at P_0 in the direction of $\vec{a}$ for

a) $P_0(1, 1)$, $\vec{a} = (3, 4)$

b) $P_0(1, 1)$, $\vec{a} = (75492, 0)$

c) $P_0(2, 2)$, $\vec{a} = (2, 2)$

d) $P_0(10, 12)$, $\vec{a} = (0, -7)$

18.3. a) Write down in complete sentences how the notions "directional derivative" and "partial derivative" are related.

b) Which of the questions of Problem 18.2 can be answered quickly by using part a?

Notation. Different books use different notation for the directional derivative of $f(x, y)$ at P_0 in the direction of $\vec{u}$. Some write $D_{\vec{u}} f \big|_{P_0}$, and others write $\dfrac{\partial f}{\partial s}$. As a rule, we do not use any notation, but spell out "directional derivative of ..." in words.

Problem

18.4. Let $f(x, y)$ be the function whose gradient at $P_0(a, b)$ is shown in the following figure.

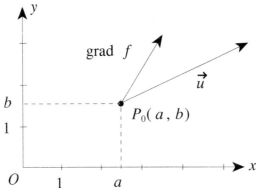

a) Find the directional derivative of f at P_0 in the direction of $\vec{u}$.

b) Draw a vector $\vec{c}$ so that the derivative in the direction of $\vec{c}$ is the highest of all directional derivatives at P_0. *Hint:* Take $\vec{u}$ and turn it slowly around P_0. As you turn, watch the scalar projection of grad f onto $\vec{u}$.

c) In b), you found the *direction* of the highest possible directional derivative at P_0. What is the *value* of this highest derivative?

Description of the gradient in terms of directional derivatives. In Problem 18.4, parts b and c, you found: The highest possible directional derivative occurs in the direction of grad f, and its value is $|\operatorname{grad} f|$. We reformulate these facts as follows:

> *The gradient of $f(x, y)$ at P_0 is a vector. Its direction is the direction in which the directional derivative of $f(x, y)$ is highest possible. Its magnitude is the value of the highest possible directional derivative at P_0.*

Problem

18.5. We do not know a formula of the function $f(x, y)$, but know the following about its directional derivatives: At $P(3, 7)$, the *lowest* directional derivative occurs in the direction of $\vec{b} = (-4, 2)$, and this lowest derivative equals -5.9. Find the gradient and the two partial derivatives of $f(x, y)$ at P.

Directional derivatives and graphs

How are the directional derivatives of a function related to its graph? We go back to the definition of the directional derivative. In Figure 18.2 on the facing page, you find the graph $z = f(x, y)$ shown above the xy-plane of Figure 18.1 on page 150. You recognize

- The point $P_0(a, b)$ (in the xy-plane) at which we take the directional derivative;

- The vector $\vec{u} = (u_1, u_2)$ in the xy-plane, the vector in the direction of which we take the directional derivative;
- The line k through the point P_0 and parallel to $\vec{u}$;
- The point $Q(x, y)$ on the line k;
- The plane α defined by the points $P_0(a, b)$, $Q(x, y)$, and $P(a, b, f(a, b))$;
- The curve L in which the plane α intersects the graph $z = f(x, y)$.

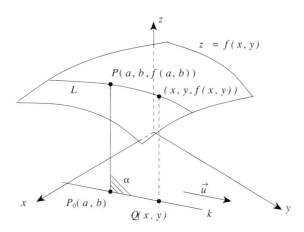

Figure 18.2

The function value $f(x, y)$ gives the height of the curve L above the line k at the point $Q(x, y)$. Since the directional derivative measures the rate of change of $f(x, y)$ with respect to distance along the line α, it gives the slope of the tangent line to the curve L, measured in the plane α.

Result: *The directional derivative of $f(x, y)$ at P_0 in the direction of $\vec{u} = (u_1, u_2)$ is the slope of the curve L in the plane formed by $\vec{u}$ and $\vec{k} = (0, 0, 1)$. In other words: You are standing at the point $P(a, b, f(a, b))$ on the graph and look straight ahead in the direction of the vector $(u_1, u_2, 0)$. The directional derivative of $f(x, y)$ at $P_0(a, b)$ in the direction of $\vec{u}$ tells you how much the surface slopes up or down under your feet. If you change $\vec{u}$ then the plane α changes, the curve L will have a different slope at P, and the directional derivative will be different.*

Problem _____

18.6. The graph of the function $z = g(x, y)$ is part of a right circular cone shown in Figure 18.3. Its axis is the z-axis, its base in the xy-plane has radius 3, and at height $z = 4$ it has radius 1. P_0 is the point in the xy-plane with coordinates $(x, y) = (2, 0.5)$.

    a) Find a vector $\vec{u} = (u_1, u_2)$ so that the directional derivative of $g(x, y)$ at P_0 in the direction of $\vec{u}$ is positive.

    b) As in part a, but with a negative directional derivative.

    c) As in part b, but with directional derivative zero.

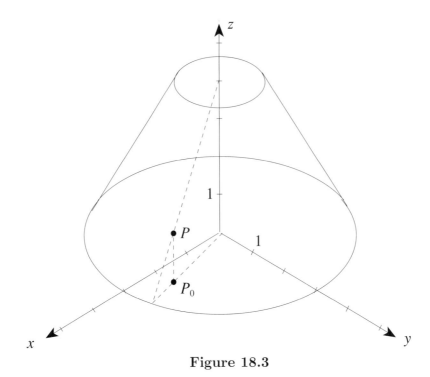

Figure 18.3

Relations among directional derivatives, gradient, and level curves

We look at a function $z = f(x, y)$ and focus at the point $P_0(a, b)$. There we have two objects connected with $f(x, y)$, namely the gradient grad f at P_0 and the level curve C which passes through P_0. The gradient and the level curve *must* be related. *How?*

The level curve C passing through $P_0(a, b)$ is the collection of all points (x, y) for which $f(x, y) = f(a, b)$. If we move through P_0 along C, the function is constant. Therefore, the directional derivative of $f(x, y)$ at $P(a, b)$ in the direction of a tangent vector $\vec{u}$ of the level curve c must be zero. The directional derivative is the dot product grad $f \cdot \frac{1}{|\vec{u}|}\vec{u}$. This dot product can be zero for two reasons:

(i) grad $f = (0, 0)$ at $P_0(a, b)$.

(ii) grad $f \neq (0, 0)$, and grad f is perpendicular to C.

Thus, the answer to our question "What is the relation between gradient and level curves?" reads as follows:

At a point P_0 in the xy-plane, the gradient of $f(x, y)$ is either zero or different from zero. If it is different from zero, it is perpendicular to the level curve of $f(x, y)$ which passes through P_0, as shown in Figure 18.4 on the facing page.

Problems _____

18.7. In the situation of the Figure 18.4 on the next page, find the directional derivative of $f(x, y)$ at P_0 in the direction of the following vectors $\vec{u}$ (use 1 cm as unit for measurements): (i) $\vec{u}$ is parallel to a tangent vector at P_0; (ii) $\vec{u}$ is parallel to grad f at P_0.

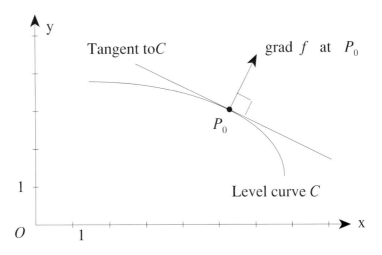

Figure 18.4

18.8. We use the fact that the gradient is normal to level curve" to find normals and tangents to curves in the plane which are given implicitly, as follows. Consider the curve C in the xy-plane given by the equation

$$F(x, y) = \text{const.}$$

We interpret C as a level curve of the function $z = F(x, y)$. We know that grad F is normal to the level curves. Therefore, grad F is normal to C. Once we have a normal of C we can quickly find a tangent vector. As an example take the curve C given by $x^4 - y^3 + 2x^2 + 6y + 1 = 9$. The point $(x, y) = (1, 1)$ lies on C.

a) Find a normal vector and a tangent vector of C at $(1, 1)$.

b) Can you find a vector $\vec{b}$ with the following property: At $(1, 1)$, the directional derivative of $z = F(x, y) = x^4 - y^3 + 2x^2 + 6y - 8$ in the direction of $\vec{b}$ is zero? If yes, find $\vec{b}$; if no, explain.

Warning: *Previously, we said that the gradient is perpendicular to level curves, and Figure 18.4 illustrates this fact. The statement is true only if we work in an xy-system where the units on both axes have the same length. If we work in a uv-system*

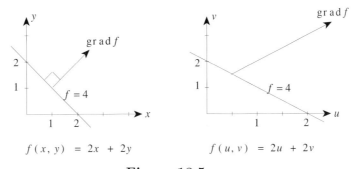

$$f(x, y) = 2x + 2y \qquad f(u, v) = 2u + 2v$$

Figure 18.5

where the units on the axes are of different lengths, the gradient will not appear at a right angle to the level curve. As an example, in Figure 18.5 on the page before, we represent the same linear function in an xy-system (same units on both axes) and in an uv-system (different units).

Additional Problems

Linear functions

18.9. With $f(x, y) = 7x + 2y + 3$:

  a) Find the directional derivative at $P_0(1, 6)$ in the direction of $\vec{u} = (4, -3)$.

  b) As in part a, but for $P_0(11, -5)$ and the same $\vec{u}$.

  c) Is there a vector $\vec{u}$ so that the directional derivative at $P_0(9, -3)$ in the direction of $\vec{u}$ is zero? If yes, find such a vector; if no, explain.

  d) Is there a point P_0 so that *all* directional derivatives at P_0 are zero? If yes, find such a point; if no, explain.

  e) Among all the directional derivatives at $P_0(-4, 8)$ find the largest and the smallest. Find vectors $\vec{u}$ and $\vec{v}$ in the direction of which the derivatives are largest and smallest.

  f) Is there a vector $\vec{u}$ so that the directional derivative at $P_0(11, 2)$ in the direction of $\vec{u}$ equals 8? If yes, find such a direction; if no, explain. *Hint:* "The directional derivative in the direction of ... is the scalar projection of ... onto ...."

  g) As in part f, but with the directional derivative equal to 2. *Hint:* What are the components of grad f?

18.10. A contour plot of a linear function $f(x, y)$ is shown at in Figure 18.6.

  a) Find the components of the gradient of f, and draw grad f into the figure.

  b) Find a formula for $f(x, y)$.

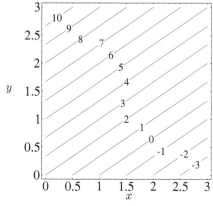

Figure 18.6

18.11. You are standing on a flat hillside with a 20% grade. In other words, the slope in the uphill direction is .20. If a due north heading takes you straight uphill, how much altitude will you gain if you head northeast for a 10 meter horizontal displacement?

Nonlinear functions

18.12. Let $f(x, y) = xy - y$.

    a) Find the directional derivative at $P_0(3, 1)$ in the direction of $\vec{u} = (1, 5)$.

    b) Is there a vector $\vec{u}$ so that the directional derivative at $P_0(3, 1)$ in the direction of $\vec{u}$ equals zero? If yes, find such a vector; if no, explain.

    c) Is there a point P_0 at which *all* directional derivatives at P_0 are zero? If yes, find such a point; if no, explain.

    d) Can you find a point $P_0(a, b)$ and a vector $\vec{u} = (u_1, u_2)$ so that the directional derivative at P_0 in the direction of $\vec{u}$ equals 327.84?

18.13. In this problem we take directional derivatives always at the same point P_0. For simplicity, we temporarily use the notation $D_{\vec{u}}f$ for the directional derivative of $f(x, y)$ in the direction of $\vec{u}$. Remember that $D_{\vec{u}}f = \operatorname{grad} f \cdot \vec{u}_0$ where $\vec{u}_0$ is the unit vector in the direction of $\vec{u}$.

    a) We take directional derivatives of $f(x, y)$, $g(x, y)$, and $f(x, y) + g(x, y)$ in the direction of $\vec{u}$. We want to compare them with each other. Is the following statement true or not? Justify your answer.

$$D_{\vec{u}}(f + g) = D_{\vec{u}}f + D_{\vec{u}}g$$

    b) We take directional derivatives of $f(x, y)$ in the direction of $\vec{u}$, $\vec{v}$, and $\vec{u} + \vec{v}$. We want to compare them with each other. Is the following statement true or not? Justify your answer.

$$D_{\vec{u}+\vec{v}}f = D_{\vec{u}}f + D_{\vec{v}}f$$

18.14. A contour plot for $f(x, y)$ is shown below.

    a) Estimate the magnitude of the gradient of f at the point $(2, 2)$, and sketch the gradient vector at that point on the plot.

    b) Use your answer to part a to estimate the partial derivatives $f_x(2, 2)$ and $f_y(2, 2)$.

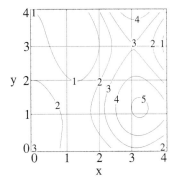

18.15. You know that $\operatorname{grad} f(P)$ has magnitude 3.

a) Find the directional derivative of f at P in the direction of grad $f(P)$.

b) Find the directional derivative of f at P in a direction forming a 45 degree angle with grad $f(P)$. (See the following figure, left.)

c) The vector v_θ forms an angle θ with grad $f(P)$. Find a formula, in terms of θ, for the directional derivative of f at P in the direction of $\vec{v}_\theta$. (See the following figure, right.)

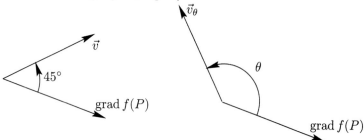

18.16. The temperature in a region in the xy-plane is given by

$$T(x,y) = e^x \cos y - y.$$

a) If you start at the origin, in what direction should you walk to experience the fastest initial rate of increase in temperature?

b) If you start at the origin and walk toward the point $(1,2)$, will you initially experience an increase, or decrease in temperature? At what rate (in temperature units per distance unit)?

19 Directional Derivatives: Functions of Three Variables

Most of the time, we can reason as we did for functions of two variables in Section 18.

The partial derivatives of $w = f(x, y, z)$ at $P_0(a, b, c)$ tell us whether $f(x, y, z)$ increases or decreases at P_0 in the directions of $\vec{\imath}$, or $\vec{\jmath}$, or $\vec{k}$ (see below left). Now we ask: *At P_0, does $f(x, y, z)$ increase or decrease in the direction of a given vector $\vec{u}$ different from $\vec{\imath}$, $\vec{\jmath}$, and $\vec{k}$ (see below right)? At which rate does $f(x, y, z)$ increase or decrease?* For the answer to these questions we need directional derivatives.

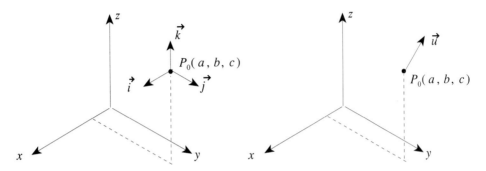

What follows is a shortened replay of the definition and computation of directional derivatives of functions of two variables in Section 18.

What are directional derivatives?

We are given three objects:

(i) A function $f(x, y, z)$;

(ii) The point $P_0(a, b, c)$ in xyz-space at which we want to take the directional derivative;

(iii) The vector $\vec{u} = (u_1, u_2, u_3)$ in the direction of which we want to take the directional derivative.

The directional derivative of $f(x, y, z)$ at $P_0(a, b, c)$ in the direction of $\vec{u} = (u_1, u_2, u_3)$ is obtained as follows. Form the "unit speed" parametrization of the line through P_0 in the direction of $\vec{u}$:

$$
\begin{aligned}
x(t) &= a + \frac{u_1}{|\vec{u}|} t \\
y(t) &= b + \frac{u_2}{|\vec{u}|} t \\
z(t) &= c + \frac{u_3}{|\vec{u}|} t.
\end{aligned}
$$

The directional derivative of $f(x, y, z)$ in the direction $\vec{u}$ at the point P_0 is then

$$
\frac{d}{dt} f(x(t), y(t), z(t)) \Big|_{t=0} . \tag{19.1}
$$

How do we compute directional derivatives?

The directional derivative is computed by using the Chain Rule to calculate the derivative is (19.1). This gives the expected result:

The directional derivative of $f(x, y, z)$ at $P_0(a, b, c)$ in the direction of a given vector $\vec{u}$ is computed as the scalar projection of the gradient of f onto the vector $\vec{u}$:

$$\operatorname{grad} f\Big|_{(x,y,z)=(a,b,c)} \cdot \frac{\vec{u}}{|\vec{u}|}.$$

Problems

19.1. The function $f(x, y, z)$ of three variables x, y, z is defined by $f(x, y, z) = x^2 z + y$.

    a) Find the directional derivative at $P_0(1, 1, 1)$ in the direction of $\vec{a} = (1, 1, -4)$.

    b) As in part a, but for $\vec{a} = (0, 0, -5)$.

    c) As in part a, but for $P_0(3, 3, 3)$ and $\vec{a} = (3, 3, 3)$.

19.2. With $g(x, y, z) = x^2 - 4x + y^2 - 9y + z^2$, is there a point Q_0 at which *all* directional derivatives of g are zero? If yes, find such a point; if no, explain.

Description of the gradient in terms of directional derivatives

At $P_0(a, b, c)$, the highest possible directional derivative is in the direction of grad f, and its value is $|\operatorname{grad} f|$. This means:

> *The gradient of $f(x, y, z)$ at P_0 is a vector. Its direction is the direction in which the directional derivative of $f(x, y, z)$ at P_0 is highest possible. Its magnitude is the value of the highest possible directional derivative at P_0.*

Problem

19.3. $f(x, y, z)$ is the function $w = g(x, y, z)$ of Problem 19.2.

    a) Find the directional derivative at $P_0(1, 1, 1)$ in the direction of grad g at P_0.

    b) Can you find a vector $\vec{u} = (u_1, u_2, u_3)$ with the following property: At $P_0(1, 1, 1)$, the directional derivative of $g(x, y, z)$ in the direction of $\vec{u}$ is zero? If yes, find such a vector, if no, explain. Read on: If your answer to part b is yes, how many such vectors are there which are not parallel to each other? If there are more than one, find two.

The tangent plane of a level surface

Remember that functions of three variables have level surfaces, and not level curves.

    Given $w = f(x, y, z)$, we consider the level surface S of level k. The equation of S is $f(x, y, z) = k$. In addition, we consider a space curve C *which lies on S* (see left in Figure 19.1 on the next page). *What does it mean for the curve C to lie on the surface S ?* If $(x, y, z) = (p(t), q(t), r(t))$ is a parametric representation of C, then the three functions $p(t)$, $q(t)$, and $r(t)$ must satisfy

$$f(p(t), q(t), r(t)) = k.$$

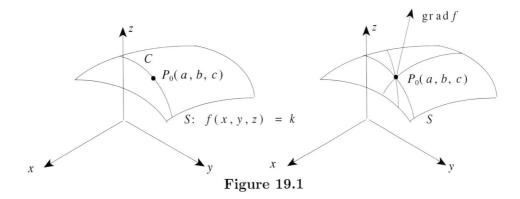

Figure 19.1

We apply the Chain Rule to both sides of this equation: $\operatorname{grad} f \cdot (p', q', r') = 0$. It means: The scalar product of the gradient of $f(x, y, z)$ at P_0 with the tangent vector (p', y', r') of C is zero. The scalar product can be zero for two reasons:

(i) $\operatorname{grad} f = (0, 0, 0)$ at $P_0(a, b, c)$.

(ii) $\operatorname{grad} f \neq (0, 0, 0)$, and $\operatorname{grad} f$ is perpendicular to C.

To discuss (ii), assume that $\operatorname{grad} f$ is different from zero at $P_0(a, b, c)$. Statement (ii) is true, *no matter what the curve C lying on S is.* It is true for *all* curves on S passing through P. For this, see Figure 19.1, right; all the curves through P_0 are meant to lie on S We conclude: The tangent lines of all curves on S passing through P_0 are perpendicular to a single vector, namely $\operatorname{grad} f$. Therefore, these tangent lines form a plane. It is called the *tangent plane* of the level surface S at P_0. A normal vector of the tangent plane at $P_0(a, b, c)$ is given by $\operatorname{grad} f$ evaluated at $P_0(a, b, c)$.

Result: *At a point P_0 in xyz-space, the gradient of $f(x, y, z)$ is either zero or different from zero. If it is different from zero, it is perpendicular to the tangent plane of the level surface which passes through P_0.*

We can describe the situation for functions f of two and three variables in one sentence:

> *The gradient at P_0 of a function f of two or three variables is either the zero vector, or it is normal to the level object of f which passes through P_0.*

Problem _____

19.4. Let $f(x, y, z) = xy^2 z + 3$.

   a) S is the level surface of $f(x, y, z)$ which passes through $P_0(1, 1, 1)$. Write down the equation of S.

   b) Give a parametric representation of the line k which passes through $P_0(1, 1, 1)$ and is normal to S.

   c) Find the equation of the tangent plane of S at $P_0(1, 1, 1)$.

   d) Find the equation of a line k which lies in the tangent plane of part c.

Application to functions of two variables

We can use the theory of functions of three variables to find the tangent plane of the graph of functions of two variables, as follows.

Given a function $g(x, y)$ of *two* variables we can define a new function $F(x, y, z)$ of *three* variables by

$$F(x, y, z) = g(x, y) - z.$$

The level surface of $F(x, y, z)$ of level zero is given by $F(x, y, z) = 0 : g(x, y) - z = 0$, that is, $z = g(x, y)$. It means:

The graph $z = g(x, y)$ is the level surface of level zero of the function $F(x, y, z) = g(x, y) - z$.

Therefore, the normal vector of the level surface $F(x, y, z) = 0$ is parallel to the normal vector of the graph $z = g(x, y)$. What is grad F?

$$\operatorname{grad} F = (F_x, F_y, F_z) = (g_x, g_y, -1).$$

This is the formula for the normal vector of the graph $z = f(x, y)$ which we found in Section 18 by going through the linear approximation of $g(x, y)$.

Note: Remember at all times:

- The gradient of a function of *two* variables is a vector in the *plane.*
- The gradient of a function of *three* variables is a vector in *space.*

Additional Problems

Directional derivatives

19.5. Let $f(x, y, z) = 4x - y + 3z + 11$.

    a) Find the average rate of change of $f(x, y, z)$ between $P_0(3, 1, 1)$ and $Q(8, 2, 5)$.

    b) P_0 and Q are the same points as in part a. Find the directional derivative of $f(x, y, z)$ at P_0 in the direction of the vector $\overrightarrow{P_0Q}$.

    c) This question is very much open-ended – do not worry if it takes some time to figure it out: You got the same answer to a) and b). Is this a coincidence? Is there a pattern? *Hint:* Go back to "How to compute directional derivatives": Look at the line k defined by P_0 and Q. Take a parametric representation of k of the form $(x, y, z) = \overrightarrow{OP_0} + t\vec{v}$, where $\vec{v}$ is a unit vector in the direction of $\overrightarrow{OP_0}$. $f(x, y, z)$ considered for points on k becomes a function $F(t)$ of t. Express the average rate of change and the directional derivative in terms of $F(t)$.

19.6. Let $f(x, y, z) = x^2 + 2y^2 - z^2 + 3yz$.

    a) At the point $P_0(2, -1, 3)$, find two nonparallel directions in which the directional derivative of f is 0.

    b) How are the vectors you found in part a related to the level surface of $f(x, y, z)$ passing through P_0?

19.7. Recall that a *radial* function $f(x, y, z)$ is one that can be expressed in the form

$$f(x, y, z) = g\left(\sqrt{x^2 + y^2 + z^2}\right)$$

where $g(t)$ is a function of one variable.

a) You discovered in Problem 13.10 that a typical level surface of a radial function f is a sphere, with center at the origin. Since grad $f(P)$ is normal to the level surface of f that passes through P, it must be perpendicular to the sphere about the origin passing through $P = P(x, y, z)$. In other words, grad $f(P)$ must be a multiple of the radial unit vector $\vec{u}$ at P, which points directly away from the origin when its tail is drawn at P. Use the Chain Rule to find a formula for grad $f(x, y, z)$ in terms of the derivative of g and the radial vector u at P. *Hint:* If you let $\vec{r} = x\vec{i} + y\vec{j} + z\vec{k}$ and $r = |\vec{r}| = \sqrt{x^2 + y^2 + z^2}$, then the radial vector $\vec{u}$ can be expressed as $\vec{u} = \frac{\vec{r}}{r}$.

b) Use the formula you found in part a to find the gradient of the Newtonian potential

$$\varphi(x, y, z) = \frac{1}{r} = \frac{1}{\sqrt{x^2 + y^2 + z^2}}$$

without calculating the partials with respect to x, y, and z.

Tangent planes of graphs and level surfaces

19.8. Let $z = f(x, y)$ be a function of two variables. We do not know a formula for $f(x, y)$, but we are told that the two curves

$$
\begin{aligned}
C : (x, y, z) &= (4 + t, 5 - t, -t^2 + t + 17) \\
K : (x, y, z) &= (u^2 + 1, 2u, 2u^3 + 2u - 3)
\end{aligned}
$$

lie on the graph $z = f(x, y)$. It is easily checked that the two curves intersect at the point $P(5, 4, 17)$. The figure below illustrates the situation; it is not to scale.

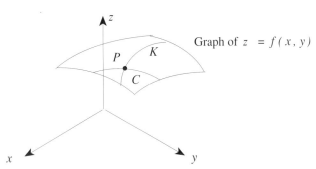

a) Do we have enough information to determine $f(5, 4)$? If yes, find $f(5, 4)$; if no, explain.

b) Do we have enough information to determine the equation of the tangent plane to the graph $z = f(x, y)$ at the point P? If yes, find the tangent plane; if no, explain.

19.9. We have the following information on the function $w = g(x, y, z)$: At $P(4, 3, 1)$ the directional derivatives in the direction of $\vec{a} = (3, 1, 2)$ and $\vec{b} = (5, 0, 2)$ are both zero.

    a) Is that enough information to determine $g(4, 3, 1)$? If yes, find $g(4, 3, 1)$; if no, explain.

    b) Is that sufficient information to determine the tangent plane of the level surface of $g(x, y, z)$ which passes through P? If yes, find the tangent plane, if no, explain.

Linear approximation

19.10. Use linear approximation to estimate $f(10^{-100}, 1 + 10^{-100}, 2 - 10^{-100})$ when $f(x, y, z) = e^{xz}(1 + \ln y)$.

19.11. You know that grad $f(P) = (1, -2, 2)$. You start at the point P, and move a short distance a in a direction at a 60 degree angle to grad $f(P)$, to a point Q. Estimate the value of $f(Q) - f(P)$. Your answer will depend on a.

19.12. The linearization of $f(x, y, z)$ at the point $(1, 2, 3)$ is $L(x, y, z) = 2x - y + 3z$. Find $f(1, 2, 3)$ and grad $f(1, 2, 3)$.

Differentials

19.13.   a) Write out the formula for the differerentials of

        (i) $W(p, q, r) = pq^3 r^7$

        (ii) $A(a, b, \theta) = ab \sin \theta$

    at the generic points (p, q, r) and (a, b, θ).

    b) Write out the formula for approximating the relative change of the two functions of a). Bring the result into the form

$$K \mathrm{rc}_1 + L \mathrm{rc}_2 + M \mathrm{rc}_3$$

    where rc_1, rc_2, and rc_3 stand for the relative changes in the first, second, and third variable, respectively.

19.14. We have the following information on the function $T(u, v, w)$: $T(2, 5, 1) = 162$, $T_u(2, 5, 1) = 4.3$, $T_v(2, 5, 1) = 12.8$, and $T_w(2, 5, 1) = -8.2$.

    a) Give the best estimate of the amount by which T will change if we change u by -0.2, v by 0.1, and w by -0.3. Would you expect the change of T to be an increase or decrease?

    b) We let u vary by at most 3%, v by at most 2.5% and w by at most 2%. Estimate by what percentage T will vary from $T(2, 5, 1)$ at most.

20   Higher-Order Approximations

In Section 16, we discussed approximation of a function of two or more variables by a linear function. Given a function $f(x, y)$ of two variables, and a point $P(x_0, y_0)$, we found that the linear function

$$L(x, y) = f(x_0, y_0) + f_x(x_0, y_0)(x - x_0) + f_y(x_0, y_0)(y - y_0)$$

is a good approximation to $f(x, y)$ when (x, y) is near (x_0, y_0). In this section, we will find better approximations by adding more terms involving powers of $(x - x_0)$ and $(y - y_0)$.

Functions of one variable

We begin by reviewing the familiar situation for functions of one variable. We consider the function $f(x)$, and we seek to approximate $f(x)$ by a simpler function when x is near some fixed number x_0. To find the best *linear* approximation, which we denote $T_1(x)$, we write $T_1(x) = a + b(x - x_0)$, where a and b are as yet unknown constants. We then require that

$$\begin{aligned} T_1(x_0) &= f(x_0) \\ T_1'(x_0) &= f'(x_0). \end{aligned}$$

Using the formula for $T_1(x)$, this gives

$$\begin{aligned} a &= f(x_0) \\ b &= f'(x_0) \end{aligned}$$

so

$$T_1(x) = f(x_0) + f'(x_0)(x - x_0).$$

This is just the linearization of f at $x = x_0$.

The function $T_1(x)$ is the *best* linear approximation to $f(x)$ for x near x_0, but we could to better with a *quadratic* approximation. To do that, we seek a quadratic function $T_2(x) = a + b(x - x_0) + c(x - x_0)^2$, and we require

$$\begin{aligned} T_2(x_0) &= f(x_0) \\ T_2'(x_0) &= f'(x_0) \\ T_2''(x_0) &= f''(x_0). \end{aligned}$$

From the formula for $T_2(x)$, this gives

$$\begin{aligned} a &= f(x_0) \\ b &= f'(x_0) \\ 2c &= f''(x_0), \end{aligned}$$

so we get

$$T_2(x) = f(x_0) + f'(x_0) + \frac{f''(x_0)}{2}(x - x_0)^2.$$

This is the best quadratic approximation to $f(x)$ for x near x_0.

Problem _____

20.1. Find the best linear and quadratic approximations to the function $f(x) = \ln x$ near $x = 1$. Use each to find an approximate value of $\ln 1.1$.

_____

Error estimates. We now ask: *In what sense is the quadratic approximation $T_2(x)$ "better" than the linear approximation $T_1(x)$?* To answer this question, we must come up with a way to measure the closeness of the approximation as x gets close to x_0. The first step is to bring in a function that measures the error in the approximation. The *error* function is just the difference between the function $f(x)$ and the approximating function. Thus, for the linear approximation $T_1(x)$, the error is

$$R_1(x) = f(x) - T_1(x).$$

For the quadratic approximation, the error is

$$R_2(x) = f(x) - T_2(x).$$

Thus we can write

$$f(x) = T_1(x) + R_1(x)$$

and

$$f(x) = T_2(x) + R_2(x).$$

We can now assess the quality of the approximations $T_1(x)$ and $T_2(x)$ by measuring the size of the error terms $R_1(x)$ and $R_2(x)$. The approximations are good if the error terms are small. The smaller the error, the better the approximation. The answer to the above question about the relative precision of the linear and quadratic approximations is therefore contained in the following "error estimates."[1]

Linear and quadratic error estimates. *If the second derivative $f''(x)$ is continuous at $x - x_0$, then*

$$|R_1(x)| \leq C_1|x - x_0|^2$$

for x near x_0.
If the third derivative $f'''(x)$ is continuous at $x = x_0$, then

$$|R_2(x)| \leq C_2|x - x_0|^3$$

for x near x_0.
Here C_1 and C_2 are constants that do not depend on x.

Since $|x - x_0|^3$ is much smaller than $|x - x_0|^2$ when x is near x_0, we have a better error estimate for the quadratic approximation than for the linear approximation.

Big O notation. There is a very convenient shorthand for expressing error estimates of the kind discussed above. For the linear approximation $T_1(x)$, we have

$$f(x) = T_1(x) + R_1(x)$$

where the error term $R_1(x)$ satisfies

$$|R_1(x)| \leq \text{const}|x - x_0|^2.$$

_____

[1] See, for example, R.C. Buck, *Advanced Calculus*, pages 74–80.

This is abbreviated by
$$f(x) = T_1(x) + O(|x - x_0|^2).$$

In this expression, the last term stands for an "error term" which has absolute value no bigger than a constant multiple of $|x - x_0|^2$ when x is near x_0. Similarly, we can write
$$f(x) = T_2(x) + O(|x - x_0|^3),$$
where the last term stands for an error term with absolute value no larger than $|x - x_0|^3$.

There is no need to stop the quest for more precise approximations with quadratics. By adding terms with higher powers of $x - x_0$, we are able to match up higher and higher derivatives at $x = x_0$. In general, the best nth degree approximation is given by the *nth Taylor polynomial*:

$$T_n(x) = f(x_0) + f'(x_0)(x - x_0) + \cdots + \frac{f^{(n)}(x_0)}{n!}(x - x_0)^n.$$

If $f^{(n+1)}(x)$ is continuous at $x = x_0$, we have[2]
$$f(x) = T_n(x) + O(|x - x_0|^{n+1}).$$

Functions of two variables

The ideas used for higher-order approximation for functions of one variable carry over in a perfectly straightforward way to functions of two (or more) variables. Here, we start with a function $f(x, y)$, which we want to approximate near a basepoint $P_0(x_0, y_0)$. The linear approximation to $f(x, y)$ is

$$T_1(x, y) = f(x_0, y_0) + f_x(x_0, y_0)(x - x_0) + f_y(x_0, y_0)(y - y_0).$$

If the second-order partial derivatives $f_{xx}(x, y)$, $f_{xy}(x, y)$, and $f_{yy}(x, y)$ are all continuous at P_0, then, with $P = P(x, y)$, the error estimate for the linear approximation can be expressed as follows[3]:

$$\begin{aligned} f(x, y) &= T_1(x, y) + O(|PP_0|^2) \\ &= T_1(x, y) + O((x - x_0)^2 + (y - y_0)^2). \end{aligned}$$

In words, this means that the error is no bigger than a constant multiple of the square of the distance between $P(x, y)$ and $P_0(x_0, y_0)$.

To find the best *quadratic* approximation $T_2(x, y)$, we write

$$T_2(x, y) = a + b(x - x_0) + c(y - y_0) + d(x - x_0)^2 + e(x - x_0)(y - y_0) + g(y - y_0)^2.$$

We want $T_2(x_0, y_0) = f(x_0, y_0)$, which forces $a = f(x_0, y_0)$. We also want the first order partial derivatives of $T_2(x, y)$ and $f(x, y)$ to agree at $(x, y) = (x_0, y_0)$.

---

[2] Again, see R.C. Buck, *Advanced Calculus*, pages 75–80.

[3] See R.C. Buck, *Advanced Calculus* pages 200–201. The astute reader may notice that the error estimate given here is sharper than the one given in the discussion of linearization in Section 16. There we only assumed that the first order partial derivatives of $f(x, y)$ were continuous. Here, we have assumed that the second-order partials are continuous. The payoff is a better error estimate.

This gives us $c = f_x(x_0, y_0)$ and $d = f_y(x_0, y_0)$. To determine the values of the remaining constants d, e, and g, we equate all of the second-order partial derivatives at $(x, y) = (x_0, y_0)$. For example, we have

$$(T_2)_{xx} = 2d,$$

so setting $(T_2)_{xx}(x_0, y_0) = f_{xx}(x_0, y_0)$ gives

$$d = \frac{f_{xx}(x_0, y_0)}{2}.$$

The values of e and g are found similarly. The final result is

$$
\begin{aligned}
T_2(x, y) &= f(x_0, y_0) + f_x(x_0, y_0)(x - x_0) + f_y(x_0, y_0)(y - y_0) \\
&+ \frac{f_{xx}(x_0, y_0)}{2}(x - x_0)^2 + f_{xy}(x_0, y_0)(x - x_0)(y - y_0) \\
&+ \frac{f_{yy}(x_0, y_0)}{2}(y - y_0)^2.
\end{aligned}
$$

And what about an error estimate? Just as in the one variable case, in going from linear to quadratic approximation, the error estimate "improves" by a factor of the distance from $P(x, y)$ to $P_0(x_0, y_0)$. Using "big O" notation, the error estimate can be expressed:

$$
\begin{aligned}
f(x, y) &= T_2(x, y) + O\left(|PP_0|^3\right) \\
&= T_2(x, y) + O\left(\left((x - x_0)^2 + (y - y_0)^2\right)^{3/2}\right).
\end{aligned}
$$

Notice that the quantity inside of the big parentheses is the cube of the distance from (x, y) to the basepoint (x_0, y_0).

Problems

20.2. Find the best linear and quadratic approximations to the function $f(x, y) = e^{x \sin y}$ near the point $(1, 0)$.

20.3.   a) Find a formula for the best cubic approximation $T_3(x, y)$ to a generic function $f(x, y)$ near the point (x_0, y_0). *Hint:* The constant, linear, and quadratic terms for $T_3(x, y)$ are the same as those for $T_2(x, y)$. However, there are four additional third order terms. Find their coefficients by matching up third order partial derivatives for $f(x, y)$ and $T_3(x, y)$ at the point (x_0, y_0).

     b) Find the best cubic approximation for the function $f(x, y) = e^{x \sin y}$ based at the point $(1, 0)$.

20.4. Let $f(x, y) = 2x^2 - xy + y^2$.

     a) Find the best quadratic approximation $T_2(x, y)$ to $f(x, y)$ based at the point $(1, 2)$.

     b) Make a brief table comparing values of $f(x, y)$ and $T_2(x, y)$ at the points $(0, 0)$, $(1, 1)$, and two other points of your own choosing. Explain why things came out the way they did.

Functions of three or more variables

We will be brief. The ideas used above for higher-order approximations of functions of one and two variables carry over immediately to functions of three or more variables. The idea is to simply choose coefficients to match up derivatives at a basepoint. For details, see, for example, R.C. Buck, *Advanced Calculus*, pages 200–201.

Problem

20.5. Find a formula for the best quadratic approximation to a generic function $f(x, y, z)$ of three variables near a basepoint (x_0, y_0, z_0). *Hint:* There are six quadratic terms, which are multiples of $(x - x_0)^2$, $(y - y_0)^2$, $(z - z_0)^2$, $(x - x_0)(y - y_0)$, $(x - x_0)(z - z_0)$, and $(y - y_0)(z - z_0)$.

Additional Problems

20.6. a) Find the best linear and quadratic approximations to $f(x, y) = e^x \cos y$ based at $(0, 0)$.

b) Use your answer to part a to estimate $e^{0.1} \cos 0.1$.

20.7. Let $f(x, y) = x^y$ for $x > 0$.

a) Find the best linear approximation $L_1(x, y)$ to $f(x, y)$ based at the point $(1, 0)$.

b) Find the best quadratic approximation $Q_1(x, y)$ to $f(x, y)$ based at the point $(1, 0)$.

c) Find the best linear approximation $L_2(x, y)$ to $f(x, y)$ based at the point $(2, 1)$.

d) Find the best quadratic approximation $Q_2(x, y)$ to $f(x, y)$ based at the point $(2, 1)$.

e) Which of the four numbers

$$L_1(2.2, 0.9), \quad Q_1(2.2, 0.9), \quad L_2(2.2, 0.9), \quad Q_2(2.2, 0.9)$$

do you expect to give the best approximation to $f(2.2, 0.9)$? Explain. You should be able to predict the result without actually calculating any of them.

Problem 20.8 makes use of an explicit formula for the remainder term when a function of one variable is approximated by a Taylor polynomial. The formula asserts that

$$h(t) = \sum_{k=0}^{n} \frac{h^{(k)}(t_0)}{k!} (t - t_0)^n + R_n(t) \tag{20.1}$$

where the "remainder term" $R_n(t)$ is given by

$$R_n(t) = \frac{h^{(n+1)}(c)}{(n+1)!} (t - t_0)^{n+1}, \tag{20.2}$$

and where c is some number between t_0 and t.

20.8. Let $f(x, y)$ be a function of two variables. For given points (x_0, y_0) and (x, y), define a function $h(t)$ of one variable t by

$$h(t) = f(x_0 + t(x - x_0), y_0 + t(y - y_0)). \tag{20.3}$$

a) Use the Chain Rule to find a formula for $h'(t)$ in terms of the partial derivatives of f.

b) Use the Chain Rule a second time to find a formula for $h''(t)$ in terms of the second-order partial derivatives of f.

c) Using your results from a) and b) above, write out the formulas (20.1) and (20.2) when $h(t)$ is given by (20.3) with $t_0 = 0$, $t = 1$, and $n = 1$.

d) Use the result of part c to find a formula for the error in a the linear approximation of $f(x, y)$ based at (x_0, y_0). Your formula will involve second-order partial derivatives of f, evaluated at a point on the line segment joining (x_0, y_0) to (x, y).

21   Local (or Relative) Extrema

A maximum or minimum value of a function is called an *extreme value* or an *extremum*. The plural of extremum is *extrema*. When we say "extrema" we mean "maximum and/or minimum values."

Finding the extrema of a function is a problem which comes up in many areas. In elementary calculus we studied extrema of functions $y = f(x)$. For such functions, there are *two* kinds of extrema, viz. local (or relative) extrema, and global (or absolute) extrema. Here we discuss extrema of functions of two variables. Sections 21–23 deal with *three* kinds of extremum problems for functions $z = f(x, y)$.

Local (or relative) extrema

Look at the graph $y = f(x)$ below left. At $x = a$, the function $f(x)$ has a peak. That is, $f(a) \geq f(x)$ for all values of x near $x = a$. At $x = a$, the function has a maximum, *but only compared to x-values nearby*. Such a maximum is called a *local* or *relative maximum*. For the part of $f(x)$ shown, there are five local maxima. Similarly, the function $y = f(x)$ has four local or relative minima.

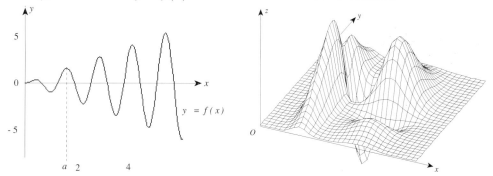

Above right is the graph $z = g(x, y)$ of a function of two variables. The graph $z = g(x, y)$ has several peaks of different heights. Each of them is called a *local* (or *relative*) *maximum* because $g(x, y)$ is highest, *but only compared to points (x, y) nearby*. For the part of $g(x, y)$ shown, there are four local maxima. The notion of local (or relative) minimum is similarly defined.

The pattern of extremum problems

Given a function of two variables, we want to find its local extrema. It means: We want to find (i) the points (x, y) where the function has extrema, and (ii) we want to know the extreme values of the function.

How do we find the local extrema of a function $y = f(x)$ of one variable? We do it in two steps.

Step 1: Find all x for which $f'(x) = 0$. These are the x-values at which $f(x)$ may – but not necessarily does – have an extremum. These x-values are called the *candidates* for extrema.

Step 2: We apply the Second Derivative Test. That is we *test the candidates*.

For a function $z = h(x, y)$ of *two* variables, the pattern is the same: step 1—find the candidates; step 2—test the candidates.

Step 1: Finding the Candidates   If $h(x, y)$ has a local extremum at $(x, y) = (a, b)$, then the tangent plane to the graph $z = h(x, y)$ must be horizontal, because otherwise the graph could neither peak out nor bottom out at (a, b). "Tangent plane horizontal" means "normal vector of the tangent plane is parallel to $\vec{k}$." The normal vector has components $(h_x, h_y, -1)$. So we conclude that both h_x and h_y must be zero at (a, b).

> If $h(x, y)$ has a local extremum at (a, b), then $\operatorname{grad} h$ is zero at (a, b).

We will see that if $\operatorname{grad} h$ at (a, b) is zero, then there may or may not be a local extremum at (a, b), exactly as in elementary calculus where "derivative zero" does not guarantee "extremum."

Terminology.   A point (a, b) is called a *critical point* of $h(x, y)$ if $\operatorname{grad} h = (0, 0)$ at (a, b). Note that a critical point is a point in the xy-plane.

Locating the critical points by means of level curves.   Below left are shown some level curves of a function $z = f(x, y)$, and below right some level curves of $z = g(x, y)$. The level curves of $g(x, y)$ are concentric circles, and $g = 3$ consists of the point which is the center of all circles.

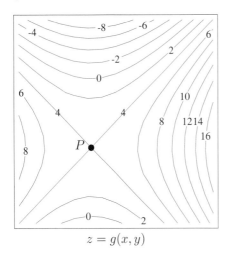

$$z = g(x, y)$$

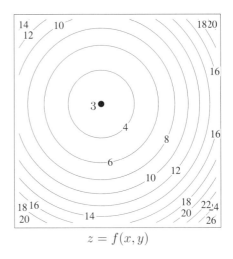

$$z = f(x, y)$$

Problem _____

21.1.   a) Find the critical point(s) of $z = f(x, y)$ on the left. *Hint:* The gradient is normal to the level curve or it is zero. Where is the shape of the level curves such that the gradient must be the zero vector?

   b) Use complete sentences to explain your answer to part a.

   c) Find the critical point(s) of $g(x, y)$ (on the right).

   d) Use complete sentences to explain your answer to part c.

Finding the critical points by computation.   We have to find the points (x, y) for which $\operatorname{grad} h = (0, 0)$. Written out in components, we get a system of two equations in two unknowns x and y: $h_x = 0$, $h_y = 0$.

Example 1. Let $h(x, y) = x^2 + xy + 5y^2 + 3x$. Here the system $h_x = 0$, $h_y = 0$ is an ordinary system of two linear equations in two unknowns:

$$\begin{aligned} 2x + \quad y &= -3 \\ x + 10y &= 0 \end{aligned}$$

■

Example 2. Let $h(x, y) = x^2 y + x^2 - y^3$. The system $h_x = 0, h_y = 0$ is a system of two <u>non</u>-linear equations in two unknowns:

$$\begin{aligned} 2xy + 2x &= 0 \\ x^2 - 3y^2 &= 0. \end{aligned}$$

How do we solve such a system? We ask:

- Is $(x, y) = (0, 0)$ a solution? Is $x = 0$ possible? What about $y = 0$?
- Can we eliminate one of the two unknowns?

In Example 1, $(0, 0)$ is a solution, and it is the only solution with $x = 0$. Therefore, we can assume $x \neq 0$, and we can divide the first equation by x to get $2y + 2 = 0$ so $y = -1$. We go with $y = -1$ into the second equation:

$$x^2 - 3(-1)^2 = 0, \quad \text{that is,} \quad x = \pm\sqrt{3}.$$

Thus, there are three critical points: $(0, 0)$, $(\sqrt{3}, -1)$, $(-\sqrt{3}, -1)$. ■

Problems ──────────────────────────────────

21.2. Find the critical points of the function of Example 1 above.

21.3. Find the critical points of $f(x, y) = y^3 - xy + x^2$.

Step 2: Testing the candidates

We first look at the functions $f(x, y)$ and $g(x, y)$ of Problem 21.1. *Are the critical points of the functions of Problem 21.1 local extrema or not?*

$f(x, y)$: The only critical point is P. If you move away from P vertically up or vertically down, the function decreases. If you move horizontally to the left or right, the function increases. P is not a local extremum. Why? Because at local extremum the function should either increase in all directions (minimum) or decrease in all directions (maximum). A point such as P is called a *saddle point* because the surface $z = f(x, y)$ looks like a saddle at P: To the north and south, the surface goes down, and to the east and west, the surface rises.

$g(x, y)$: The point $g = 3$ on the right is a local minimum because the surface rises in all directions once you move away from that point.

Views of the graphs $z = f(x, y)$ and $z = g(x, y)$ are shown in Problem 21.5. The upshot of our examples $f(x, y)$ and $g(x, y)$ is that at a critical point, a function may or may not have a local extremum.

The Second Derivative Test. For a function $z = h(x, y)$ given by a formula, there is a Second Derivative Test. It requires more computations than the one-variable test, and as in the case of the one-variable test, it may be inconclusive. Although we will not do so here, the test can be deduced from the quadratic approximation formula discussed in Section 20.[1]

Start with a critical point (a, b) for $h(x, y)$. The test makes use of the number

$$D(a, b) = h_{xx}(a, b)h_{yy}(a, b) - (h_{xy}(a, b))^2,$$

which can be easily remembered as the determinant of the matrix of second order partial derivatives of h:

$$D(a, b) = \begin{vmatrix} h_{xx}(a, b) & h_{xy}(a, b) \\ h_{xy}(a, b) & h_{yy}(a, b) \end{vmatrix}.$$

The rules for the Second Derivative Test are as follows:

> (i) If $D(a, b) > 0$ and $h_{xx}(a, b) > 0$, then $h(x, y)$ has a local minimum at (a, b).
>
> (ii) If $D(a, b) > 0$ and $h_{xx}(a, b) < 0$, then $h(x, y)$ has a local maximum at (a, b).
>
> (iii) If $D(a, b) < 0$, then $z = h(x, y)$ has a saddle point at (a, b).
>
> (iv) If $D(a, b) = 0$, the test is inconclusive. We cannot say anything at all.

Problems

21.4. The functions discussed in Problem 21.1 have the formulas $f(x, y) = x^2 - y^2 + 4$ and $g(x, y) = x^2 + y^2 + 3$. Go with these and a third function $h(x, y) = -x^2 - y^2 + 7$. Apply the Second Derivative Test to all three functions.

21.5. Four surfaces are shown below. The dotted grid underneath represents a horizontal plane. Could any of these surfaces be the graph of $f(x, y)$, $g(x, y)$, and $h(x, y)$ of Problem 21.4? Explain why each surface could or could not be the graph of $f(x, y)$, $g(x, y)$, $h(x, y)$.

(a)   (b)

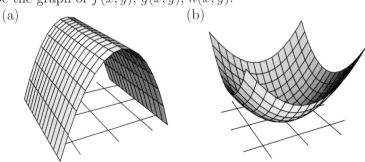

---

[1] See, for example, R.C. Buck, *Advanced Calculus*, pages 288-290.

(c)   (d)

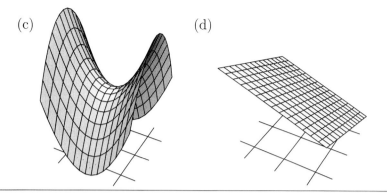

Additional Problems

21.6. $g(x, y) = xy + \frac{8}{x} + \frac{1}{y}$

    a) Find the critical points of $g(x, y)$ by computation.

    b) Use the Second Derivative Test to classify the critical points.

21.7. Let $f(x, y) = y(y - x^2)$.

    a) Find the critical points and apply the Second Derivative Test. What does it tell you?

    b) Draw the level curve of level zero in an xy-system. *Hint:* $f(x, y)$ is the product of two functions (which are they?). If a product of scalars is zero, at least one factor must be zero.

    c) The different branches of the level curve of level zero subdivide the plane into different parts. Hatch those parts where $f(x, y) > 0$ with lines of slope $+1$, and those where $f(x, y) < 0$ with lines of slope -1. Look at the picture—does $f(x, y)$ have a local extremum? Use complete sentences to explain your answer.

21.8. We consider the function $f(x, y) = x^2 + y^3 - 3y$.

    a) At which of the following points can $f(x, y)$ possibly have a local extremum:
$(1, 0)$, $(-1, 0)$, $(0, 1)$, $(0, -1)$, $(1, 1)$, $(-1, -1)$

    b) Find the critical points and apply the Second Derivative Test.

21.9. Let $f(x, y) = x^4 - 16y^4$.

    a) Find all critical points for $f(x, y)$, and apply the Second Derivative Test to each.

    b) Sketch the level curve $f(x, y) = 0$, and mark the regions in the plane where $f(x, y)$ is positive, and those where it is negative.

    c) Classify each critical point as a local maximum, local minimum, or saddle point. Explain your answer.

21.10. Find all critical points for $f(x, y) = x^4 + 2y^8$, and classify each.

21.11. The function $f(x, y)$ is defined by $f(x, y) = -x^3 - y^3 + 18xy$. Find the critical points and classify them.

21.12. We have the following information on a function $f(x, y)$: grad $f = (x^2 - 9, y - 2)$. Is that enough information to find the critical points? If yes, find them and apply to them the Second Derivative Test.

22 Constrained Extrema

Below are shown some level curves of a function $z = f(x, y)$ together with six curves J, K, L, M, N, and Q. Of these, J, K, L and M are finite stretches with endpoints marked by bullets. The curve Q consists of a line segment and a circular arc, and it has two corners. The curves Q and N have no endpoints. A curve that forms a loop without endpoints is called a *closed curve*.

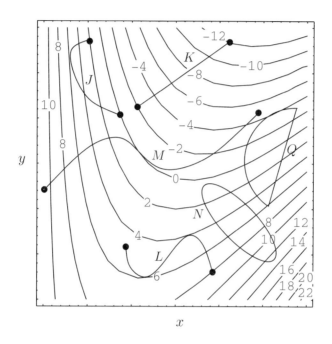

Figure 22.1

The key question

What are the highest and lowest values the function $f(x, y)$ reaches for points (x, y) which lie on the curve J? What about K and the other curves?

Problem _____

22.1. a) On each curve in Figure 22.1, mark with "AA" the point where $f(x, y)$ is highest in comparison to all other points on the curve. Mark with "aa" the point where $f(x, y)$ is lowest in comparison to all other points on the curve.

b) Estimate the highest and the lowest value of the function $f(x, y)$ along each curve.

_____

Terminology. A highest or lowest value of a function $f(x, y)$ on a curve C is called an *extremum* of $f(x, y)$ on C. Such an extremum is called a *constrained extremum*. The word "constrained" is used because the independent variable cannot vary freely, but is restricted or constrained to a curve C. The curve, or an equation that defines it, is called the *constraint*. The constraint is either a finite stretch of a curve or a closed curve, as in Figure 22.1.

How do we find constrained extrema?

The routine resembles the routine for relative extrema. We proceed in two steps. The first step is to find the points on the constraint curve where extrema could occur. We call these points on the constraint curve *candidates* The second step is to test the candidates to determine which ones yield extrema. We will discuss the strategy we will follow for each step before proceeding to techniques for carrying out the strategies.

Step 1 – Finding the candidates. We go back to Figure 22.1 and ask: *What features distinguish the points where the extrema occur?* The extrema occur at points of the following kinds:

  (i) Endpoints;

 (ii) Corners;

(iii) Points where the constraint curve is tangent to the level curve of the function which passes through that point. Such a point P is shown below. At P, the directional derivative of $f(x, y)$ in the direction of the tangent of C is zero. Why? The tangents of C and the level curve are the same. But the directional derivative in the direction of the tangent of the level curve is zero.

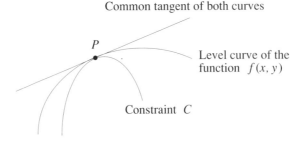

In addition to the three types of extrema listed above, there is a fourth type that is not illustrated in Figure 22.1:

 (iv) Points on the constraint curve where the gradient of $f(x, y)$ is the zero vector.

A point P on the constraint curve of type (iii) or (iv) is called a *critical point* of the function $f(x, y)$ along C.[1]

Problem ⎯⎯⎯⎯⎯⎯⎯⎯⎯⎯⎯⎯⎯⎯⎯⎯⎯⎯⎯⎯⎯⎯⎯⎯⎯⎯⎯⎯⎯⎯

22.2. We consider the function $z = f(x, y)$ and the six curves of Figure 22.1. For each curve determine the kind of points where the extrema occur (endpoint, corner, or critical point).

Step 2 – Testing the candidates. Once we have the candidates $P, \ldots$ we evaluate the function at the candidate points $P, \ldots$. The highest of these values is the maximum on the curve, and the lowest is the minimum. Thus, Step 2 is easy *once we have found the candidates.*

⎯⎯⎯⎯⎯⎯⎯⎯⎯⎯

[1]Note that the term *critical point* was used in Section 21 to mean a point at which the gradient vanishes. The meaning here is different, but the purpose is the same: to identify candidates for extrema. Despite the different usages of this term, the meaning should always be clear from context.

Techniques for finding the candidates

There are different techniques, depending on how the constraint curve is given. We consider four cases.

Case 1: The curve is given as a graph of $y = f(x)$, a function of one variable.

Case 2: The curve is given by a parametric representation $(x, y) = (a(t), b(t))$.

Case 3: The constraint is given in implicit form $F(x, y) = 0$ (i.e., C is a level curve of a function of two variables).

Case 4: The constraint curve is given in graphical form, by a drawing.

For each case we look at an example of a curve C, describe the work that has to be done, and do part of it. In the four examples we take the same function

$$g(x, y) = xy - x + 1.$$

Example for case 1: *C given by $y = f(x)$.* We take the constraint C to be the parabola $y = \frac{1}{3}x^2$ from $x = 0$ to $x = 2$. It is shown below left.

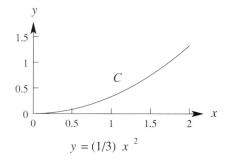

$y = (1/3)\ x^{\,2}$

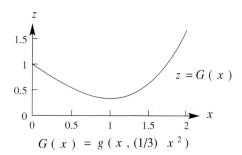

$G(\,x\,)\ =\ g\,(\,x\,,(1/3)\ x^{\,2}\,)$

A point (x, y) of C has coordinates

$$(x, f(x)) = \left(x, \frac{1}{3}x^2\right).$$

For points on C, the function $z = g(x, y)$ becomes a function of one variable x:

$$z = g\left(x, \frac{1}{3}x^2\right) = x\left(\frac{1}{3}x^2\right) - x + 1 = \frac{1}{3}x^3 - x + 1.$$

We write $G(x)$ for this new function. The graph of $z = G(x)$ is shown above right. Now the problem of finding the extrema of $g(x, y) = xy - x + 1$ on the curve C reads as follows: *Find the extrema of $G(x) = \frac{1}{3}x^3 - x + 1$ for x in the interval $0 \le x \le 2$.* This is a problem of elementary calculus: Find the absolute extrema of a function of *one* variable. To solve it, you first find the values of x in the interval $0 \le x \le 2$ such that $G'(x) = 0$. You then evaluate $G(x)$ for each such x, and at the endpoints $x = 0$ and $x = 2$, and see which give the largest and smallest values.

Problem

22.3. Carry out the remaining steps for our example constraint $y = \frac{1}{3}x^2$.

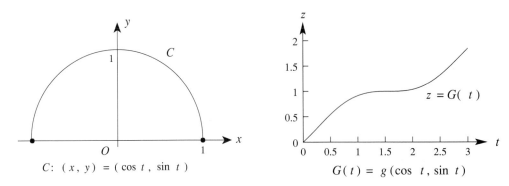

Figure 22.2

Example for case 2: *C given by* $(x, y) = (a(t), b(t))$. As C we take the half-circle $(x, y) = (\cos t, \sin t)$, $0 \le t \le \pi$. It is shown at left in Figure 22.2. *Remember:* $g(x, y) = xy - x + 1$. We work as in case 1 above: On the curve C, the function $z = g(x, y)$ becomes a function $z = G(t) = g(\cos t, \sin t) = \cos t \sin t - \cos t + 1$ of one variable (see Figure 22.2, right). We have to find the absolute extrema of $G(t)$ in the interval $0 \le t \le \pi$.

Problem _____

22.4. Carry out the computations to find the extrema of $g(x, y)$ on the half-circle C. *Hint:* You will have to solve the equation $\cos^2 t - \sin^2 t + \sin t = 0$ for t. To do this, rewrite the equation in terms of $\sin t$ alone. Then set $\sin t = u$ and find u.

_____

Case 3: *C given by* $F(x, y) = 0$. We take the constraint curve C to be the ellipse $x^2 + 4y^2 = 4$, and the function is still $g(x, y) = xy - x + 1$.[2] Figure 22.3 on the facing page shows the ellipse and some level curves of $g(x, y)$. Our main task is to find the candidates for extrema. Each will be of one of the four types listed under the heading *Finding the candidates* on page 178. Our curve C has neither endpoints nor corners, so there are no candidates of type (i) or (ii). As for (iv), the only point where grad g vanishes is the point $(0, 1)$ which does lie on C. Therefore, we have found one candidate, namely, the point $(0, 1)$.

The critical points of type (iii) are more difficult to find. Let us first use ruler and pencil to get an estimate.

Problem _____

22.5. Use ruler and pencil to find the type (iii) critical points of $g(x, y)$ on C. Draw the missing level curves as well as you can and estimate their level.

_____

[2]It may occur to you that the constraint curve considered here (an ellipse) is quite similar to the one considered in the case 2 example (a half-circle). This might lead you to suspect that the extrema can be found by the same technique used in case 2. That is in fact the case. Just parametrize the ellipse by $x = 2 \cos t$ and $y = \sin t$.

There are no hard-and-fast rules to tell you which technique to use for a given problem.

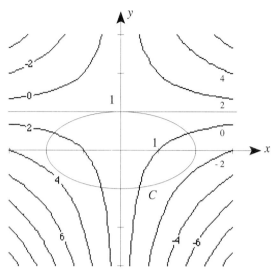

Figure 22.3

Now we use formulas to find the critical points along C, as follows. At a type (iii) critical point of $g(x, y)$ along C the tangent of C is parallel to the tangent of the level curve which passes through the point. If the tangents are parallel, so are the normals. We are going to use the normals. First we raise three questions.

Question 1: *How do we find a normal vector of the constraint curve C?*
Answer: We consider C as the level curve of level zero of the function $F(x, y) = x^2 + 4y^2 - 4$. The normal of a level curve has the direction of the gradient. Therefore $\operatorname{grad} F = (2x, 8y)$ is a normal vector of C.

Question 2: *How do we find a normal vector of the level curve of the function $g(x, y)$?*
Answer: We reason as in question 1, but apply it to a level curve of $g(x, y)$. Thus, $\operatorname{grad} g = (y - 1, x)$ is a normal vector of the level curve of our $g(x, y) = xy - x + 1$.

Question 3: *What does it mean to say that the vectors $\operatorname{grad} F$ and $\operatorname{grad} g$ are parallel?*
Answer: We must have $\operatorname{grad} g = \lambda \operatorname{grad} F$ for a scalar λ. The scalar λ (Greek letter, pronounced "lambda") is called a *Lagrange multiplier*.

Now we put the answers to questions 1 to 3 together: At a type (iii) critical point of $g(x, y)$ along the curve $F(x, y) = constant$ we must have

$$\operatorname{grad} g = \lambda \operatorname{grad} F. \tag{22.1}$$

In fact, this equation is also satisfied by type (iv) critical points, with $\lambda = 0$, so at *every* critical point (22.1) must hold for *some* scalar value λ. Equation (22.1) is called the *Lagrange condition*.

In our case $g(x,y) = xy - x + 1$, and C is given by $F(x,y) = x^2 + 4y^2 = 4$. The Lagrange condition gives us

$$(y - 1, x) = \lambda (2x, 8y).$$

Written out in components, this equation reads

$$\begin{aligned} y - 1 &= \lambda\, 2x \\ x &= \lambda\, 8y. \end{aligned}$$

These equalities must hold at a critical point of $g(x,y)$ along C. We want to use them to determine x and y. However, λ is not known. We need a third equation. As third equation we take the equation of C. The problem of finding the critical points of $g(x,y)$ along C has become the problem of solving a system of three equations for x and y:

$$\begin{aligned} y - 1 &= \lambda\, 2x \\ x &= \lambda\, 8y \\ x^2 + 4y^2 &= 4 \qquad \text{(the constraint equation)} \end{aligned}$$

It is a system of nonlinear equations. We dealt with systems of nonlinear equations in Section 21 when we were looking for the points where the gradient vanishes. There the approach was to ask two questions:

- *Is $(x,y) = (0,0)$ a solution? Is $x = 0$ possible? What about $y = 0$?*
- *Can we eliminate one of the two unknowns, preferably λ?*

Now we start working the specific system under discussion. If x is zero, then $y = 1$ because of the first equation. The point $(x,y) = (0,1)$ satisfies also the second equation (with $\lambda = 0$) and the constraint equation. Therefore $(0,1)$ is a critical point on C.

We have dealt with the case $x = 0$. So we can assume $x \neq 0$. From the first equation we get

$$\lambda = \frac{y - 1}{2x}.$$

We plug that into the second equation and find

$$x = 8 \frac{y - 1}{2x} y.$$

This simplifies to

$$x^2 = 4y^2 - 4y.$$

Now we use the constraint to replace x^2 by $4 - 4y^2$. After simplifying, the result is

$$2y^2 - y - 1 = 0.$$

This equation has the solutions $y = 1$ (which we already found) and $y = -\frac{1}{2}$. Therefore there are three critical points of $g(x,y)$ along C:

$$(0, 1), \qquad (\sqrt{3}, -\frac{1}{2}), \qquad (-\sqrt{3}, -\frac{1}{2}).$$

What happened to the other unknown λ? We are not interested in the value of λ, but we had to work with it in order to find x and y.

Problem _____

22.6. Complete the work for the case 3 example: Make a list of all candidates for extrema and test the candidates by calculating the value of $g(x, y)$ at each. Compare with your answer to Problem 22.5.

_____

Remarks on Lagrange multipliers

1. **Remembering the formula.** When the constraint curve is given by the equation $F(x, y) = 0$, there is a simple trick to remember the equations for the critical points. Form the *Lagrange function*

$$L(x, y, \lambda) = g(x, y) - \lambda F(x, y).$$

The equations for the critical points are obtained by setting all three partial derivatives, with respect to x, y, and λ, equal to 0:

$$\frac{\partial L}{\partial x} = 0 \qquad \frac{\partial L}{\partial y} = 0 \qquad \frac{\partial L}{\partial \lambda} = 0.$$

The first two of these equations give the components of the vector equation $\operatorname{grad} g = \lambda \operatorname{grad} F$, which is the Lagrange condition. The last is just the constraint equation $F(x, y) = 0$.

2. **Terminology.** Some books use less geometrical language. Instead of saying "find the extrema of $f(x, y)$ on a curve C" they say "find the extrema of $g(x, y)$ under the constraint $F(x, y) = 0$." In this language, a critical point of $g(x, y)$ along the curve C is called "a critical point of $g(x, y)$ under the constraint $F(x, y) = 0$." The method of finding the points is still the same.

3. **A confusion to avoid.** Whether or not the problem is formulated in geometric language, one always works with two formulas: The function whose extrema we want, and the formula describing the constraint curve. Do not confuse the two. To avoid confusion, some books call the function whose extrema you want the *objective function* and the formula for the constraint curve the *constraint*.

4. **Unclear situations.** If the curve C is given in implicit form, it is often not clear whether it is a curve of the kind we consider (a finite stretch of a curve, or a closed curve). You can still apply the Lagrange method to find the critical points under the constraint, but if the curve has infinite extent, it may be impossible to decide whether they are extrema without more advanced methods.

5. **Other uses of Lagrange Multipliers.** The method of Lagrange multipliers works in many more general situations (e.g., see R.C. Buck, *Advanced Calculus*, pages 295–298).

Case 3 : *C given graphically.* In this case, you superimpose the constraint curve on a contour plot of the function $g(x, y)$, and work with ruler, pencil, and eyeballs to locate the four types of candidates for extrema.

Problem _____

22.7. We are interested in the extrema of $g(x, y) = xy - x + 1$ on the curve C shown below. The curve C overlays a contour plot of $g(x, y)$.

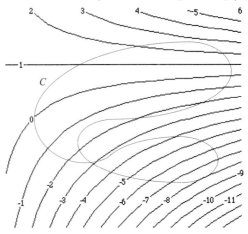

a) Mark the endpoints of C (candidates of type (i)). Label them $D, E \ldots$.

b) Mark the corners of C (candidates of type (ii)). Label them $K, L \ldots$.

c) Mark the type (iii) critical points of $g(x, y)$ along C. Label them $Q, R \ldots$.

d) Mark the points of C where the gradient of $g(x, y)$ vanishes (critical points of type (iv)). Label them $M, N \ldots$.

e) For each candidate you found in a) to d), indicate whether at this point $g(x, y)$ has the maximum on C, or the minimum on C, or a local maximum on C, or a local minimum on C, or no extremum at all.

_____

Additional Problems

22.8. Write your own summary "How to find constrained extrema."

22.9. Go back to the example for case 1 on page 179. There you used substitution and avoided finding the four types of candidates. Now go through the procedure of finding the candidates for extrema. That is:

a) Find the coordinates of the endpoints of C (candidates of type (i)).

b) Find the coordinates of the corners of C (candidates of type (ii)).

c) Find the coordinates of the type (iii) critical points of $g(x, y)$ along C.

d) Find the coordinates of the points of C where grad g is zero (critical points of type (iv)).

22.10. Find the critical points for the function $f(x, y) = xy$ subject to each of the following constraints. Decide whether each is an extremum, and if so, say which type.

a) $x = 1$ b) $x + y = 2$ c) $x^2 + 4y^2 = 8$
d) $x = 2y$ e) $x = 4, \ y = t^2$ f) $x = t, \ y = t^2$

22.11. A contour plot for $f(x,y)$ is shown below, along with a constraint curve C.

    a) Mark all critical points along the curve, and label them A, B, etc.

    b) The critical points you marked partition the curve C into several pieces. As the curve is traversed from left to right, say whether the function $f(x,y)$ increases or decreases on each of these pieces.

    c) Which critical points give the maximum and minimum values of $f(x,y)$ along C?

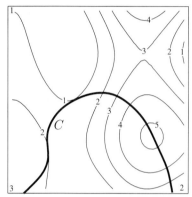

22.12. Find the extrema of $f(x,y) = x^2 + y^2$ under the constraint $x^2 + xy + y^2 = 1$. The constraint curve is an ellipse. An ellipse is a closed curve and has no endpoints.

22.13. Find all critical points for the function $f(x,y) = x^3 + y^2$ along the circle $x^2 + y^2 = 1$, and decide which give the maximum and minimum of $f(x,y)$ along this circle.

22.14. Estimate the maximum and minimum values of the function $f(x,y) = 1 - x + 2y$ along the curve below.

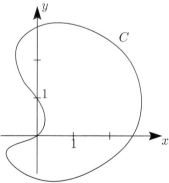

22.15. Find the points on the ellipse $x^2 + 4y^2 = 4$ that are closest to the point $(1,0)$. *Hint:* Write out a formula for the square of the distance from (x,y) to $(1,0)$, and minimize it over the ellipse.

22.16. A cylindrical can with a top and bottom is to have a volume of 100 cm^3. Find the dimensions which will minimize the surface area.

22.17. Throughout all parts of this problem we work with the two functions

$$p(x,y) = x - 2y + 1; \quad H(x,y) = (x-6)^2 + (y-1)^2.$$

Note that $H(x,y)$ is the square of the distance between the points (x,y) and $(6,1)$. The following problems $\mathcal{A}$ and $\mathcal{B}$ will be discussed:

> Problem $\mathcal{A}$: Find the critical points of $p(x,y)$ under the constraint $H(x,y) = 1$.

> Problem $\mathcal{B}$: Find the critical points of $H(x,y)$ under the constraint $p(x,y) = 1$.

a) Which method is better for Problem $\mathcal{A}$, substitution or Lagrange multipliers? Solve $\mathcal{A}$.

b) Which method is better for Problem $\mathcal{B}$, substitution or Lagrange multipliers? Solve $\mathcal{B}$.

c) Use the grid in Figure 22.4 to solve $\mathcal{A}$ by ruler and pencil (you may or may not need the circle drawn). Are the critical points you found extrema under the constraint, or local extrema, or neither?

d) Use the grid in Figure 22.4 to solve $\mathcal{B}$ by ruler and pencil (you may or may not need the circle drawn). Are the critical points you found extrema under the constraint, or local extrema, or neither?

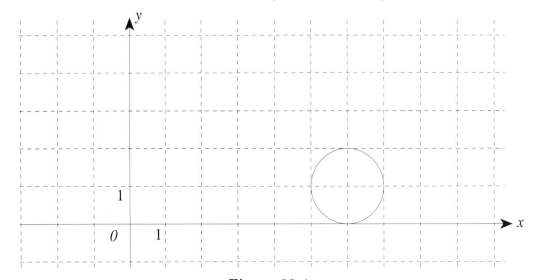

Figure 22.4

23   Global (or Absolute) Extrema

By a filled-in circle we mean a circle together with all points in its interior. A filled-in triangle consists of a triangle together with all points in its interior, and similarly for other figures in the plane. Often we show filled-in figures with shading.

Filled-in figures are called *regions* or *domains*.

An example and the key question

Below are shown some level curves of a function $z = f(x, y)$ and a triangular region T. We are interested in the values of the function only for points in the region T.

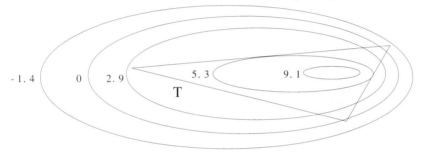

Figure 23.1

We ask:

What are the highest and lowest values which $f(x, y)$ has for points (x, y) of the region T?

The highest and lowest values of $f(x, y)$ for (x, y) in T are called the *global (or absolute) extrema* of $f(x, y)$ in the region T.

Problem ———————————————————————————————————————

23.1. The function $f(x, y)$ and the region T are shown in Figure 23.1 above. Find the global extrema of $f(x, y)$ in the region T.

———

The key question of this section is the general version of Problem 23.1:

Key question: *Given a function $g(x, y)$ and a region D. Find the global extrema of $g(x, y)$ in D.*

Before we start work on the key question we have to look more closely at regions.

A closer look at regions. A region D in the plane is typically bounded by one or more curves which form the dividing line between the region and the rest of the plane. We will say:

- The *boundary* of a region D consists of the curve(s) which separate it from the rest of the plane.

- The *interior points* of a region D are the points of the region which do not lie on the boundary.

Example 1. A filled-in circle is often called a *disk*. A disk D is shown below. It has center M and radius r. It consists of all points whose distance from M is less or equal to r.

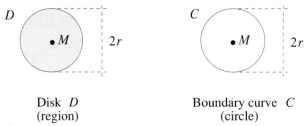

Disk D
(region)

Boundary curve C
(circle)

The boundary of the disk D is the circle C of radius r and center M. The interior points of D are those points whose distance from M is strictly less than r. ■

Example 2. The triangular region T is shown below. Its boundary curve consists of the three line segments AB, BC, CA:

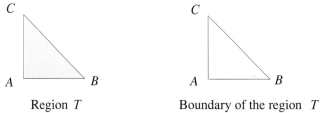

Region T Boundary of the region T

The interior of T is made up of all points of the filled-in triangle which do not lie on any of the line segments AB, BC, CA. ■

Problems

23.2. The triangular region T in the xy-plane is the one of Example 2 above. The corners have coordinates $A(2,0)$, $B(6,0)$, $C(2,4)$.

    a) Give the coordinates of two points R, S of the interior of T.

    b) Give the coordinates of two points U, V of the boundary of T.

23.3. Go back to your answer to Problem 23.1 and choose the correct combination of the words in italics:

The function $f(x,y)$ has the global minimum

in the interior of T   *on the boundary of T*.

The function $f(x,y)$ has the global maximum

in the interior of T   *on the boundary of T*.

How to find the global extrema in a region D

We have the usual two steps of an extremum problem: step 1—find the candidates, step 2—test the candidates. We test the candidates by evaluating the function. So the problem is to find the candidates.

Any point in a region is either an interior point, or it lies on the boundary. Therefore we can say:

> A global extremum of $g(x,y)$ in D occurs either in the interior of D or on the boundary of D.

We conduct separate searches for candidates in the interior and on the boundary of D:

What are the candidates in the interior of D?
If a global extremum of $g(x, y)$ occurs at a point $Q(a, b)$ in the interior of D, then the graph $z = g(x, y)$ must have a horizontal tangent plane at the point $(a, b, g(a, b))$. Therefore $Q(a, b)$ must be a critical point of $g(x, y)$. Thus, the candidates in the interior of D are those critical points of $g(x, y)$ which lie in the interior of D.

What are the candidates on the boundary of D?
They are the candidates for extrema on the boundary curve C. It means: We have a curve C (the boundary of D) and a function $z = g(x, y)$, and we need the candidates for the extrema of $g(x, y)$ on C. *This is a problem on constrained extrema, discussed in Section 22.* We found in Section 22: If C is given in a parametric representation, the candidates for extrema on C are found by substitution. In case C is given by an equation $F(x, y) = 0$ we may have to use Lagrange multipliers to find the critical points of $g(x, y)$ along C.

Summary

To find the global extrema of $g(x, y)$ in a region D, do the following.

Step 1: Finding the candidates. There are two kinds of candidates:

(i) The critical points of $g(x, y)$ which lie in the interior of D. We find them by solving the equation $\operatorname{grad} g = (0, 0)$, and we disregard any solutions which are not in the interior of D.

(ii) The candidates for extrema of $g(x, y)$ on the boundary curve of D. We use the methods developed in Section 22.

Step 2: Testing the candidates. Evaluate the $g(x, y)$ for all candidates (x, y), and the global extrema appear as the highest and lowest values.

Example 3. We are going to outline the steps for finding the absolute extrema of $g(x, y) = xy - y$ in the triangular region T shown below.

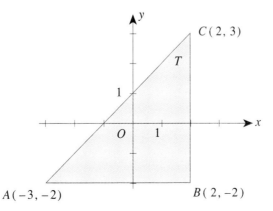

Step 1: Find the candidates (x, y). There are *two* kinds of candidates:

(i) *The critical points of $g(x,y)$ in the interior of T*

You find them by solving the equation $\operatorname{grad} g = (0,0)$ for (x,y).

(ii) *The candidates for extrema of $g(x,y)$ on the boundary curve*

The boundary curve consists of the three line segments AB, BC, and CA. Each can be described by a parametric representation and has to be worked through separately. For example, look at AB:

- On AB we have $y = -2$. For points on AB, the function $g(x,y) = xy - y$ becomes a function $G_1(x)$:

$$G_1(x) = g(x, -2) = -2x + 2.$$

  The candidates for extrema on AB are the x-values for which $G_1'(x) = 0$ (there are none because $G_1' = -2$) and the endpoints A, B.

- On BC we have $x = 2$, and $g(x,y)$ becomes a function $G_2(y)$. You treat $G_2(y)$ on BC the way you treated $G_1(x)$ on AB.

- On AC we have $y = x + 1$, and $g(x,y)$ becomes a function $G_3(x)$. You repeat the process for $G_3(x)$.

Step 2: Evaluate the function for all candidates. Then pick out the global extrema.

∎

Problem ───────────────────────────────────────

23.4. Carry out the computations of Example 3.

Example 4. A ruler and pencil problem. We consider polar coordinates (r, θ) (for a review of polar coordinates, see the beginning of Section 26). Any point P in the xy-plane has Cartesian coordinates (x,y) and polar coordinates (r, θ), as shown below.

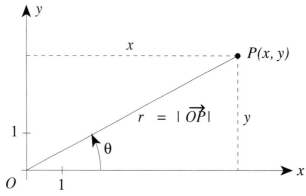

We take θ between 0 and 2π.

Now we define two functions $z = f_1(x,y)$ and $f_2(x,y)$ as follows:

$f_1(x,y) = r$ where r is the r of polar coordinates (r, θ) of the point $P(x,y)$;

$f_2(x,y) = \theta$ where θ is the θ of polar coordinates (r, θ) of the point $P(x,y)$.

D is the region shown in Figure 23.2 on the next page.   ∎

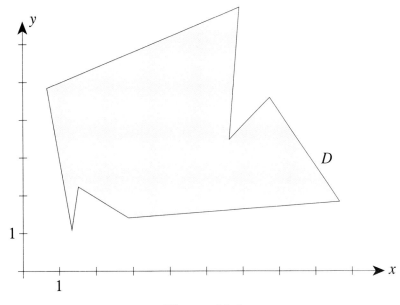

Figure 23.2

Problem _____

23.5. a) Use ruler and pencil to find the global extrema of the function $z = f_1(x, y) = r$ in the region D shown in Figure 23.2.

b) As a), but for the function $z = f_2(x, y) = \theta$.

Additional Problems

23.6. With $g(u, v) = 2u - v + 5$, find the points in the circular region (filled-in circle) shown below where $g(u, v)$ has the global minimum and maximum, and find the value of $g(u, v)$ at these points. *Hints:* This is a ruler-and-pencil problem. Use level curves.

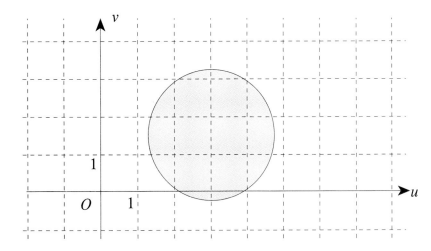

23.7. We consider the function $f(x, y) = x^3 + y^3 - 12x - 3y + 5$.

  a) Find the global extrema of $f(x, y)$ in the filled-in triangle OAB below.

  b) Same with the filled-in rectangle $OCDE$ below.

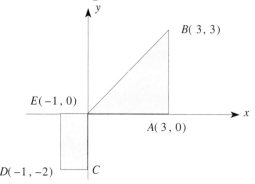

23.8. Find the points at which the function $f(x, y) = \cos x + \sin y$ has its global extrema in the square region shown below.

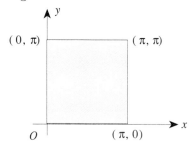

23.9. Estimate the maximum and minimum values of

$$\varphi(x, y) = \frac{1}{\sqrt{x^2 + y^2}}$$

over the region shown below. *Hint:* You may want to revisit Problem 19.7.

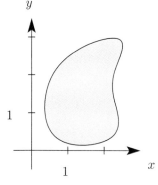

23.10. A steel plate covers the region inside the ellipse $x^2 + 4y^2 = 4$. The temperature at any point (x, y) on the plate is given by $f(x, y) = xy$.

  a) Find the maximum and minimum temperature along the rim of the plate, *i.e.* along the boundary ellipse.

  b) Find the maximum and minimum temperature on the whole plate (interior and boundary combined).

23.11. You sell your wares in two different markets. You are free to set separate prices in the two markets. In both markets, demand decreases with price, but at different rates. In the first market, if you charge p dollars per unit, you will be able to sell about $1000 - 2.3p$ units. In the jargon of economics, this means that the *demand* in the first market is approximately

$$1000 - 2.3p.$$

Demand in the second market is approximated by

$$1400 - 4.1p.$$

On the other hand, the cost to produce and bring to market x items per week increases linearly with x, according to the formula

$$\text{cost} = 50000 + 70x.$$

a) Your weekly revenue is the amount of money you take in each week by selling your wares. You set your unit price in the first market at p_1 dollars, and in the second market at p_2 dollars. Find a formula for your weekly revenue in terms of p_1 and p_2.

b) Each week, you must produce and bring to market enough units to satisfy the total demand in the two markets. Find a formula for your weekly cost in terms of the prices p_1 and p_2 in the two markets.

c) Your weekly profit is the difference between revenue and cost. How should you set your selling price in each market to maximize your profit? How many units should you ship to each market?

Chapter 3

Integration

24 Double Integrals I—Description and Properties

What is a double integral?

Before coming to grips with this question, we review the meaning of ordinary integrals of functions of one variable.

Review of one variable integrals. If $f(x)$ is a function defined on an interval $I : a \leq x \leq b$, the *integral* $\int_a^b f(x)\,dx$ is defined by a limiting process which may be described as follows.

1. Partition the interval I into small intervals $I_1, I_2, \ldots$.

2. For each of the small intervals I_j, sample the function value at some point x_j in the interval, obtaining the value $f(x_j)$.

3. Multiply each sampled value $f(x_j)$ by the length Δx_j of the interval I_j from which it was sampled, and add up the results, obtaining

$$f(x_1)\,\Delta x_1 + f(x_2)\,\Delta x_2 + \cdots = \sum_j f(x_j)\,\Delta x_j.$$

The integral $\int_a^b f(x)\,dx$ is defined to be the number that is approximated by the above sums. In other words, the integral can be approximated to any precision you choose by making the small intervals $I_1, I_2, \ldots$ short enough. Depending on the physical units of x and $f(x)$, the integral may measure area, displacement, mass, work, etc.

Double Integrals. Double integrals follow the same general pattern, except that the interval I is replaced by a two-dimensional region D. The region D is partitioned into small pieces $D_1, D_2, \ldots$ from which function values are sampled, and approximating sums are formed as in the one-dimensional case considered above, with the lengths Δx_j replaced by areas of the small regions D_j. Before describing the general situation, we illustrate the process with two concrete examples.

Example 1. Measuring pollutants. A toxic substance from a factory smokestack has been settling on a 16 square kilometer plot of land for years. We are charged with measuring the total amount of the substance that contaminates this region.

We cannot measure this total directly, but we can take soil samples at various points in the region. Measuring the amount of the substance in our samples gives us an estimate for the "mass density" of the substance at various points in the region. The mass density is a function $f(x, y)$ that estimates the number of grams of the substance per square meter of ground at any point (x, y) in the region. Thus, for example, to say $f(a, b) = 7$ means that in a small portion D of the region containing the point (a, b), the number of grams of the toxic substance is approximately 7 times the area of D in square meters. Because we are dealing with a relatively large tract of land, it will be more convenient to measure distance in kilometers and mass in metric tons. Since each square kilometer measures one million square meters, and each metric ton consists of one million grams, we may view the units for the mass density $f(x, y)$ as metric tons per square kilometer. The mass density $f(x, y)$ is shown in the contour plot below.

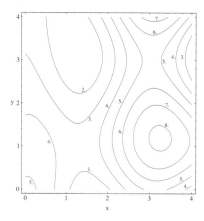

We ask: Now that we have a representation for the mass density how can we estimate the total amount, in metric tons, of pollutant in the region? To get started, we observe the following: We can get a crude lower bound by observing that throughout the region, the mass density is at least 1 metric ton/km^2, so the total number of metric tons is at least

$$
\begin{aligned}
(1 \text{ metric ton/km}^2)(\text{area of region}) &= (1 \text{ metric ton/km}^2)(16 \text{ km}^2) \\
&= 16 \text{ metric tons.}
\end{aligned}
$$

You can get an upper bound in the same way. Since the mass density is at most 9 metric tons/km^2, the total number of metric tons is at most

$$
\begin{aligned}
(9 \text{ metric tons/km}^2)(\text{area of region}) &= (9 \text{ metric tons/km}^2)(16\text{km}^2) \\
&= 144 \text{ metric tons,}
\end{aligned}
$$

So you can be sure that the number of metric tons is between 9 and 144.

You can refine these bounds by laying a rectangular grid over the contour plot, and estimating the number of metric tons in each rectangle of the grid, as shown in Figure 24.1 on the next page. In the upper left hand rectangle of the left figure, the mass density is between 1 and 4 metric tons/km^2 and the area of the rectangle is 1 km^2, so you can be sure that the number of metric tons in the upper left rectangle

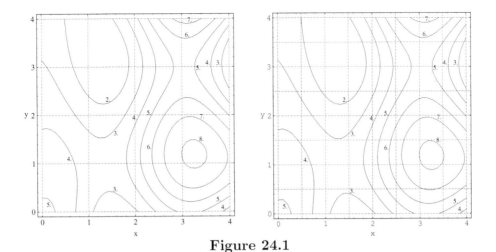

Figure 24.1

is between 1 and 4. You can repeat this procedure for each rectangle in the grid, and sum the results to estimate the total mass. The pair of tables below gives lower and upper estimates for the mass density in each square:

Lower density estimates				Upper density estimates			
1	1	2	3	4	4	8	8
1	1	2	3	4	4	8	8
2	2	3	6	5	5	8	9
3	2	3	3	6	5	8	9

To obtain a lower bound for the number of metric tons in the region, multiply each lower density estimate by the area of the corresponding small square (1 km^2) to get the number of metric tons in the small square, and add up the results to get

$$\text{total number of metric tons} \geq 37.$$

Do the same with the upper density estimates to get

$$\text{total number of metric tons} \leq 103.$$

Combining these two estimates, we conclude that the total mass of the pollutant in the region is between 37 and 103 metric tons. This is better than our original estimate, but still pretty crude. You could improve the estimate by refining the grid, using, say, 64 small squares, each with edge 0.5 km, as shown at left in Figure 24.1. ∎

Problem _____

24.1. Obtain improved upper and lower bounds for the mass of pollutant in the above region by using a grid consisting of squares with edge 0.5 km. *Tip:* You will have to deal with sums having 64 terms, which seems like a lot, but many terms will be repeated. Count up the number of times each value occurs, and multiply.

Example 2. The volume under a graph. Figure 24.2 represents a three-dimensional solid consisting of all points above a portion of the xy-plane and below a curved surface. The top of the pictured solid is the graph of the function

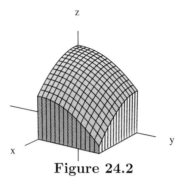

Figure 24.2

$$f(x,y) = 1.1 - \frac{1}{2}(x^2 + y^2),$$

the base is the square in the xy-plane defined by $0 \le x \le 1$ and $0 \le y \le 1$, and the sides are vertical planes. Now we ask: *What is its volume? How can we calculate it?* Instead of calculating it directly, we will attempt to *estimate* the volume by chopping the solid into small pieces, estimating the volume of each piece, and adding those estimates to get an estimate for the volume of the whole solid. We start by overlaying the base with a grid which partitions it into small squares of size 0.2×0.2 units, as shown in Figure 24.3. Now take one of the small squares, say the one with lower

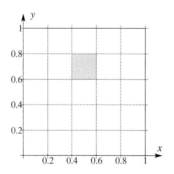

Figure 24.3

left corner $(x,y) = (0.4, 0.6)$, and look at the part of the solid that sits over that square, as shown in the following figure, left.

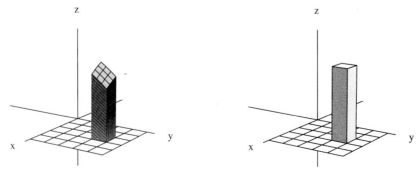

You can estimate the volume of this piece of the solid by replacing it with a rectangular parallelepiped with the same base, but with height equal to the height of, say, the back corner of the the column, which is

$$\text{back corner height} = f(0.4, 0.6) = 0.84.$$

The approximating solid is shown in the right figure above. The volume of the approximating solid is

$$\text{volume} = (\text{area of base}) \cdot \text{height} = (0.2)^2 \cdot f(0.4, 0.6) = 0.336.$$

This is the *exact* volume of the figure shown on the right above, which gives a good *estimate* for the volume of the one on the left.

We now repeat the above performance for each of the small squares in Figure 24.3, and sum up the results, obtaining

$$\sum f(P_j) \cdot (0.2)^2 = 0.86.$$

Here the points P_j run over the lower left corners of the small squares shown in Figure 24.3, giving the volume of the solid shown below.

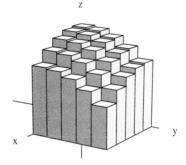

Since the above solid closely approximates the original solid, the calculated value of 0.86 gives a good estimate for the volume of the original solid. At the expense of greater computational effort, we could get a better estimate by using a finer grid. ∎

Problem

24.2. a) The estimate 0.86 for the solid of Example 2 is an overestimate of the exact volume of the given solid. Explain why.

  b) Using the same grid used in the example, give your best underestimate for the volume. *Hint:* Sample function values at different points in the small squares.

We are now ready to answer the question posed at the beginning of this section: *What is a double integral?* Look closely at the procedure used in Example 1 to estimate the total mass of pollutant, and in Example 2 to estimate the volume under a graph. In each case, we had a function $f(x, y)$ defined on a plane region. We estimated some quantity (mass or volume) by partitioning the region, sampling function values times area, and summing. The *procedure* is the same in the two examples, even though the *interpretation* of the result is different. Roughly speaking, a double integral is the number we estimate by the procedure used in Examples 1 and 2. To evaluate a double integral, you must have

- A region D in the xy-plane;
- A function $f(x, y)$ defined on the region D.

The *double integral* of $f(x, y)$ over the region D, which is denoted by the symbol $\iint_D f(x, y) \, dA$, is the *number* that is estimated by the following procedure.

1. Partition D into small[1] regions $D_1, D_2, \ldots$, which don't overlap and which fill up D.

2. For each of the small regions D_j, sample the function value in D_j by choosing a point P_j in D_j and calculating or estimating $f(P_j)$.

3. Multiply the sampled values $f(P_j)$ by the area $A(D_j)$ of the corresponding small region, and add up the results. The sum approximates the value of the double integral:

$$\iint_D f(x, y) \, dA \approx f(P_1)A(D_1) + f(P_2)A(D_2) + \cdots = \sum_j f(P_j)A(D_j).$$

The small regions D_j used in the approximation scheme described above are often taken to be rectangles, as was the case in Examples 1 and 2, but they don't have to be. If D is not itself a rectangle, it may not be possible to partition D into small rectangles, so that some of the D_j's must be taken to be of a different shape.[2]

Problem

24.3. Returning to Example 2, let D be the region inside the triangle with vertices $(0, 0)$, $(1, 0)$, and $(0, 1)$. Estimate $\iint_D f(x, y) \, dA$ by partitioning D into ten of the small squares shown in Figure 24.3, and five small triangles.

What does a double integral measure?

Double integrals can measure many different things, depending on what the integrand measures.

Densities. In Example 1, we estimated the integral of a function that measured mass density, in number of metric tons per unit area. The estimate was obtained by summing terms obtained by multiplying mass density (in metric tons/unit area) by area, so the units for the integral are metric tons, which is what we wanted. In general a *density* is a function $\rho(x, y)$ which measures some physical quantity per unit area. Specific examples include mass densities (mass per unit area), charge densities (electrical charge per unit area), population densities (number of organisms per unit area), pressure (force per unit area), and temperature (heat energy per unit area). The reasoning used in Example 1 shows that if $\rho(x, y)$ is a density that measures

---

[1] Here "small" means that the maximum distance between any two points in D_j is small. Having small area is not enough. Thus, a long thin rectangle would not be considered small, even though it may have small area.

[2] Actually, it is not necessary to fill up every bit of the region D with the small regions D_j. It is ok to omit a portion of D with small area. Thus, you *can* use small rectangles if you wish, as long as you make them small enough and use enough of them to fill up almost all of D.

some quantity per unit area then integrating $\rho(x, y)$ over a region D gives the *total amount* of that quantity in the region D. Thus, if a flat plate covers a region D in the xy-plane, and has mass density $\rho(x, y)$ at the point (x, y), then $\iint_D \rho(x, y)\, dA$ gives the total mass of the plate. If x and y are measured in meters and ρ is in kg/m^2, then the units for $\iint_D \rho(x, y)\, dA$ are kilograms. Similarly, integrating a charge density gives total charge, integrating pressure (force per unit area) gives total force, and integrating temperature (heat energy per unit area) gives total heat energy.

Volumes. We saw in Example 2 that the approximation procedure that defines double integrals can be used to estimate volumes. In general, if $f(x, y)$ is a positive function, then $\iint_D f(x, y)\, dA$ measures the volume of the prism like solid shown in the following figure, right.

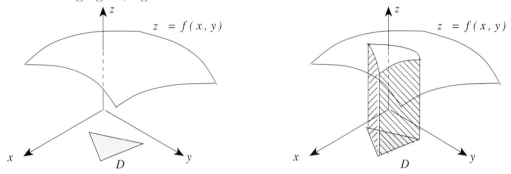

Example 3. Let $f(x, y)$ be the constant function $f(x, y) = c = \text{constant}$. Then the solid bounded by D and the plane $z = c$ is a prism or cylinder, as shown in Figure 24.4. Its volume is given by

$$\text{volume} = (\text{height } c) \cdot (\text{area of } D).$$

Therefore,

$$\iint c\, dA = c \cdot (\text{area of } D).$$

■ From now on we write $A(D)$ for the area of D.

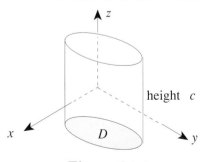

Figure 24.4

Problem _____

24.4. Let $f(x, y)$ be the constant function $f(x, y) = 644$, and let E be the circular region given by

$$(x - 3.1 \cdot 10^6)^2 + (y - 8.5 \cdot 10^8)^2 = 4.$$

Find $\iint_E f(x,y)\,dA$.

Example 4. In Example 3, we take $c = 1$. Then

$$\iint_D 1\,dA = 1 \cdot A(D) = A(D).$$

In this case, the double integral is the area of D. Here, you should think of the integrand 1 as a dimensionless quantity, so that the units for the integral are area units. The integral of $f(x,y) = 1$ is written $\iint_D dA$. ∎

Problem _____

24.5. For every statement in Example 4, there is a corresponding statement for functions of one variable. Rewrite Example 4 for functions of one variable, and make the drawing corresponding to Figure 24.4. Your rewritten Example 4 must make sense to somebody who has had only one-variable calculus.

_____

Example 5. In Figure 24.5 below, R is a rectangle in the xy-plane with corners $(0,0,0)$, $K(7,0,0)$, $L(7,4,0)$, and $M(0,4,0)$. Also, part of the graph $z = f(x,y) = -y^2 + 4y$ is shown. The formula for $f(x,y)$ does not contain x. Therefore, the graph $z = f(x,y)$ is formed as follows: First, draw the parabola $z = -y^2 + 4y$ in the yz-plane. Then, at all points of the parabola, attach lines parallel to $\vec{\imath}$. ∎

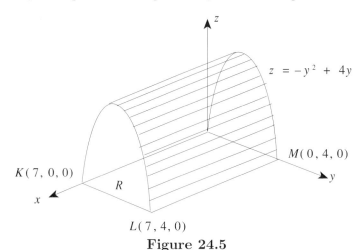

Figure 24.5

Problem _____

24.6. Evaluate the double integral $\iint_R (-y^2 + 4y)\,dA$ where R is the rectangle $OKLM$ of Figure 24.5 above. *Hint:* We have to compute the volume of a solid cylinder. The walls are perpendicular to the yz-plane, and the base in the yz-plane is the filled-in parabola $z = -y^2 + 4y$ from $y = 0$ to $y = 4$. The volume of such a cylinder is given by *(height) · (area of the base)*. In our case, the height equals 7, and you find the area of the base by ordinary integration in the yz-plane.

Bounds and averages

Suppose you have two functions $f(x, y)$ and $g(x, y)$ that are both defined on the *same* region D. Then the integrals $\iint_D f(x, y)\, dA$ and $\iint_D g(x, y)\, dA$ are both meaningful. Now suppose we know that $f(x, y) \leq g(x, y)$ for *every* (x, y) in D. This means that the graph of $f(x, y)$ is always below the graph of $g(x, y)$. *What can you say about how the integrals compare?* Both integrals are *numbers* that can be estimated by the general approximation procedure that defines a double integral. The procedure involves sampling values of f and g, and if you sample each at the same point, then *the sampled values of f will always be less than or equal to the sampled values of g.* That means that estimates for $\iint f(x, y)\, dA$ will always be less than or equal to the corresponding estimates for $\iint g(x, y)\, dA$. Consequently, the same inequality will persist for the numbers being approximated, which are the double integrals. This gives the

Comparison Principle for double integrals. *If we have* $f(x, y) \leq g(x, y)$ *for every point* (x, y) *in* D, *then*

$$\iint_D f(x, y)\, dA \leq \iint_D g(x, y)\, dA.$$

Problem

24.7.   a) Assume that $f(x, y) \leq b$ (a constant) for every (x, y) in the region D. Give a convincing argument based on the Comparison Principle that

$$\iint_D f(x, y)\, dA \leq b \cdot A(D).$$

Here $A(D)$ denotes the area of D. *Hint:* Example 3 on integrals of constant functions may be useful.

b) Assume that $f(x, y) \geq a$ (a constant) for every (x, y) in D. Give a convincing argument that

$$\iint_D f(x, y)\, dA \geq a \cdot A(D).$$

Upper and lower bounds for double integrals. In Problem 24.7, you saw the following. If a function $f(x, y)$ is trapped between two constant values a and b, then the integral $\iint_D f(x, y)\, dA$ is trapped between the numbers $a \cdot A(D)$ and $b \cdot A(D)$. This gives you a way of estimating or *bounding* the integral above and below if you can find upper or lower bounds for the integrand $f(x, y)$. The best possible estimates are obtained when the constants a and b are taken to be the smallest and largest values of $f(x, y)$ on D, which we denote by $\min f$ and $\max f$, respectively. We get the estimates

$$\min f \cdot A(D) \leq \iint_D f(x, y)\, dA \leq \max f \cdot A(D). \tag{24.1}$$

Problem
24.8. The function $f(x, y)$ is defined as the distance of the point (x, y) from the origin $(0,0)$. Let D be the circular region shown below. Estimate $\iint_D f(x, y)\, dA$ above and below. *Hint:* Apply formula (24.1). Use ruler and pencil to find the global extrema of $f(x, y)$ on D.

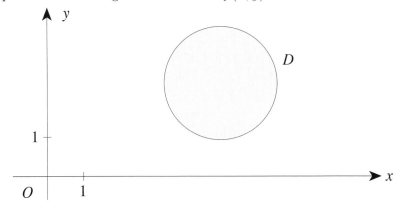

Averages. The double integral lies between $\min f \cdot A(D)$ and $\max f \cdot A(D)$. Therefore, there *must* be a number f^* between $\max f$ and $\min f$ such that the double integral equals $f^* \cdot A(D)$:

$$\iint_D f(x, y)\, dA = f^* \cdot A(D)$$

The number f^* is the average of $f(x, y)$ over D. It represents all the different values of $f(x, y)$ on D in such a way that its product with $A(D)$ gives the double integral.

The Average Value. *The average of $f(x, y)$ over the region D is the number f^* given by*

$$f^* = \frac{\iint_D f(x, y)\, dA}{A(D)}$$

Problem
24.9. This is about the function $z = f(x, y) = -y^2 + 4y$ and the rectangular region R of Problem 24.6.

    a) Find the average of $f(x, y)$ over the region R. *Hint:* Use the boxed formula above.

    b) Find a number d with the following property: The volume of the rectangular box with base R and height d is the same as the volume above R and under the graph $z = f(x, y) = -y^2 + 4y$.

The basic rules for double integrals. They are as follows.

(i) $\iint_D \big[f(x, y) + g(x, y)\big]\, dA = \iint_D f(x, y)\, dA + \iint_D g(x, y)\, dA.$

(ii) Let k be a constant. Then $\iint_D k\, f(x, y)\, dA = k \iint_D f(x, y)\, dA.$

(iii) The region D is subdivided by a curve in two subregions D_1 and D_2. For example, D could be like any of the following:

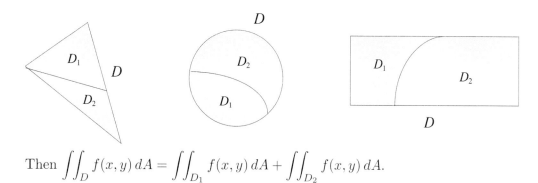

Then $\iint_D f(x,y)\,dA = \iint_{D_1} f(x,y)\,dA + \iint_{D_2} f(x,y)\,dA.$

Problem

24.10. Your friend says: "Rules (i) and (iii) are the same. They say that you can add double integrals." What do you say?

Additional Problems

24.11. T is the triangular region OAB in the xy-plane (see below). The triangle ABC is part of the graph of $z = p(x,y) = 5(-\frac{x}{4} - \frac{y}{3} + 1)$, as shown below.

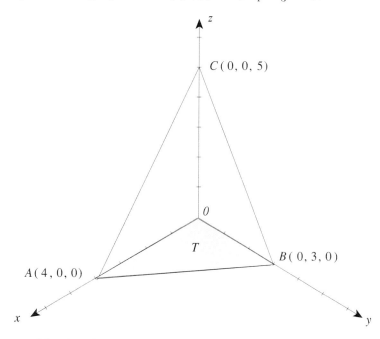

a) Find $\iint_T p(x,y)\,dA$. *Hint:* The volume of a pyramid is given by $\frac{1}{3}$(area of base) $\cdot$ (height).

b) We write p^* for the average of $p(x,y)$ on the region T. Find p^*. *Hint:* Use the fact (double integral) = (average) $\cdot$ (area).

c) We keep the region T in the xy-plane, but take a new graph $z = k\,p(x,y) = k[5(-\frac{x}{4} - \frac{y}{3} + 1)]$ where k is a scalar. In the figure below, k is approximately $\frac{7}{10}$.

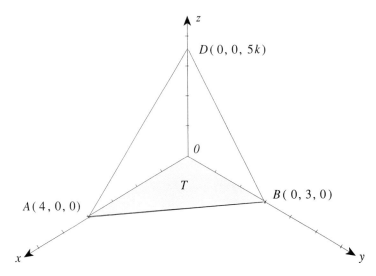

(i) Find $\iint_T k\,p(x,y)\,dA$. Your answer will contain k.

(ii) Which of the following three statements is true:

$(1)\quad \iint_T p(x,y)\,dA = \iint_T k\,p(x,y)\,dA$

$(2)\quad \iint_T k\,p(x,y)\,dA = k\iint_T p(x,y)\,dA$

$(3)\quad \frac{1}{k}\iint_T p(x,y)\,dA = \iint_T k\,p(x,y)\,dA$

24.12. In the following figure, left is shown part of the graph $z = -x^3 + 2$; the figure is not to scale. On the right is shown a rectangular region D_a in the xy-plane. Its width is 2. Its length is $a + 1$ where $a \geq -1$ is a real number. We integrate $f(x,y) = -x^3 + 2$ over D_a for different values of a.

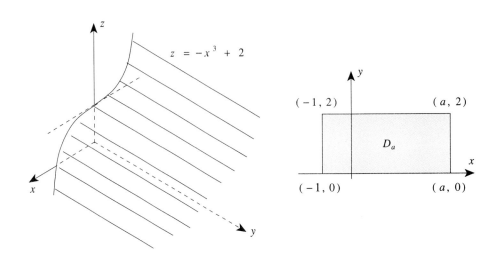

To answer the questions, look where the graph is above or below the xy-plane.

   a) Find a such that the double integral of $f(x, y)$ over D_a is zero.

   b) Find a such that double integral of $f(x, y)$ over D_a is negative.

   c) Find a such that double integral of $f(x, y)$ over D_a is positive.

24.13. The triangular region T is shown in Figure 24.6. The function $q(x, y)$ is defined by $q(x, y) = x - y$. Estimate $\iint_T (x - y) \, dA$. That is, find constants k and K such that

$$k \leq \iint_T (x - y) \, dA \leq K.$$

Hint: Find the global extrema of $q(x, y)$ by means of level curves. Use ruler and pencil.

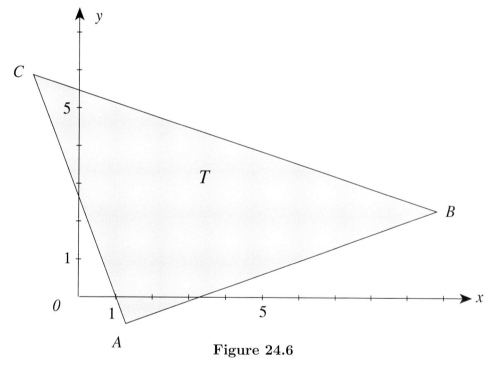

Figure 24.6

24.14. Go back to "The basic rules for double integrals" just before Problem 24.10 on page 204. Look at the three rules. Are there corresponding rules for ordinary definite integrals of elementary calculus? What are they?

24.15. A one meter by one meter flat plate is made of material of variable density. Its mass density is proportional to the distance from one edge and is equal to 2 kg/m² along the opposite edge. Find the total mass of the plate. *Hint:* Choose coordinates so that the plate covers a convenient region in the xy-plane, and work out a formula for the mass density. The total mass of the plate is given by a certain double integral, which may also be interpreted as the volume of a certain solid. Calculate the volume of that solid by elementary geometry.

24.16. A rectangular viewing window at the Pittsburgh Zoo sea lion tank measures 15 ft horizontally and 5 ft vertically. The fluid pressure s ft below the top edge is given by

$$\text{pressure} = 62.4 \cdot s \ \text{lb/ft}^2.$$

Calculate the total force that the water exerts on the window. *Hint:* Choose coordinates so that the window covers a region of the xy-plane. The force is measured by an integral, which can also be interpreted as a volume. Calculate the volume by elementary geometry.

24.17. Circle City is a perfect disk with radius 10 km. The population density in the center of town is 500 people per square kilometer, and the density falls off linearly as a function of the distance from the town center, to a density of 0 at the edge of town. What's the total population? *Hint:* Interpret the total population as an integral, then interpret the integral as a volume.

24.18. Figure 24.7 shows part of a topographical map. The curves shown are contour lines. They are the level curves of the function $e(x, y)$ defined as follows. The point $P(x, y)$ on the map represents a point P_1 out there in the real world. Then $e(x, y)$ is defined as the elevation above sea level of the real point P_1. A contour line w of the map represents a path w_1 in the real world. The real path w_1 is completely level. The map covers a piece of land L of the shape of a square. The levels of the level curves shown are in meters above sea level. The scale of the map is $1 : 10,000$ so that 1 cm on the map corresponds to 100 m in reality.

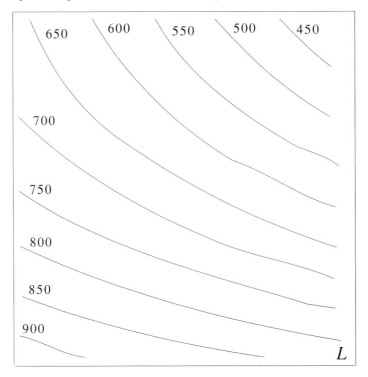

Figure 24.7

a) By just looking at the level curves, estimate the average elevation e^* of the piece of land L. *Hint:* It will be a very rough guess, but you can do better than picking a number at random.

b) Give the best estimate you can for $\int\int_L e(P)\, dA$, the double integral of the elevation function $e(P)$ over the region L. *Hints:* (i) Use the fact (double integral) = (average) $\cdot$ (area). (ii) Remember that 1 cm on the map are 100 m in reality.

25 Double Integrals II—Evaluation in Cartesian Coordinates

In Section 24, we discussed what a double integral *is*. In this section, we discuss how to *evaluate* a double integral. We work with a region D in the xy-plane, and a function $f(x, y)$ defined on D. Our task is to evaluate the double integral $\iint_D f \, dA$.

We illustrate the technique using as D a triangular region KLM in the xy-plane. Below left D is shown as it lies in the xy-plane, and on the right D is shown in its true shape in the xy-plane.

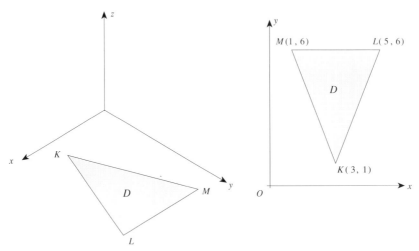

Figure 25.1

The idea. We saw in Section 24 that one interpretation of a double integral is as the volume under a surface. We will use this interpretation to calculate $\iint_d f \, dA$ by measuring a volume. We have to find the volume of the solid between D and the graph $z = f(x, y)$. First, we compute the volume of a thin slice. Such a slice is shown at left in Figure 25.2. It has thickness Δy. The walls are parallel to the xz-plane. The back wall faces the xz-plane, and the front wall faces us, the viewers. Once we have the volume of a slice, we add up all of them by integrating over y.

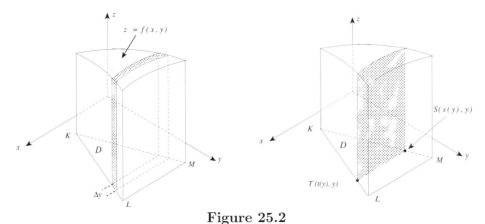

Figure 25.2

The volume of a slice.   The right side of Figure 25.2 shows the back wall of the slice. Its baseline is the line segment ST in the xy-plane. We compute the volume of a slice in three steps.

Step 1: The baseline ST of the back wall is parallel to the x-axis. The x-coordinates of the endpoints S and T depend on their y-coordinates. *Why?* If you change the position of the slice to another position y, the x-coordinates of S and T will change. The x-coordinate of S is a function $s(y)$ of y, and the x-coordinate of T is a function $t(y)$. Thus, the coordinates of S and T are of the form

$$S(s(y), y), \quad T(t(y), y).$$

Similarly, the area of the back wall is a function of y. We write $A(y)$ for the area of the back wall at position y.

Problem _____

    25.1. a) We increase y from the value as shown at right in Figure 25.2. Which of the functions $A(y)$, $s(y)$, $t(y)$ will increase? Stay constant? Decrease?

           b) Read off the xy-coordinates of K, L, and M from Figure 25.1 on the facing page. Find the following: $A(1)$, $s(3.5)$, $t(6)$.

_____

Step 2: The area $A(y)$ of the back wall is the area under the graph $z = f(x, y)$ from $x = s(y)$ to $x = t(y)$ in a plane $y = $ constant. This plane is parallel to the xz-plane. Therefore we can compute $A(y)$ as an ordinary integral in the xz-plane:

$$A(y) = \int_{x=s(y)}^{x=t(y)} f(x, y) \, dx.$$

In this integration with respect to x, the letter y represents a constant and is not affected by integration. The limits of integration are formulas rather than numbers. The evaluated integral will contain y because y occurs in the function $f(x, y)$ and in the limits of integration.

Step 3: Now we look at the slice, and not only at the back wall. The slice has thickness Δy, and its back wall is at position y. The volume of the slice is approximately the volume of a prism of whose base is the back wall and whose height is Δy:

$$\text{volume of slice} \approx A(y) \cdot \Delta y.$$

Problem _____

    25.2. a) Explain why in general the *exact* volume of the slice is different from $A(y) \cdot \Delta y$.

           b) For the slice with back wall and thickness as shown at left in Figure 25.2, would you think that the exact volume of the slice is less than, or equal to, or greater than the approximation $A(y) \cdot \Delta y$? Explain your answer.

Adding up the volume of all slices. Now we slice up the whole solid. The back walls of the many slices are at different positions y, with y varying between 1 and 6.

We start by subdividing the y-interval from 1 to 6 into smaller pieces, for example into 489 pieces of length $\Delta y = \frac{5}{489}$ each. We number the points which subdivide the y-interval from 1 to 6:

$$
\begin{array}{ccccc}
1 & & y_{211} \quad y_{212} \quad y_{213} & & 6 \\
\end{array}
$$

The first point y_1 is $y = 1$. We add up the approximate volumes of the 489 slices:

$$A(y_1)\Delta y + A(y_2)\Delta y + \ldots + A(y_{489})\Delta y. \tag{25.1}$$

This number is an approximation for the total volume, that is, for the double integral.

A careful look at the sum (25.1) reveals that it is an approximating sum for an ordinary one-variable integral of the function $A(y)$:

$$A(y_1)\Delta y + A(y_2)\Delta y + \ldots + A(y_{489})\Delta y \approx \int_{y=1}^{y=6} A(y)\,dy. \tag{25.2}$$

Further subdivision of the interval $1 \leq y \leq 6$ gives both a better and better approximation to the volume we are after, *and* a better and better approximation to the integral on the right of (25.2). The inevitable conclusion is that the integral on the right of (25.2) gives the exact value of the volume.

We have computed the double integral by means of two ordinary integrations:

$$\iint_D f(x,y)\,dA = \int_{y=1}^{y=6} A(y)\,dy = \int_{y=1}^{y=6} \left(\int_{x=s(y)}^{x=t(y)} f(x,y)\,dx \right) dy.$$

In words: *To evaluate $\iint_D f(x,y)\,dA$ we first integrate $f(x,y)$ with respect to x from $x = s(y)$ to $x = t(y)$ where we have to find the functions $s(y)$ and $t(y)$ by looking at D. After this first integration with respect to x we are left with an expression which is a function of y. Then we integrate this function of y between limits which again we find by looking at D.*

Terminology. We have used slices whose walls are planes $y = $ constant. For this reason, such slices are called *y-slices*.

In the following summary, D will denote any region, and not necessarily the triangular region KLM we discussed in detail.

Evaluating a double integral with slices

Using y-slices

Step 1: Template

Write down the template for the integral. It shows the limits of integration which you have to find:

$$\int_{y=\ldots\ldots}^{y=\ldots\ldots} \left(\int_{x=\ldots\ldots}^{x=\ldots\ldots} f(x,y)\,dx \right) dy.$$

Step 2: Sketch

Make a *big* sketch of D—the bigger the sketch, the easier and safer the work.

Step 3: Limits

Limits for x

Pick a y, say $y = c$, and draw the line $y = c$.

Walk on this line in the positive direction, starting with a very low value of x. Call S the point at which you enter D. The coordinates of S are $(s(y), y)$. Find the formula for $s(y)$.

Call T the point at which you leave D. The coordinates of T are $(t(y), y)$. Find the formula for $t(y)$.

Insert the x-limits into the template: $s(y)$ is the lower limit and $t(y)$ is the upper limit.

Limits for y

Determine the range of y-values so as to cover all of D. Call y_1 the lowest and y_2 the highest.

Insert the y-limits into the template: y_1 is the lower limit and y_2 is the upper limit.

Step 4: Setting up the integral

At this point, you have "set up the integral." It means that the integral can be evaluated by two *ordinary* integrations:

$$\iint_D f(x, y)\, dA = \int_{y=y_1}^{y=y_2} \left(\int_{x=s(y)}^{x=t(y)} f(x, y)\, dx \right) dy.$$

Step 5: Integration

Integrate from inside out: First, integrate with respect to x. Then, integrate with respect to y.

Problem _____

25.3. Let D be the triangle KLM shown in Figure 25.1 at the beginning of this section. Set up $\iint_D xy\, dA$. *Hint:* You need the equation of the lines KM and KL. Here they are:

$$KM : x + 0.4y - 3.4 = 0; \qquad KL : x - 0.4y - 2.6 = 0.$$

_____

Notation. With y-slices, the first integration is with respect to x and the second with respect to y. To indicate the order of integration, one writes the double integral in the form

$$\iint_D f(x, y)\, dx\, dy \quad \text{or} \quad \int_{y=y_1}^{y=y_2} \int_{x=s(y)}^{x=t(y)} f(x, y)\, dx\, dy.$$

The method of x-slices

Instead of using y-slices parallel to the xz-plane, we can take x-slices parallel to the yz-plane. With x-slices we get the same result (this fact is known as one of the theorems of Fubini). *Why would one need two methods?* For some regions y-slices work better, for others x-slices. This issue is taken up at the end of this section.

Problem _____

25.4.   a) Rewrite the summary "Evaluating a double ..." for x-slices. With x-slices you will walk through the region in the y-direction. Write A for the point where you enter the region, and B for the point where you leave.

   b) D is the triangular region of Figure 25.3. Set up $\iint_D x^2 y^3 \, dA$ with x-slices.

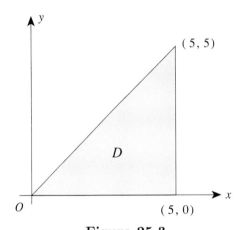

Figure 25.3

_____

Notation (continued).   If a double integral is evaluated with x-slices, one uses $dy \, dx$ and writes

$$\iint_D f(x,y) \, dy \, dx \quad \text{or} \quad \int_{x=x_1}^{x=x_2} \int_{y=a(x)}^{y=b(x)} f(x,y) \, dy \, dx.$$

On the right, the line $x = c$ enters D at $A(x, a(x))$ and leaves D at $B(x, b(x))$.

Since dA is replaced by $dx \, dy$ or $dy \, dx$ in order to evaluate a double integral using Cartesian coordinates, we will write

$$dA = dx \, dy = dy \, dx.$$

The expressions $dx \, dy$ and $dy \, dx$ are often called the *area element in Cartesian coordinates*. It is often convenient to think of $dx \, dy$ as the area of an "infinitesimally small" rectangle with side lengths dx and dy, and to think of the double integral $\iint_D f(x,y) \, dx \, dy$ as summing the "infinitesimal" quantities $f(x,y) \, dx \, dy$ over a partition of D into an infinite number of infinitesimally small rectangles.

Should we use y-slices or x-slices? At the beginning of the section we used y-slices to evaluate the integral over the triangular region KLM (see Figure 25.1). What about using x-slices? With x-slices you would have to split the integral into two because the formulas for the entrance and exit points depend whether you are between $x = 1$ and $x = 3$ or between $x = 3$ and $x = 5$: If you use y-slices you

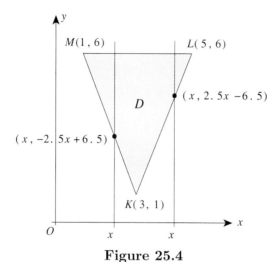

Figure 25.4

have to set up only one integral. This is a typical example where one kind of slices is more convenient than the other. When choosing slices, ask yourself: Can I avoid setting up two integrals? Are the formulas for the boundary curves of the region as simple as possible?

Problem _____

25.5. We have to integrate $f(x, y)$ over each of the four regions D_1, ..., D_4 shown below. Where would you use y-slices, and where x-slices?

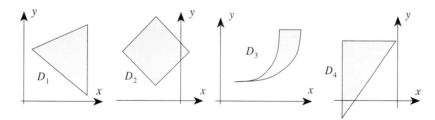

_____

Note: Double integrals require caution. In particular:

- Be aware of the order of integration, and stick to the rule "$dx\,dy$ means y-slices, and $dy\,dx$ means x-slices."

- When you set up an integral write not only the limits of integration, but also the name x or y of the variable. For example,

$$\text{write} \quad \int_{y=3}^{y=5} \int_{x=6-y}^{x=21-4y} e^{xy}\,dx\,dy \quad \text{and not only} \quad \int_{3}^{5} \int_{6-y}^{21-4y} e^{xy}\,dx\,dy.$$

Your instructor may insist on writing double integrals in this way.

Additional Problems

Please read. Problems 25.6 and 25.8–25.12 are about setting up and evaluating double integrals. Before getting into 25.6 we summarize the two ways of setting up a double integral in Cartesian coordinates, as follows.

R is the region shown below. If you want to set up $\iint_R f(x,y)dA$ by means of x-slices, you have to find the formulas for the y-coordinate $a(x)$ of A and $b(x)$ of B shown at left. For y-slices, you need the x-coordinates $s(y)$ of S and $t(y)$ of T shown at right.

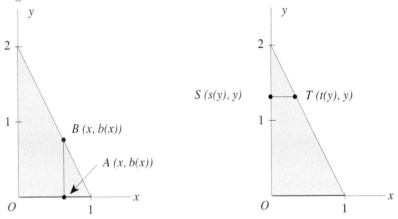

For our R we have $a(x) = 0$ and $s(y) = 0$. Here are the corresponding integrals set up:

$$\int_{x=0}^{x=1} \int_{y=0}^{y=b(x)} f(x,y)\, dy\, dx \qquad\qquad \int_{y=0}^{y=2} \int_{x=0}^{x=t(y)} f(x,y)\, dx\, dy$$

In short: "$dx\, dy$" means "y-slices," that is, you keep y constant and integrate first with respect to x." Similarly, "$dy\, dx$" means "x-slices," that is, you keep x constant and integrate first with respect to y."

25.6.   a) Below you find two rectangular regions D_1, D_2. Do the following:

        (i) Set up $\iint_{D_1} e^{x+y}\, dx\, dy$ and $\iint_{D_1} e^{x+y}\, dy\, dx$.

        (ii) Evaluate $\iint_{D_2} (28 + m^2)\, dA$, where m is a constant.

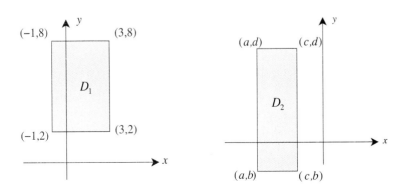

> **Watch out for the limits of double integrals.** Say you have to integrate $z = f(x, y)$ over a region D of the xy-plane. If D is a rectangle (such as D_1, D_2 just discussed) then the limits are easy to write down. *If the region is not a rectangle, then finding the limits usually requires detail work.*

b) In the following figures, you find two regions D_3, D_4. Do the following:

   (i) D_3 is a right triangle, and $f(x, y)$ is a function of two variables the formula of which we do not know. Set up $\iint_{D_3} f(x, y)dxdy$. Then set up $\iint_{D_3} f(x, y)dydx$.

   (ii) D_4 is again a right triangle. – Evaluate $\iint_{D_4} x \, dydx$. Then evaluate $\iint_{D_4} x \, dxdy$.

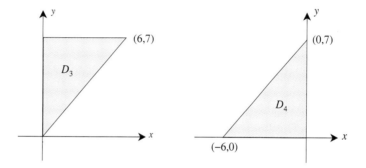

c) Below you find two regions D_5, D_6. Do the following:

   (i) D_5 is bounded by part of a graph $y = g(x)$ and the x-axis. Decide whether to use x-slices or y-slices. Then set up $\iint_{D_5} f(x, y) \, dA$ with the slices you have chosen.

   (ii) A friend sees what you are doing in (i) and says: "I do not get it. Isn't the integral in question just the ordinary integral $\int_u^v g(x)dx$ of elementary calculus?" Write down a response.

   (iii) D_6 is a triangle standing on its head. If you have to integrate over D_6 will you use x-slices or y-slices? Set up and evaluate $\iint_{D_6} (y + 2) \, dA$.

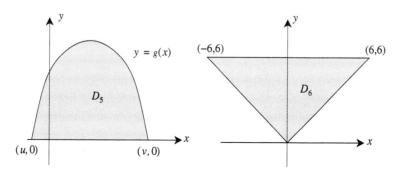

   d) Below you find two regions D_7, D_8. Do the following:

      (i) D_7 is a parallelogram. – x-slices or y-slices? Evaluate $\iint_{D_7} y \, dA$.

      (ii) D_8 looks like a fat letter "ell." Evaluate $\iint_{D_8} x \, dA$.

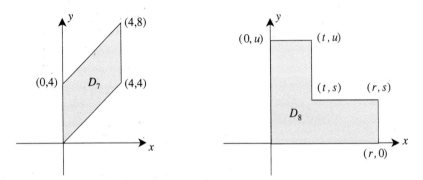

25.7. Evaluate the following ordinary integrals. *Do not simplify.*

$$\text{a) } \int_1^3 xy^2 \, dx \quad\quad \text{b) } \int_{x-1}^{x+1} xy^2 \, dy \quad\quad \text{c) } \int_{3p+4}^{q^2} xy^2 \, dw \quad\quad \text{d) } \int_{e^y}^{y} (x + 2y) \, dx$$

25.8. Let D be the region bounded by the parabola $y = x^2 - 1$ and the x-axis.

   a) Set up a double integral that gives the area of D.

   b) Find the area of D by evaluating the double integral of part a).

25.9. D is the half-disk (filled-in half-circle) of radius 3 and center $M(4,2)$, as shown below.

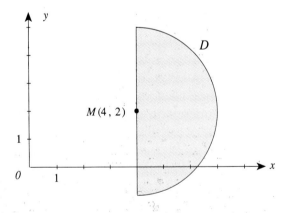

   a) Set up the integral $\iint_D xy \, dy \, dx$. Observe the given order of integration.

   b) Carry out the first (inside) integration. Do not simplify. Your answer looks like $\int_{x=\ldots}^{x=\cdots} \ldots dx$.

25.10. T is the triangular region shown in Figure 25.5. Set up the integral $\iint_T (x + y + 1) \, dx \, dy$. *Hints:* (1) Draw a horizontal line to subdivide T into two triangles T_1 and T_2. (2) The equations of the lines MK, KL, LM are as follows:

$$MK : x + y = 6; \quad KL : x - 2y = 3; \quad LM : x + 4y = 21.$$

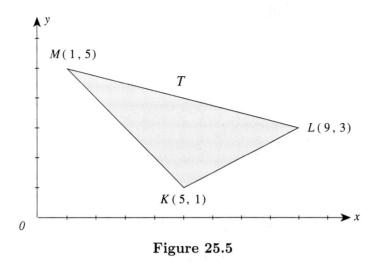

Figure 25.5

25.11. Return to the situation described in Problem 24.15. There, you set up a double integral to measure the mass of a flat plate and used elementary geometry to evaluate the integral. Do it again, this time using the methods of this section to evaluate the integral.

25.12. Return to the situation described in Problem 24.16. There, you set up a double integral to measure the force on a window. Calculate the force by using the methods of this section to evaluate the integral.

26   Double Integrals in Polar Coordinates

The region D shown below looks like part of a flat washer. To set up a double integral over D will be cumbersome, no matter which order of integration you use.

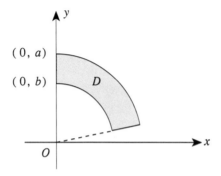

Figure 26.1

For regions such as D, there are a better integration procedures. They use *polar* coordinates instead of Cartesian xy-coordinates.

Review of polar coordinates.   We describe a point P in the plane by

(i) its distance $r = |\overrightarrow{OP}|$ from the origin, and

(ii) the angle θ ("theta") which $\overrightarrow{OP}$ forms with $\vec{\imath}$, counted in counterclockwise direction from $\vec{\imath}$ on, as shown in Figure 26.2.

The numbers (r, θ) are called the *polar coordinates* of P.

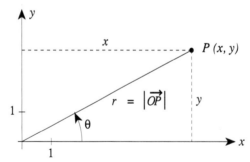

Figure 26.2

Note: From now on, we have to make clear which coordinates we are using. For instance, we will say "P has coordinates $(x, y) = (3, 2)$" or "Q is the point $(r, \theta) = (2, 1)$."

Problem _____

26.1.   a) Plot the points $Q_1, \ldots$ with polar coordinates as follows: $Q_1(2, \frac{3}{2}\pi)$, $Q_2(3, 0)$, $Q_3(4, \frac{5}{4}\pi)$, $Q_4(1, \pi)$, $Q_5(1, 455\pi)$.

b) Find the polar coordinates (r, θ) of the points whose xy-coordinates are as follows: $T_1(11, -11)$; $T_2(1, \sqrt{3})$, $T_3(-328, 0)$.

Pointers for the use of polar coordinates

1. $(r, \theta) = (2, 257)$ and $(r, \theta) = (2, 257 + 2\pi)$ represent the same point. Why? θ and $\theta + 2\pi$ differ by one full rotation of 2π and look the same in the geometric representation. Similarly, $(r, \theta) = (2, 257)$ and $(r, \theta) = (2, 257 + 266\pi)$ represent the same point. Therefore: *Given the point P, its r is uniquely determined, but its θ is not.* On the other hand: *Given (r, θ), the point P is uniquely determined.*

2. $r = |\overrightarrow{OP}|$ is always greater than or equal to zero.

3. At the origin we have: $r = 0$, θ is undetermined.

4. We have two different coordinate systems in the plane, namely, the xy-system and the $r\theta$-system. In such a situation, how do we move from one coordinate system to the other? There are *two* questions:

   Question 1: *Given the polar coordinates (r, θ) of a point P, how do we find its Cartesian coordinates?*
   Answer: $x = r \cos \theta$, $y = r \sin \theta$.

   Question 2: *Given the Cartesian coordinates (x, y) of a point P. How do we find its polar coordinates?*
   Answer: $r = \sqrt{x^2 + y^2}$, θ determined by $\tan \theta = \frac{y}{x}$.[1] The formula for θ shows again that the θ of a point P is not uniquely determined. If θ is the angle of P, then so is $\theta \pm 2\pi, 4\pi, 6\pi \ldots$.

5. As an example, we consider the following function $f(P)$ for points in the plane: $f(P)$ equals the distance of P from the origin. In xy-coordinates, the formula for the function is $\sqrt{x^2 + y^2}$. In polar coordinates, the formula for the same function is r. It means: *One and the same function has different formulas in different coordinate systems.*

   Problem ──────────────────────────────

   26.2. The function $f(x, y)$ is given by $f(x, y) = 4x^2 y + y^3$. Express $f(x, y)$ in polar coordinates. That is, write out the expression $f(r \cos \theta, r \sin \theta)$ as a formula $F(r, \theta)$ in r and θ.

   ──────────────────────────────

6. The points (x, y) for which $x = $ constant form lines in the plane, and similarly for $y = $ constant. These lines are called the *coordinate curves* (or *coordinate lines*) of the xy-system. What can we say about the points (r, θ) for which $r = $ constant? They form circles centered at the origin. What about $\theta = $ constant? These points form a ray (half-line) issued from the origin. It means: *The coordinate curves of polar coordinates are circles and rays.*

   Problem ──────────────────────────────

   26.3. Draw the following parts of coordinate curves of polar coordinates:
   (a) $r = 3$, $0 \le \theta \le \pi$ (b) $\theta = \frac{\pi}{4}$, $0 \le r \le 5$
   (c) $r = 4$, $37.5\pi \le \theta \le 38\pi$

   ──────────────────────────────

---

[1] If $x = 0$ then θ equals $\frac{\pi}{2}$ or $\frac{3\pi}{2} \pm 2\pi, 4\pi, 6\pi \ldots$.

Evaluating a double integral in polar coordinates

Working out the idea

The idea for polar coordinates is similar to the idea for xy-coordinates, but the details work out differently. We interpret the integral as the volume under the graph. In xy-coordinates, the method of y-slices works as follows:

1. We use planes $y = $ constant to cut up the solid under the graph $z = f(x, y)$ into slices.

2. We add up the volume of the slices by ordinary integration.

In polar coordinates, the solid under the graph is cut up in a different way. We look at all points (x, y, z) in space which have the following property: Their shadow $(x, y, 0)$ lies on a ray $\theta = c = $ constant. These space points form a *half-plane*. The half-plane is normal to the xy-plane. It is the polar version of a plane $y = $ constant or $x = $ constant. Now we can describe the idea for evaluating double integrals in polar coordinates.

1. We use half-planes $\theta = $ constant to cut up the solid under the graph $z = f(x, y)$ into wedge-shaped pieces. *We call these pieces θ-wedges.*

2. We add up the volume of all θ-wedges by ordinary integration.

We work through this method for the washer-shaped region D shown in Figure 26.1 on page 220. Figure 26.3 shows the washer D in the xy-plane, the part of the graph $z = f(x, y)$ above D, and the θ-wedge. The sidewalls of the θ-wedge are parts of two half-planes $\theta = $ constant. The angle between them is $\Delta\theta$. To simplify the picture we have taken the graph $z = f(x, y)$ to be a plane.

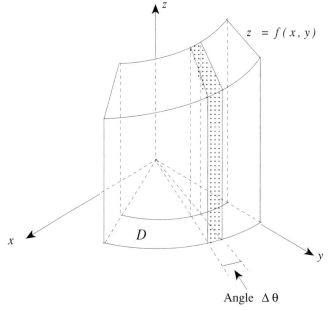

Figure 26.3

Now we look at the base of the θ-wedge in the xy-plane, shown in Figure 26.4. The base consists of several pieces like the shaded one. We approximate the shaded

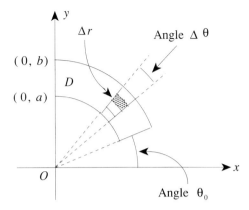

Figure 26.4

piece by a rectangle of length Δr, width $r\,\Delta\theta$, and area $\Delta r \cdot r\Delta\theta$. The volume of the solid above the shaded piece is approximately

$$\text{(area of shaded piece)} \cdot \text{(height)} = (\Delta r \cdot r\Delta\theta) \cdot f(x,y)$$
$$= (\Delta r \cdot r\Delta\theta) \cdot f(r\cos\theta, r\sin\theta)$$
$$= F(r,\theta)\, r\, \Delta r\, \Delta\theta$$

In the last term, $F(r,\theta)$ is the function $f(x,y)$ rewritten in polar coordinates. The value of $F(r,\theta)$ is taken at a point in the shaded piece.

Now we add up the volume above all shaded pieces by integration. Then we obtain the approximate volume of the θ-wedge:

$$\text{(volume of } \theta\text{-wedge of angle } \Delta\theta) \;\approx\; \left(\int_{r=a}^{r=b} F(r,\theta)\, r\, dr \right) \Delta\theta$$

To get the total and exact volume above D, we add up the volumes of all θ-wedges by integrating over θ from $\theta = \theta_0$ to $\theta = \frac{\pi}{2}$:

$$\text{total volume} \;=\; \iint_D f(x,y)dA = \int_{\theta=\theta_0}^{\theta=\frac{\pi}{2}} \left(\int_{r=a}^{r=b} F(r,\theta)\, r\, dr \right) d\theta$$

We have evaluated a double integral in polar coordinates. The evaluation procedure differs from the procedure for xy-coordinates in two ways:

1. *The function $f(x,y)$ has to be converted to polar coordinates and becomes a different expression $F(r,\theta)$.*

2. *In the first integration with respect to r there is an additional r. We have to integrate $F(r,\theta)\, r$, and not only $F(r,\theta)$.*

Problem ────────────────────────────────

26.4. D is the washer we have been using (see Figure 26.4). Find $\iint_D xy\, dA$. *Hints:* (1) Convert the function $f(x,y)$ to polar coordinates. (2) Use the formula *total volume* above. (3) Remember to insert the additional r. (4) You will need the fact $\int \cos\theta \sin\theta\, d\theta = \frac{1}{2}\sin^2\theta + C$. (5) Your answer will contain the letters a, b, and θ_0.

Since dA is replaced by $r\,dr\,d\theta$ in order to evaluate a double integral in polar coordinates, we write

$$dA = r\,dr\,d\theta.$$

The expression on the right is called the *area element in polar coordinates.*

Doing the calculations

We want to evaluate a double integral over a region D using polar coordinates. For example, D may look as shown below.

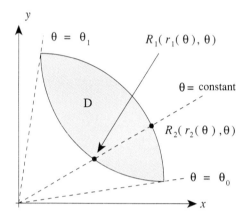

Step 1: Conversion and Template

Write $f(x, y)$ in polar coordinates:

$$f(r\cos\theta, r\sin\theta) = F(r, \theta).$$

and write out a template for the integration:

$$\int_{\theta=\ldots\ldots}^{\theta=\ldots\ldots} \left(\int_{r=\ldots\ldots}^{r=\ldots\ldots} F(r,\theta)\,r\,dr \right) d\theta.$$

Don't forget the extra factor of r.

Step 2: Sketch

Make a *big* sketch of D. The bigger the sketch, the easier and safer the work.

Step 3: Limits

Limits for r

- Pick a θ, say $\theta = c$, and draw the ray $\theta = c$.
- Walk on this ray away from the origin. Call R_1 the point at which you enter D. The polar coordinates of R_1 are $(r_1(\theta), \theta)$. Find the formula for $r_1(\theta)$.
- Call R_2 the point at which you leave D. The polar coordinates of R_2 are $(r_2(\theta), \theta)$. Find the formula for $r_2(\theta)$.
- $r_1(\theta)$ is the lower limit for r and $r_2(\theta)$ is the upper limit.

Limits for θ

- Determine the range of θ-values so as to cover all of D. Call θ_1 the lowest and θ_2 the highest. θ_1 is the lower limit of θ and θ_2 is the upper limit.

Step 4: Setting up the Integral

Add the limits of integration to your template:

$$\iint_D f(x, y)\, dA = \int_{\theta=\theta_1}^{\theta=\theta_2} \left(\int_{r=r_1(\theta)}^{r=r_2(\theta)} F(r, \theta)\, r\, dr \right) d\theta.$$

Step 5: Integration

Integrate from inside out: First with respect to r, then with respect to θ.

Area in polar coordinates

To get the area $A(D)$ of D, you integrate the function $f(x, y) = 1$ (see Example 4 of Section 24). In polar coordinates, this means:

$$A(D) = \iint_D dA = \iint_D r\, dr\, d\theta.$$

Frequently asked questions about polar coordinates

1. *In polar coordinates, why is there an additional r in the integration with respect to r?* The evaluation of a double integral in Cartesian coordinates rests on subdividing the region D into rectangles of area $\Delta x \Delta y$ (below left), while in polar coordinates segments of washers are used. The area of a washer segment is approximately $r\, \Delta\theta \cdot \Delta r = r\, \Delta r\, \Delta\theta$ (below right).

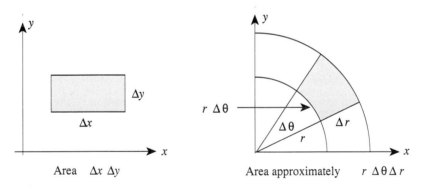

Area $\Delta x\, \Delta y$ Area approximately $r\, \Delta\theta \Delta r$

The additional r of polar coordinates comes from the r in the approximate area $r\, \Delta r\, \Delta\theta$ of the shaded piece above right.

2. *When should I use Cartesian coordinates, and when polar coordinates?* There is no rule, only a general guideline: The more the boundary of D resembles the coordinate curves, the simpler it will be to find the limits of integration (see Section 23 for "boundary"). Thus, if D is bounded by straight lines (triangle, rectangle, etc), try xy-coordinates. If D is bounded by circles, or part of circles, and radii of these circles, try polar coordinates (for example our fourth of a washer). If the curve is given or easily expressible in polar coordinates, use polar coordinates.

Additional Problems

26.5. a) Rewrite the equation $x^2 + y^2 = 16$ in polar coordinates.

b) As a), but for $2x + y = 0$.

c) The line m is given by $(x, y) = (4, 2) + t(3, 1)$. Write the equation of m as one formula $p(r, \theta) = 0$.

26.6. In the following figures, you find two regions. D_1 is a half-circle of radius 5, and D_2 has the shape of a washer. Do the following:

(i) Set up $\iint_{D_1} e^{x+y}\, dA$ in polar coordinates.

(ii) Evaluate $\iint_{D_2} (x^2 + y^2)\, dA$ in polar coordinates.

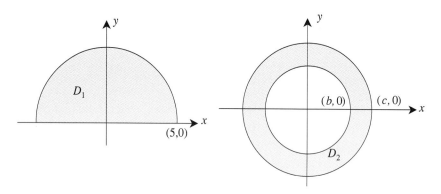

26.7. The equation $r = \theta$ describes a curve C in the xy-plane.

a) Sketch C for $0 \le \theta \le 2\pi$. *Hint:* Consider θ as the independent variable and make a table of r for different values of θ.

b) Find a parametric representation $(x, y) = (x(t), y(t))$ of C. *Hint:* Use θ as parameter.

c) Draw the half-line $\theta = \frac{3}{4}\pi$. Shade the region bounded by C and the line you just drew. Find the area of this region.

26.8. D is the region bounded by the curve $r = 2\theta$, $0 \le \theta \le \pi$ and the x-axis, as shown below.

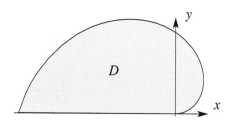

The function $f(P)$ of points P in the xy-plane is defined as follows: $f(P)$ is the distance of P from the straight line $x = -411$.

a) Set up $\displaystyle\iint_D f(P)\, dA$.

b) Evaluate the integral, giving a symbolic answer (involving symbols like π), not a decimal approximation. The second integration, with respect to θ is somewhat tedious to do by hand. You may use a table of integrals or a computer algebra system to get the answer, but you should cite your source if you do.

26.9. The function $f(r, \theta)$ is defined as the area of the shaded circle sector shown in Figure 26.5 below. Use a double integral to find the formula for $f(r, \theta)$.

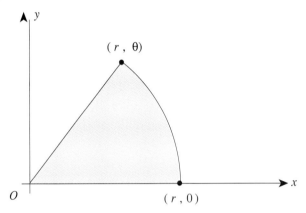

Figure 26.5

Hint: We do not want to confuse the upper corner (r, θ) of the sector with the r, θ over which you have to integrate to find the area. Therefore, use (r_0, θ_0) for the coordinates of the upper corner and integrate. The area will be an expression in terms of r_0 and θ_0. If you write r instead of r_0 and θ instead of θ_0 you have the formula for the function $f(r, \theta)$.

26.10. Return to the situation of Problem 24.17 on page 208. Evaluate the integral that gives the total population using the methods of this section.

26.11. Figure 26.6 on the next page shows two regions D_1 and D_2. On the left, D_1 is part of a circle of radius m and center at $L(m, 0)$. The line segment OP forms the angle α with the positive x-axis. On the right, D_2 is the region between a circle of radius 1 (with center at the origin) and the line segment which connects $(1, 0)$ and $(0, 1)$.

    a) Do the following with D_1 on the left:

        (i) Set up the double integral which gives the area of the region D_1. *Hint:* The equation of the circle in polar coordinates is $r = 2m\cos\theta$.

        (ii) Evaluate the integral. Your answer will contain m and α.

    b) Do the following with D_2 on the right:

        (i) Set up $\iint (x^2 + y)\, dx\, dy$. That is, use Cartesian coordinates.

        (ii) Set up $\iint (x^2 + y)\, dA$ in polar coordinates.

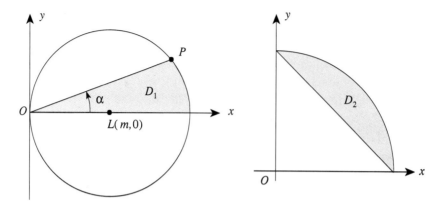

Figure 26.6

27   Triple Integrals

What is a triple integral?

Triple integrals are defined in much the same way as double integrals, except that the domain of integration is a three-dimensional solid instead of a two-dimensional planar region. We begin with a physical example.

Example 1. Mass density. We are given a block of material. At any point P within the material, we let $\rho(P)$ denote the *mass density* at the point P, measured in kilograms per cubic meter. This means that if we take a small chunk of the solid containing P, then the mass of that chunk, in kilograms, is approximately $\rho(P)$ times the volume of the chunk in cubic meters. The material is not assumed to be of uniform consistency, so the mass density may vary from point to point. We now ask: *If we know the mass density of the solid at every point, how can we calculate the total mass of the solid?* Before attacking this problem, you should notice the similarity with the problem addressed in Example 1 of Section 24. There, we knew a mass density, measured in metric tons per square kilometer, and we wanted to find the total mass. We will attempt to estimate the total mass of our solid by a method which closely parallels the one we used earlier to estimate the total mass in a plane region in Example 1. To estimate the total mass of the material, we partition it into small chunks $S_1, S_2, \ldots$, and sample the density at a point P_j of each small chunk S_j. The mass of the small chunk S_j is approximately

$$(\text{mass density at } P_j)(\text{volume of } S_j) = \rho(P_j)V(S_j),$$

where $V(S_j)$ stands for the volume of the small chunk S_j. We can now estimate the total mass my summing the masses of the small chunks:

$$\text{total mass} \approx \sum_j \rho(P_j)V(S_j).$$

■

Problems _____

27.1. A cube of length 2 meters on each edge has constant mass density 50 kg/m^3. What is its total mass? *Hint:* No matter how you partition the cube, you'll always get the same "estimate" for the total mass. What is it?

27.2. The cube defined by $0 \le x \le 1$, $0 \le y \le 1$, $0 \le z \le 1$ has mass density $\rho(x, y, z) = xyz$. Find upper and lower bounds for its total mass by partitioning the cube into eight smaller cubes, estimating the mass of each small cube, and summing the results.

_____

We now answer the question posed at the beginning of this section: *What is a triple integral?* Roughly speaking, a triple integral is the number that is a approximated by the procedure used in Example 1. To evaluate a triple integral, you must have

- A three-dimensional solid K in xyz-space;
- A function $f(x, y, z)$ defined on the solid K.

The *triple integral* of the function $f(x, y, z)$ over the solid K, which is denoted by the symbol $\iiint_K f(x, y, z)\, dV$, is the number which is estimated by the following procedure.

1. Partition K into small solids $K_1, K_2, \ldots$ which fill up the solid K and do not overlap.

2. For each small solid K_j, sample the function value by choosing a point P_j in K_j and calculating or estimating $f(P_j)$.

3. Multiply the sampled values $f(P_j)$ by the volumes $V(K_j)$ of the corresponding small solids, and add up the results. The sum approximates the triple integral:

$$\iiint_K f(x, y, z)\, dV \approx \sum_j f(P_j) V(K_j). \tag{27.1}$$

Problems _____

27.3. Find upper and lower estimates for $\iiint_K xyz\, dV$ when K is the cube defined by $0 \le x \le 1$, $0 \le y \le 1$, and $0 \le z \le 1$ by partitioning K into eight small cubes of the same size. *Hint:* You have already done it! See Problem 27.2.

27.4. Evaluate $\iiint_B 2\, dV$ when B is the ball defined by $x^2 + y^2 + z^2 \le 1$. *Hint:* No matter how you partition the ball, the approximating sum (27.1) will give the same result. What is it?

Properties of triple integrals

The properties of double integrals listed in Section 24 all carry over in a straightforward way to triple integrals. It is only necessary to replace all references to "area" by analogous references to volume. Thus, for example, if c is any constant, then

$$\iiint_K c\, dV = c \cdot V(K). \tag{27.2}$$

where $V(K)$ stands for the volume of K.

Evaluating triple integrals

As with double integrals, triple integrals can be evaluated "one variable at a time." The simplest case is integration over a box with edges parallel to the coordinate axes. We illustrate this case in Example 2. Later on, in Example 3, we consider a solid tetrahedron (pyramid). For such a solid, finding the limits requires detail work.

Example 2. Integration over boxes. We want to evaluate $\iiint_B xy^2\, dV$, where B is the box defined by $0 \le x \le 2$, $0 \le y \le 1$, and $1 \le z \le 3$. Inserting the limits for x, y, and z gives

$$\iiint_B xy^2\, dV = \int_{z=1}^{z=3} \int_{y=0}^{y=1} \int_{x=0}^{x=2} xy^2\, dx\, dy\, dz.$$

We now evaluate, starting with the inside integral and working our way out:

$$\begin{aligned} \iiint_B xy^2\, dV &= \int_{z=1}^{z=3} \int_{y=0}^{y=1} \left(\int_{x=0}^{x=2} xy^2\, dx \right) dy\, dz \\ &= \int_{z=1}^{z=3} \int_{y=0}^{y=1} 2y^2\, dy\, dz = \int_{z=1}^{z=3} \frac{2}{3}\, dz = \frac{4}{3}. \end{aligned}$$

The order in which the three integrations are performed is not important, provided that the limits on the integral signs are matched up with the correct variable. There are six different orders in which the integration may be performed. ∎

Problems ————————————————————————————

27.5. Repeat the calculation of Example 2 using the order of integration represented by $dz\,dy\,dx$. You should get the same answer in the end.

27.6. In Problem 27.2, you obtained estimates for the total mass of a cube of material from its mass density. Find the *exact* value for the total mass.

——

In Example 3, we integrate over a solid tetrahedron (pyramid) K. For such a solid, integration can still be performed one variable at a time, but finding the limits requires more work than for a box.

Example 3. Integration over a tetrahedron. The tetrahedron K shown below is bounded by the three coordinate planes and the plane $6x + 3y + 2z = 6$.

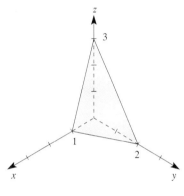

We want to evaluate

$$\iiint_K z\,dV.$$

We first choose an order of integration. There are six possible choices. We use the order of integration represented by $dz\,dy\,dx$. Our main task is to fill in the limits of integration in the template

$$\iiint_K z\,dV = \int_{x=?}^{x=?} \int_{y=?}^{y=?} \int_{z=?}^{z=?} z\,dz\,dy\,dx.$$

Once this is done, the mechanics of evaluating the integral are straightforward. We work from the inside to the outside. The innermost integral is taken with respect to z, with x and y held fixed, and so *the limits of integration on the inner integral may depend on x and y*. To get the lower limit for the inner integral, look at the above figure. The solid is bounded below by the plane $z = 0$, so 0 is the lower limit for the z integral. The upper limit is a little trickier. The solid is bounded above by the plane $6x + 3y + 2z = 1$, so on the top surface, $z = 3 - 3x - \frac{3}{2}y$, which gives the upper limit of $3 - 3x - \frac{3}{2}y$ for the inner integral. Here is where we are:

$$\iiint_K z\,dV = \int_{x=?}^{x=?} \int_{y=?}^{y=?} \int_{z=0}^{z=3-3x-(3/2)y} z\,dz\,dy\,dx.$$

We still need to find the limits for the two outer integrals. For this, look at the shadow of the solid on the xy-plane, which we call K_{xy}:

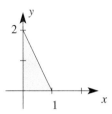

Since (x, y) varies over this region, we get

$$\iiint_K z \, dV = \iint_{K_{xy}} \left(\int_{z=0}^{z=3-3x-(3/2)y} z \, dz \right) dA_{xy}.$$

Here, dA_{xy} indicates that the outer integrations are treated as a double integral in the xy-plane. The limits for the outer integrals are determined in the usual way for double integrals. The result is

$$\iiint_K z \, dV = \int_{x=0}^{x=1} \int_{y=0}^{y=2-2x} \int_{z=0}^{z=3-3x-(3/2)y} z \, dz \, dy \, dx.$$

Finally, we evaluate the integral working from the inside to the outside:

$$\begin{aligned}
\iiint_K z \, dV &= \int_{x=0}^{x=1} \int_{y=0}^{y=2-2x} \left(\int_{z=0}^{z=3-3x-(3/2)y} dz \right) dy \, dx \\
&= \int_{x=0}^{x=1} \int_{y=0}^{y=2-2x} \frac{1}{2}\left(1 - 3x - \frac{3}{2}y\right)^2 dy \, dx \\
&= \int_{x=0}^{x=1} \frac{1}{3}(3 - 3x)^3 \, dx = \frac{3}{4}.
\end{aligned}$$

∎

In working out the limits of integration over the tetrahedron K of Example 3, we used the "shadow" K_{xy} of K on the xy-plane. The xy-shadow of K is obtained by shifting all the points of the solid K to the xy plane by moving them along lines parallel to the z-axis. In Example 3, the xy-shadow was used to obtain the limits for the two outer integrals. We now ask: *What happens if we change the order of integration?* If the innermost integral is with respect to y, then we use the "xz-shadow" K_{xz} to find the limits for the outer integrals. It is obtained by shifting all the points of K to the xz-plane along lines parallel to the y-axis. Similarly, if the innermost integral is with respect to x, we use the yz-shadow K_{yz}. It is obtained by shifting the points of K into the yz plane along lines parallel to the x-axis. For the tetrahedron of Example 3, the xz and yz-shadows are shown below.

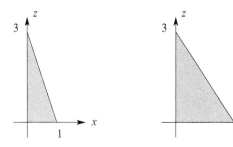

Problem

27.7. a) With K as in Example 3, fill in the limits of integration in the formula below:

$$\iiint_K z \, dV = \int_{z=?}^{z=?} \int_{y=?}^{y=?} \int_{x=?}^{x=?} z \, dx \, dy \, dz.$$

Hint: You will need to look at the shadow of K on the yz-plane.

b) Evaluate the integral using the limits of integration you found in part a. Make sure you get the same answer that was obtained in Example 3.

Measuring volume

Taking $c = 1$ in (27.2) gives a formula for calculating volume as a triple integral:

$$V(K) = \iiint_K 1 \, dV.$$

Problem

27.8. a) Set up a triple integral to measure the volume of the tetrahedron of Example 3.

b) Calculate the volume of the tetrahedron by evaluating the integral.

The volume element. Because the symbol dV is replaced by $dx \, dy \, dz$ to evaluate a triple integral in Cartesian coordinates, we write

$$dV = dx \, dy \, dz.$$

The symbol dV is called the *volume element,* and the right side of the above equation expresses the volume element in Cartesian coordinates. It is often useful to think of the $dx \, dy \, dz$ as the volume of an "infinitesimal box" with dimensions dx, dy, and dz, and to think of $\iiint + K f(x, y, z) \, dx \, dy \, dz$ as summing the "infinitesimal" quantities $f(x, y, z) \, dx \, dy \, dz$ over a partition of K into infinitely many infinitesimal boxes.

Summary

You are given a three-dimensional solid K, and a function $f(x, y, z)$ defined on K, and you want to evaluate $\iiint_K f(x, y, z) \, dV$. Proceed as follows.

Step 1: Let K_{xy} denote the shadow of K on the xy-plane.

Step 2: Try to find two functions $\varphi(x, y)$ and $\psi(x, y)$ on K_{xy} so that K consists of all points above the surface $z = \varphi(x, y)$ and below the surface $\psi(x, y)$.

Step 3: Use the functions you found in Step 2 as the z-limits on the inner integral:

$$\iiint_K f(x, y, z) \, dV = \iint_{K_{xy}} \int_{z=\varphi(x,y)}^{z=\psi(x,y)} f(x, y, z) \, dz \, dA_{xy}.$$

Step 4: Find the limits of the outer integral may be obtained by the usual method for double integrals, as in Section 25.

The method outlined above works well when the solid K can be conveniently described as in Step 2. In other cases, it may be necessary to interchange the roles of the variables x, y, and z as discussed after Example 3 on page 232. For yet another approach, see Problem 27.13.

Additional Problems

Setting up triple integrals

27.9. K is the triangular solid shown in Figure 27.1 on the facing page. Find
$\iiint_K (z+3)dV$.

27.10. K is the triangular solid in Figure 27.1 on the next page, and Q is the point with coordinates $(5b, 0, a)$.

We use the point Q to define a function $f(P)$ of points P in space as follows:

$$f(P) = |\overrightarrow{QP}|^2.$$

Estimate $\iiint_K f(P)dV$. That is, find constants e and E such that

$$e \leq \iiint_K f(P)dV \leq E.$$

It is not necessary to partition K here. Find the maximum and minimum values of the integrand on the solid K, and use them to estimate the value of the integral.

27.11. K is the three-dimensional solid in Figure 27.2 on the facing page. It looks like a box where some of the angles are not right angles. Two faces of K are parallel to the xz-plane, two are parallel to yz-plane, and two are parallel to the plane $x - z = 0$. Go to part a to figure out how K looks and how it sits in the xyz-system.

   a) In Figure 27.3, you have three views of an xyz-system. On the left, you look down onto the xy-plane, and the z-axis sticks out toward you. In the middle, you look at the yz-plane, and the x-axis sticks toward you. On the right, you look at the xz-plane, and the y-axis points toward you. Draw into each system what you see of K. Take b to be $\frac{1}{2}$.

   b) Evaluate $\iiint_K y\,dV$. *Hint:* Use the order of integration $dz\,dy\,dx$ and operate as in Example 3. Your answer will contain b.

27.12. K is the solid of the preceding Problem 27.11. We define a function $f(P)$ for points in space as follows: Write $d(P)$ for the distance of P from the x-axis. Then

$$f(P) = 2 + d(P).$$

Give an estimate for $\iiint_K f(P)\,dV$. That is, find constants e and E such that

$$e \leq \iiint_K f(P)\,dV \leq E.$$

Hint: The volume of K equals $b(1)(1) = b$.

Please read. Up to now, we have set up triple integrals using only xyz-coordinate systems. In Problem 27.13, you will use polar coordinates at some point.

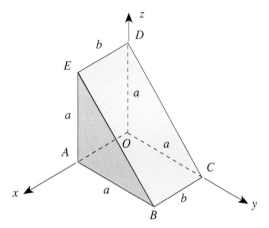

Figure 27.1

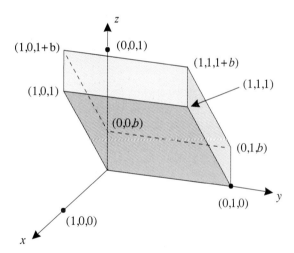

Figure 27.2

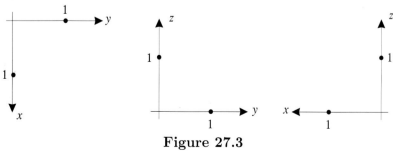

Figure 27.3

27.13. The solid K shown below is a solid circular cone of radius b and height h.

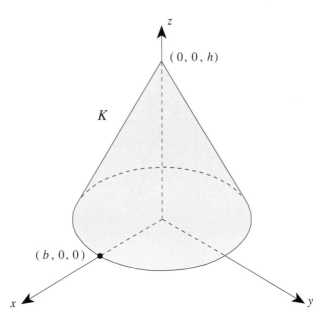

You will set up $\iiint_K xyz\, dV$ by a different method. You will do this in steps.

a) Intersect K with a plane $z = c$ where c is a constant. Write $W(c)$ for the intersection of K with that plane. Describe the shape and size of $W(c)$. The letter c will figure in your description. When you have written out your description, replace the constant c by the letter z.

b) We want to integrate with respect to z last. That is, set up the integral in the form

$$\iiint_K xyz\, dV = \int \left[\iint_{W(z)} xyz\, dA \right] dz$$

$W(z)$ is a filled-in circle (a disk). Therefore, we use polar coordinates (r, θ) for the integration over $W(z)$. Set up the double integral over $W(z)$ in polar coordinates. Do it in such a way that the first integration is with respect to r, and the second with respect to θ.

c) Write down $\iiint_K xyz\, dV$ with all the limits shown (remember—give also the names of the variables).

Different ways of setting up triple integrals

- What did you just do in Problem 27.13? You set up the integral over a cone. How did you do that? You took a plane $z = constant$. It is parallel to the xy-plane, and it intersects the cone in a region $W(z)$. Then you set up a double integral over $W(z)$. Finally you integrated over z. We call this method *method of planes* because we used a plane.

- At the beginning of this section, you set up an integral over a tetrahedron (pyramid). How did you do that? You took a *line* where x and y are *both* constant. The line is parallel to the z-axis, and it intersects the tetrahedron in an interval $I(x, y)$. You set up the inner integral with respect to z over the interval $I(x, y)$. Finally, you found the limits for the outer two integrals by letting x and y vary over the shadow of the tetrahedron in the xy-plane. We call this the *method of lines*.

For the method of planes you used a plane parallel to the xy-plane. You could have equally taken a plane parallel to the yz-plane or xz-plane. For the cone, a plane parallel to the xy-plane works best.

For the method of lines in Example 3, you used lines parallel to the z-axis and the shadow of K in the xy-plane. You could have equally taken lines parallel to the x-axis and the shadow in the yz-plane or or lines parallel to the y-axis and the shadow on the xz-plane.

Before you set up a triple integral, ask first what is likely to work better, shadows or plane. Then decide which shadow or plane to use.

27.14. B is one half of a solid three-dimensional ball, given by the equations

$$x^2 + y^2 + z^2 \leq 1, \qquad z \geq 0.$$

We also consider the generic function $w = f(x, y, z)$ of three variables. You will be asked to set up the integral

$$\iiint_B f \, dV$$

in three different orders of integration. In the answers you have to name the variables in the limits of integration. For example, you have to write $\int_{y=4x+1}^{y=5}$. Writing only $\int_{4x+1}^{5}$ is not enough.

Below you find three sketches of B, one for each integral. Drawing into the sketches may help you finding the limits.

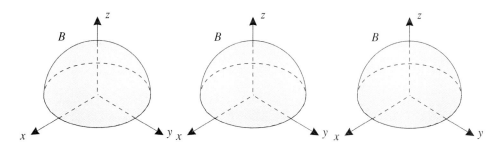

a) Use the order of integration $dz \, dy \, dx$. It is the order used for the tetrahedron K of Example 3.

b) In part a, you asked first: "what are the limits for the innermost integral with respect to z?" You found these limits by asking "for a given x and y, how does z vary?" Now we want to integrate in the order $dx \, dy \, dz$. The innermost integration is with respect to x. So we ask "for a given y and z, how does x vary?"

c) Use the same kind of reasoning to do it in the order $dy \, dz \, dx$.

27.15. K is a solid tetrahedron (a pyramid) lying on its side, as shown below.

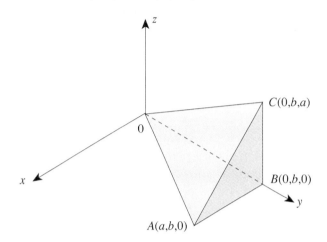

a) $H(x, y, z)$ is a generic function of three variables. Set up the integral $\iiint_K H(x, y, z)\, dx\, dz\, dy$ in the given order of integration.

b) Now we take for $H(x, y, z)$ the function $H(x, y, z) = xz$. Imagine you have carried out the first integration of the triple integral which you have set up in part a. At this point, you will have to evaluate a double integral of a function of the following variable(s)—choose one:

only x only y only z

only x and y only x and z only y and z x, and y, and z

27.16. K is the solid sphere (the ball) of radius 1 and center at the point $M(4, 6, 12)$. Estimate $\iiint_K \dfrac{1}{|\overrightarrow{OP}|}\, dV$. That is, find numbers e and E such that

$$ e \ \leq \ \iiint_K \frac{1}{|\overrightarrow{OP}|}\, dV \ \leq E. $$

28   Triple Integrals in Cylindrical Coordinates

Polar coordinates simplify many double integrals which would be cumbersome to set up in Cartesian coordinates. In a similar way, cylindrical coordinates simplify many triple integrals.

Cylindrical coordinates

The position of a point P in space can be described either by Cartesian coordinates (x, y, z) or by

(i) the polar coordinates (r, θ) of the point's shadow $P_0(x, y, 0)$ in the xy-plane, and

(ii) the point's elevation z above (or below) the xy-plane, as shown below.

The *cylindrical coordinates* of P are (r, θ, z). Thus, cylindrical coordinates are "polar coordinates plus z."

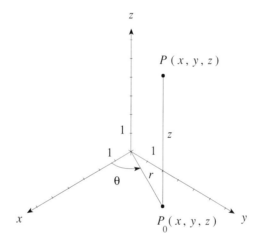

Problem

28.1. Into a standard coordinate system, draw the points with the following cylindrical coordinates (r, θ, z): $Q_1(3, \pi, 1)$, $Q_2(4, 0, 3)$, $Q_3(5, \frac{5}{2}\pi, -1)$.

Things to keep in mind when using cylindrical coordinates:

1. In cylindrical coordinates, r is the distance of P from the z-axis, and *not* the distance of P from the origin. The value of r is always greater than or equal to zero.

2. The cylindrical coordinates of the origin are $r = 0$, θ undetermined, $z = 0$.

3. We have two different coordinates systems in space, namely the xyz-system and the $r\theta z$-system. *How do we convert from one to the other system?* We have two questions:

   Question 1: Given the cylindrical coordinates (r, θ, z) of a point P. How do we find its Cartesian coordinates?

   Answer: $x = r\cos\theta$, $y = r\sin\theta$, $z = z$.

   Question 2: Given the Cartesian coordinates (x, y, z) of a point P. How do we find its cylindrical coordinates?

Answer: $r = \sqrt{x^2 + y^2}$, $\theta = \tan^{-1}(\frac{y}{x})$, $z = z$. Note that again the θ of a point P is not uniquely determined, but only up to a multiple of 2π.

Problem _____

28.2. The function $f(x, y, z)$ is given by $f(x, y, z) = xyz$. Express $f(x, y, z)$ in polar coordinates. That is, write out $f(r \cos \theta, r \sin \theta, z)$ as a formula in terms of r, θ, and z.

_____

4. The planes $x = $ constant, $y = $ constant, and $z = $ constant are called the *coordinate planes* or *coordinate surfaces* of the xyz-system. What are the coordinate surfaces of the $r\theta z$-system of cylindrical coordinates?

Problem _____

28.3. Draw the following cylindrical coordinate surfaces:
(a) $r = 3$, $0 \le \theta \le \pi$, $z \ge 0$ (b) $\theta = \frac{\pi}{2}$, $4 \le r \le 5$, $z \ge 0$
(c) $z = -3$, $0 \le \theta \le \frac{1}{2}\pi$

_____

Evaluating a triple integral in cylindrical coordinates

The evaluation of a triple integral in cylindrical coordinates rests on subdividing the solid K into solid pieces bounded by coordinate surfaces $r = $ constant, $\theta = $ constant, and $z = $ constant. Such a piece is shown below.

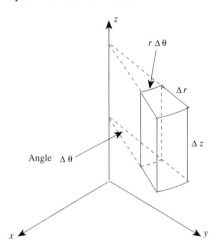

The base of the piece is a washer segment of polar coordinates, except that it does not necessarily lie in the xy-plane. Its volume is approximated by

$$\text{volume} \approx r \, \Delta r \, \Delta \theta. \tag{28.1}$$

How can we use this to evaluate a triple integral $\iiint_K f \, dV$? We partition K into small solids K_j using cylindrical coordinate surfaces as described above. Then the approximating sums for the triple integral take the form

$$\iiint_K f \, dV \approx \sum f(P_j) V(K_j) \approx \sum f(P_j) r \, dr \, d\theta \, dz.$$

The last sum approximates the integral $\int\int\int f\, r\, dr\, d\theta\, dz$, so

$$\iiint_K f\, dV = \iiint f\, r\, d\theta\, dz,$$

where the limits for ρ, ϕ, and θ are chosen to describe the solid K in cylindrical coordinates. For this reason, we write

$$dV = r\, dr\, d\theta\, dz$$

and refer to the right side as the *volume element in cylindrical coordinates.*

Example. K is the right circular cone shown below (radius a, height h), and $f(x, y, z)$ is defined as the distance of $P(x, y, z)$ from the plane $y = -6$. We are going to set up $\iiint_K f(x, y, z)\, dV$ in cylindrical coordinates. Why are we using cylindrical coordinates? If you intersect K with a plane $z = $ constant, the figure $W(z)$ is a solid circle over which we have to integrate, and a double integral over a circle is simpler in polar coordinates—remember: cylindrical coordinates are polar coordinates plus z.

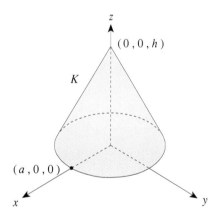

Step 0: Rewrite $f(x, y, z)$ in cylindrical coordinates:

$$f(x, y, z) = f(r\cos\theta, r\sin\theta, z) = F(r, \theta, z).$$

In our case, $f(x, y, z) = y + 6 = r\sin\theta + 6$.

Step 1: Pick a plane $z = $ constant. The plane intersects K in the region $W(z)$. Make a separate, *big* drawing of $W(z)$.

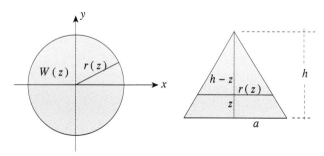

The region $W(z)$ is a disk (above left). Its radius r depends on z: $r = r(z)$. To determine $r(z)$ we use the side view of the cone (above right). Similarity of triangles tells us

$$\frac{r(z)}{h-z} = \frac{a}{h},$$

from which we get

$$r(z) = \frac{a}{h}(h-z).$$

Step 2: Consider $f(x, y, z)$ as a function of the two variables x, y and set up the double integral $\iint_{W(z)} f \, dA$ over $W(z)$ in polar coordinates. *Do not forget the additional r for the integration with respect to r.*

In our case:

$$\iint_{W(z)} f \, dA = \int_{\theta=0}^{\theta=2\pi} \left[\int_{r=0}^{r=\frac{a}{h}(h-z)} [r\sin\theta + 6] \, r \, dr \right] d\theta.$$

Step 3: Integrate the integral obtained in Step 2 with respect to z, and you have set up the triple integral of $f(x, y, z)$ over K in cylindrical coordinates. The limits of z are to be chosen such that you sweep out K with your horizontal slice $W(z)$.

In our case:

$$\iiint_K (y+6) \, dV = \int_{z=0}^{z=h} \left[\int_{\theta=0}^{\theta=2\pi} \left[\int_{r=0}^{r=\frac{a}{h}(h-z)} [r\sin\theta + 6] \, r \, dr \right] d\theta \right] dz.$$

∎

Volume in cylindrical coordinates. To get the volume $V(K)$ of K, you integrate the function $f(x, y, z) = 1$ (see Section 27, page 233). In cylindrical coordinates, this means:

$$V(K) = \iiint_K dV = \iiint_K r \, dr \, d\theta \, dz.$$

Cartesian coordinates and cylindrical coordinates compared

1. *In cylindrical coordinates, why is there an additional r in the volume element $dV = r \, dr \, d\theta \, dz$?* The additional r is the r of approximate volume $r \, \Delta r \, \Delta \theta \, \Delta z$ of the pieces into which one subdivides the solid K, and the base of these pieces are the washer segments of polar coordinates.

2. *When to use Cartesian coordinates, and when cylindrical coordinates?* There is no hard-and-fast rule, only a general guideline:

   If K is bounded by planes, try Cartesian coordinates. If K is in any way "cylinder-like," try cylindrical coordinates. By a "cylinder-like" solid is meant a solid K which is bounded by a surface of revolution (for example, a sphere flattened at the top, a heavy-wall pipe), or part of such a surface (for example, one fourth of a bagel cut along two small circles, one half of a round cake). The more the surface which bounds K resembles the coordinate surfaces of cylindrical coordinates, the simpler it will be to find the limits of integration. If the surfaces that bound the solid are given in cylindrical coordinates, use cylindrical coordinates.

Additional Problems

28.4. Figure 28.1 on the next page shows two solid circular cylinders K_1, K_2, and two solid circular cones K_3, K_4. The axes of the cylinders and cones are along the axes of the xyz-system.

a) The function $f(P)$ for points P in space is defined as follows: $f(P) = |\overrightarrow{OP}|^4$. Set up $\iiint_{K_1} f(P) \, dV$.

b) Keep the function $f(P)$ of a) and set up $\iiint_{K_2} f(P) \, dV$.

   Hint: In cylindrical coordinates, we have polar coordinates (r, θ) in the xy-plane, and the z-axis is perpendicular to the (r, θ)-plane. The ordinary cylindrical coordinates (r, θ, z) will not work well for K_2 because the axis of K_2 is the y-axis. Therefore, we define cylindrical coordinates for which the xz-plane carries polar coordinates and the y-axis plays the role of z-axis. We do this as follows. Write (r^*, θ^*) for polar coordinates in the zx-plane. Count θ^* in such a way that θ^* turns counterclockwise if you look at the zx-plane in the direction of $-(0, 1, 0)$. Then $z = r^* \cos \theta^*$, $x = r^* \sin \theta^*$, $y = y$ are cylindrical coordinates which are based in the zx-plane in exactly the same way the ordinary cylindrical coordinates are based in the xy-plane. Use (r^*, θ^*, y) to set up the integral.

c) The function $g(P)$ for points in space is defined as follows: $g(P)$ is the square of the distance of P from the x-axis. Set up $\iiint_{K_3} g(P) \, dV$.

d) We keep the function $g(P)$ of c). – Set up $\iiint_{K_4} g(P) \, dV$.

   Hint: Adapt the method developed for K_2 in b) to the x-axis. *Watch out:* In which way do you have to count θ^* positively so that the x-axis plays the same role as the z-axis in ordinary cylindrical coordinates? It does not make a difference for K_4 because we make one full turn around the x-axis.

28.5. a) On the left in Figure 28.2 on the following page is shown a solid K. It is one-half of a piece of solid pipe. The inner radius of the pipe equals a, the outer radius equals b, and the length of the pipe is h. Set up

$$\iiint_K (4x^2 + 4y^2 + z^2) \, dV.$$

b) On the right in Figure 28.2 on the next page, you have a view of the yz-plane so that the x-axis sticks out toward you. C is a filled-in circle (a disk) of center $(0, m, 0)$ and radius d. Rotate C around the z-axis. You obtain a solid T. It has the shape of a bagel, and mathematically it is called a torus. The function $f(P)$ for points in space is defined as follows: $f(P)$ is the distance of P from the plane $z = 600b$ where b is the radius of C. Set up $\iiint_T f(P) \, dV$.

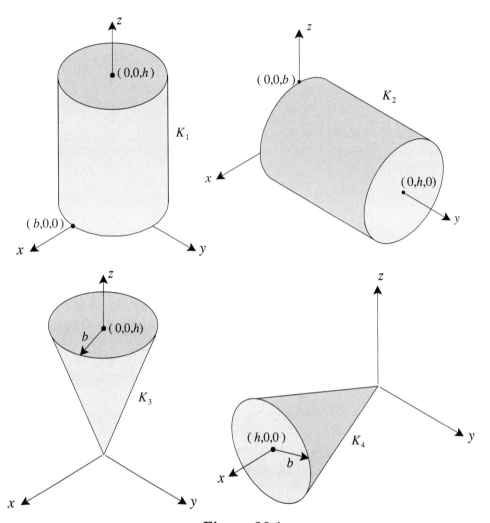

Figure 28.1

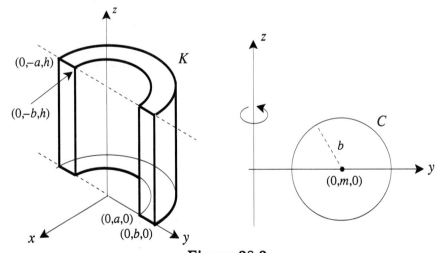

Figure 28.2

28.6.   a) Below are two views of the xz-plane so that the x-axis points to the right. Will the y-axis point towards you or will it point away from you?

b) On the left, M is a filled-in half-circle (one half of a disk) in the xz-plane. It is centered at the origin, and its radius equals b. Rotate the half-circle around the z-axis. You obtain a solid sphere B. Use a triple integral to find the volume of B. *The volume of a solid sphere is well known, and you might even remember it. The point of the exercise is to practice once more triple integrals in cylindrical coordinates.*

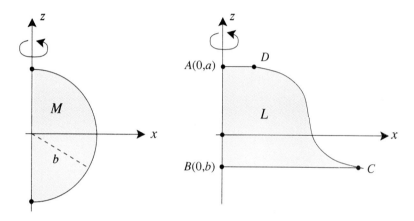

c) On the right we have a region L in the xz-plane. It is bound by three line segments AD, AB, BC, and by a curve AC. For the curve BC, we have two equations: $z = f(x)$, $x = g(z)$. Rotate the region L around the z-axis and you obtain a solid K. Set up the integral that gives the volume of K. Your answer will contain the letters a and b and one or both of the functions $f(x)$, $g(z)$.

29   Triple Integrals in Spherical Coordinates

Cylindrical coordinates were introduced because they simplify triple integrals for certain types of solids. Another system of coordinates which similarly works for other solids is the system of "spherical coordinates."

Spherical coordinates

The position of a point P in space can be described either by Cartesian coordinates (x, y, z) or by the following quantities:

(i) The distance $\sqrt{x^2 + y^2 + z^2}$ of P from the origin. This distance is written as ρ (read as "roe");

(ii) The θ of the cylindrical coordinates of P;

(iii) The angle ϕ ("fi") which $\overrightarrow{OP}$ forms with $\vec{k}$.

The three numbers (ρ, ϕ, θ) are called the *spherical coordinates* of P.

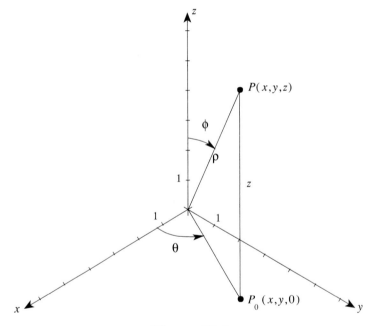

Figure 29.1

Problem

29.1.   a) Into the system in Figure 29.1, draw the point Q with spherical coordinates $(\rho, \phi, \theta) = (3, \frac{\pi}{2}, \pi)$.

     b) T is the point with spherical coordinates $(2, \frac{5\pi}{6}, \frac{7\pi}{4})$. What can you say about the Cartesian coordinates (x, y, z) of T—is x positive, or zero, or negative? What about y and z? Looking at the figure might help you.

     c) The point U has Cartesian coordinates $(x, y, z) = (3, 5, -2)$. Its spherical coordinates are (ρ, ϕ, θ).

        (i) Compute ρ.

(ii) Estimate θ as follows: If it is equal to one of the numbers, circle that number; if it is between two of them draw a circle between them:

$$0 \qquad \frac{\pi}{2} \qquad \pi$$

(iii) Do the same for ϕ:

$$0 \qquad \frac{\pi}{2} \qquad \pi$$

Remarks on the use of spherical coordinates

1. The number $\rho = |\overrightarrow{OP}|$ is the distance of P from the origin. It is always greater than or equal to zero. Do not confuse the ρ of spherical coordinates with the r of cylindrical coordinates: In cylindrical coordinates, r is the distance of P from the z-axis.

2. The angle ϕ is counted from the z-axis *down* and not from the xy-plane up. Thus, ϕ is not geographical latitude, but $\frac{\pi}{2} - latitude$.

3. *The angle ϕ ranges from zero to π only.* Why? Consider the point P with spherical coordinates $(3, \phi, 0)$. If you increase ϕ starting at $\phi = \frac{\pi}{4}$, P moves in the xz-plane. For $\phi = \pi$, P has Cartesian coordinates $(-3, 0, 0)$. As an experiment, increase ϕ to $\pi + \frac{\pi}{4}$. Now you are the point with spherical coordinates $(3, \frac{3\pi}{4}, \pi)$. That is, the angle ϕ of this point is less than π. So it is not necessary to consider values of ϕ greater than π.

4. The spherical coordinates of points on the positive z-axis are $\rho = z$, θ undetermined, $\phi = 0$. Points on the negative z-axis have spherical coordinates $\rho = -z$, θ undetermined, $\phi = \pi$.

5. By now, we have three different coordinates systems in three-dimensional space, namely the Cartesian, cylindrical and spherical coordinates. *How do we convert from one to the other system?* We have six questions of the kind "given the coordinates of P in one system, find its coordinates in the other system." Here is the answer to one such question.

   Transformation from spherical to Cartesian coordinates. Given the spherical coordinates (ρ, ϕ, θ) of a point P, the Cartesian coordinates are given by

   $$x = \rho \sin \phi \cos \theta \qquad y = \rho \sin \phi \sin \theta \qquad z = \rho \cos \phi.$$

   How does one get these equations? In Figure 29.1 above, the segment OP_0 equals $\rho \sin \phi$.

Problem

29.2. Given the spherical coordinates (ρ, ϕ, θ) of P, find formulas for its cylindrical coordinates (r, θ, z). *Hint:* Use a sketch, e.g., Figure 29.1 on the facing page.

The coordinate surfaces of spherical coordinates. The coordinate surface $\rho = 5$ is a sphere (spherical shell, not a solid sphere) of radius 5, centered at the origin. The surface $\theta = \frac{\pi}{4}$ is the same for spherical and cylindrical coordinates; it is a half-plane given by $x - y = 0$, $x \geq 0$. *What is the shape of the surface $\phi = \frac{\pi}{6}$?* It is the collection of all points P so that the angle between $\overrightarrow{OP}$ and $\vec{k}$ equals 30°. It is a circular cone, opened upwards, as shown in Figure 29.2.

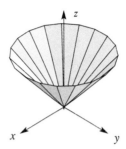

Figure 29.2

Problems

29.3. Find the equation of the coordinate surface $\phi = \frac{\pi}{6}$ (shown above) in Cartesian coordinates. *Hint:* Express the fact that the angle between $\overrightarrow{OP} = (x, y, z)$ and $\vec{k}$ equals $\frac{\pi}{6}$.

29.4. The coordinate surface $\phi = \frac{\pi}{6}$ above intersects the coordinate surface $\rho = a$ in a curve C.

    a) Describe the shape of C. What can you say about the position of C in space? *Hint:* Try to "see" it using Figure 29.2.

    b) Write down a parametric representation $(x, y, z) = (x(t), y(t), z(t))$ of the curve C. *Hint:* Make a "side view" of the sphere and the cone. The cone will appear as two legs of an equilateral triangle.

29.5. Write down the equation of the coordinate surface $\phi = \frac{\pi}{2}$ in (i) Cartesian coordinates (x, y, z), and (ii) in cylindrical coordinates (r, θ, z).

Triple integrals in spherical coordinates

In order to evaluate triple integrals using spherical coordinates, it is necessary to work out the formula for the volume element dV in in terms of ρ, ϕ, and θ. The pattern is the same as that used in Section 28 to work out the volume element in cylindrical coordinates. We subdivide the solid K into pieces bounded by coordinates surfaces $\rho = $ constant, $\theta = $ constant, and $\phi = $ constant. How do these pieces look? They are bounded by planes, spheres, and cones. We figure out their shape in two steps.

The left portion of Figure 29.3 on the next page shows the coordinate surface $\rho = \rho_0$, a sphere of radius ρ_0. The shaded patch $ABCD$ on the sphere is bounded by four circular arcs. The arcs AB and DC are parts of half-circles of radius ρ_0 in half-planes $\theta = $ constant. The arcs AD and BC are parts of circles in which cones $\phi = $ constant intersect the sphere. The spherical coordinates (ρ, ϕ, θ) of the four points are as follows:

    $A(\rho_0, \phi_0, \theta_0)$;
    $B(\rho_0, \phi_0 + \Delta\phi, \theta_0)$ (remember that ϕ is measured from the top down);
    $C(\rho_0, \phi_0 + \Delta\phi, \theta_0 + \Delta\theta)$;
    $D(\rho_0, \phi_0, \theta_0 + \Delta\theta)$.

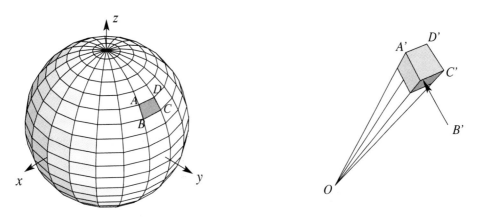

Figure 29.3

The length of the arc AB is $\rho_0 \cdot \Delta\phi$. The length of the arc AD is $\rho_0 \cdot \sin\phi_0 \cdot \Delta\theta$ because $\rho_0 \cdot \sin\phi_0$ is the radius of the circle of which AD is a part. As approximation for the surface area of the curved patch $ABCD$, we take the product of the lengths of the arcs AB and AD:

$$\text{area of patch} \approx (\rho_0 \Delta\phi) \cdot (\rho_0 \sin\phi_0 \Delta\theta) = \rho_0^2 \sin\phi_0 \Delta\theta \Delta\phi.$$

Now we can describe the shape of the subdivision pieces. Imagine a second sphere of radius $\rho_0 + \Delta\rho$, also centered at the origin. On that sphere, we look at the patch $A'B'C'D'$ which corresponds to the patch $ABCD$ on the smaller sphere. The spherical coordinates of A', B', C', D' are the same as those of A, B, C, D, except that ρ has been increased by $\Delta\rho$. As subdivision piece, we take the solid of thickness $\Delta\rho$ between the patches $ABCD$ and $A'B'C'D'$. The right portion of Figure 29.3 is a sketch of such a piece. The walls $ABA'B'$ and $CDC'D'$ are planes because they are part of coordinate surfaces $\theta = $ constant. The walls $BCC'B'$ and $ADD'A'$ are curved because they are parts of cones $\phi = $ constant. The patches $ABCD$ and $A'B'C'D'$ are parts of spheres $\rho = $ constant. The volume of a subdivision piece is approximately

$$\text{volume} \approx \text{surface area of base } ABCD \times \text{height} \approx \rho^2 \sin\phi_0 \, \Delta\rho \, \Delta\theta \, \Delta\phi.$$

Thus, the *volume element in spherical coordinates* is

$$dV = \rho^2 \sin\phi \, d\rho \, d\phi \, d\theta.$$

Problem _____

29.6. Look at the subdivision piece $ABCDA'B'C'D'$ of Figure 29.3. Its approximate volume is a function of the spherical coordinates (ρ, ϕ, θ) of A and of the increments $\Delta\rho$, $\Delta\phi$, and $\Delta\theta$.

   a) We increase θ and keep the rest of the variables the same. Will the approximate volume increase, or stay the same, or decrease? *Hint:* Look at Figure 29.3.

   b) Now we increase ϕ and keep everything else the same. Will the approximate volume increase, stay the same, or decrease?

Evaluation of a triple integral in spherical coordinates. The shaded region in the xz-plane shown in Figure 29.4 is bounded by a quarter circle of radius a, a line of slope -1 and the coordinates axes. We rotate the region around the z-axis and obtain a solid K. We are going to set up the triple integral of $f(x, y, z)$ over K in spherical coordinates.

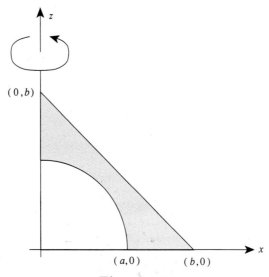

Figure 29.4

Problem

29.7. Describe in words the shape of the solid K.

Evaluating a triple integral in spherical coordinates We will outline the steps to set up a triple integral using spherical coordinates. We illustrate the steps with the solid K obtained by rotating the region in Figure 29.4 about the z-axis.

Step 0: Rewrite $f(x, y, z)$ in spherical coordinates:

$$f(x, y, z) = f(\rho \cos \theta \sin \phi, \rho \sin \theta \sin \phi, \rho \cos \phi) = F(\rho, \phi, \theta).$$

Step 1: Pick a half-plane $\theta = $ constant. The half-plane intersects K in the region $W(\theta)$. Make a separate, *big* drawing $W(\theta)$.

For our K, all the $W(\theta)$ look as shown in Figure 29.5 on the facing page.

Step 2: In $W(\theta)$ draw a ray $\phi = $ constant. Walk on this ray, starting at the origin. You enter $W(\theta)$ at $\rho = \rho_1$, and you leave at $\rho = \rho_2$.

For our K, $\rho_1 = a$. To find ρ_2, note that the equation of the slanted line is $\rho \sin \phi_\rho \cos \phi = b$ (see Figure 29.5). Therefore,

$$\rho_2 = \frac{b}{\sin \phi + \cos \phi}.$$

In order to cover all of $W(\theta)$ with such a ray, you have to start with a lowest ϕ_1 and end with a highest ϕ_2.

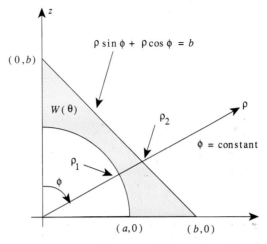

Figure 29.5

For our K, $\phi_1 = 0$ (ray is vertically up) and $\phi_2 = \frac{\pi}{2}$ (ray is in horizontal position).

Determine the range of θ so as to cover all of K: θ has to range from $\theta = \theta_1$ to $\theta = \theta_2$.

For our K, we have $\theta_1 = 0$ and $\theta_2 = 2\pi$.

Step 3: Set up the integral with limits ρ_1, ρ_2 for ρ, ϕ_1, ϕ_2 for ϕ, and θ_1, θ_2 for θ, and do not forget to insert the "correction factor" discussed above:

$$\int_{\theta=\theta_1}^{\theta=\theta_2} \left[\int_{\phi=\phi_1}^{\phi=\phi_2} \left[\int_{\rho=\rho_1}^{\rho=\rho_2} \rho^2 \sin\phi \, F(\rho,\phi,\theta)\, d\rho \right] d\phi \right] d\theta.$$

You integrate first with respect to ρ, then ϕ, then θ.

For our K we get

$$\int_{\theta=0}^{\theta=2\pi} \left[\int_{\phi=0}^{\phi=\frac{\pi}{2}} \left[\int_{\rho=a}^{\rho=\frac{b}{\sin\phi+\cos\phi}} \rho^2 \sin\phi \, F(\rho,\phi,\theta)\, d\rho \right] d\phi \right] d\theta.$$

Volume in spherical coordinates. To get the volume $V(K)$ of K, you integrate the function $f(x,y,z) = 1$ (see Section 27). In spherical coordinates, this means:

$$V(K) = \iiint_K dV = \iiint_K \rho^2 \sin\phi \, d\rho d\phi d\theta.$$

When to use spherical coordinates. If K is in any way "sphere-like," try spherical coordinates. By a "sphere-like" solid is meant a solid K which is bounded by a sphere or part of a sphere, or one which is symmetric with respect to the origin. The more the surface which bounds K resembles the coordinate surfaces of spherical coordinates, the simpler it will be to find the limits of integration. You have to experiment. If the boundary surfaces are easily represented in terms of (ρ, ϕ, θ), use spherical coordinates.

Additional Problems

29.8. Below are shown two regions D_1, D_2 in the yz-plane. Both are bounded by quarter-circles, D_1 by one and D_2 by two.

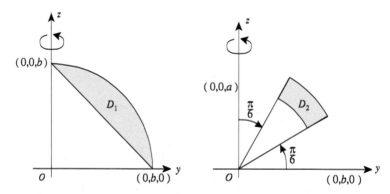

a) In this part, we work with D_1 on the left. We rotate D_1 around the z-axis in space. We obtain a solid K_1.

   (i) Describe the shape of K_1 in such a way that a fellow student will understand you.

   (ii) For a point P in space, we write $f(P)$ for the distance of P from the origin. Use spherical coordinates to set up the triple integral of $f(P)$ over K_1.

   (iii) Use cylindrical coordinates to set up the same integral.

b) In this part we work with D_2 on the right. – The right half of the yz-plane shown is the plane $\theta = \frac{\pi}{2}$. We rotate D_2 around the z-axis, *but only until D_2 lies in the half-plane $\theta = \frac{3}{2}\pi$.* We obtain a solid that we call K_2.

   (i) Describe K_2 in words in the same way you described K_1. *Hint:* Take the surface of the earth and operate with latitude and longitude. Then thicken the strip limited by parallels and meridians.

   (ii) Consider a point P in space. It has spherical coordinates (ρ, ϕ, θ). Find the conditions which ρ, ϕ, θ have to satisfy if P is to be a point of K_2. You will get three inequalities involving ρ, θ, and ϕ.

   (iii) Use an integral to find the volume of K_2.

   (iv) Your friend says: "I am confused. Take Cartesian coordinates. What is the volume of the solid defined by $0 \le x \le a$, $0 \le y \le b$, $0 \le z \le c$? It is of course abc. In (ii) we found that K_2 is given by $a \le \rho \le b$, $(\pi/2) \le \theta \le (3\pi/2)$, $(\pi/6) \le \phi \le (\pi/3)$. It is the same kind of formulas. So the volume should be $(b-a)(\pi)(\pi/6)$, right? But in (iii) we found something totally different. There must be something wrong somewhere. I always thought that formulas are formulas, and that they always work the same way. What is going on here?" Write down a response.

29.9. D is the region in the yz-plane shown at left in Figure 29.6. It is bounded by parts of a circle of radius b and by a horizontal line segment. When we rotate D around the z-axis we get a solid S_1 called a spherical cap.

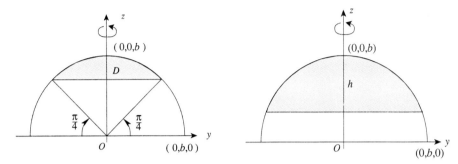

Figure 29.6

a) Use spherical coordinates to set up $\iint\int_S (x + y + z)\, dV$.

b) S_2 is another spherical cap of the same solid sphere. The cap has height h, as shown at right in Figure 29.6. Use spherical coordinates to set up the integral for the volume of S_2.

c) Use your answer to b) to find again the volume of a solid sphere of radius b.

29.10. K is the solid obtained as follows. Below left you have a view of the yz-plane. The region D in the yz-plane is part of the region between two circles. We rotate D around the z-axis and get a solid K. The solid K is shown in the perspective view on the right.

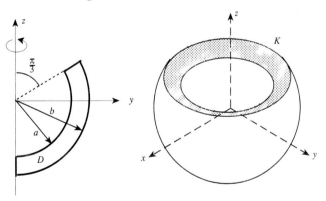

We look at $\iint\int_K f(x, y, z)\, dV$ for different functions $f(x, y, z)$.

a) The function $f(x, y, z)$ is defined as the distance of $P(x, y, z)$ from the point $Q(100(a + b), 23(a + b), -68(a + b))$. Would you expect the integral to be positive, or zero, or negative, or is it impossible to predict anything of this sort? Explain your answer.

b) As a), but for $f(x, y, z) = z$. Would you expect the integral to be positive, or zero, or negative, or is it impossible to predict anything of this sort? Explain your answer.

c) Set up and evaluate the integral for $f(x, y, z) = z$.

29.11. The mass density of the earth's atmosphere, in kg/m^3, at an altitude of y meters above sea level is approximately

$$\delta(y) = \delta_0 e^{-ay}$$

where $\delta_0 = 1.20\,\text{kg/m}^3$ is the atmospheric density at sea level, and $a = 0.000116$. The radius of the earth is 6.87×10^6 meters. Estimate the total mass of the earth's atmosphere. *Hint:* Express the density in terms of the distance ρ from the earth's center, and set up an integral using spherical coordinates. The integration can be carried out by hand but is somewhat tedious. You may want to consult a table of integrals or use a calculator or computer. If you do so, you should say how you got your answer.

Chapter 4

Vector Fields

30 Vector Fields in the Plane

A vector field attaches a vector to each point in a portion of a plane or three-dimensional space. Vector fields play an important role in many aspects of the physical sciences, including the study of gravitation, electricity and magnetism, and motion of fluids.

In this section, we will introduce the idea of a vector field in the plane. Later, in Section 32, we will extend the idea to three-dimensional space.

☞ **What is a vector field in the plane?**

Example 1. Rigid rotation of the plane. In the plane, we pick a point O, and call it the origin. We let the plane rotate around O in the counterclockwise direction with constant angular velocity ω. That is, the plane turns at a constant $\frac{\omega}{2\pi}$ revolutions per second. Now we take a picture of the plane at time $t = t_0$.

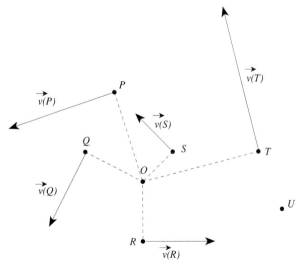

Figure 30.1

At this moment, a generic point P is in the course of moving along a circle with velocity vector $\vec{v}(P)$. The vector $\vec{v}(P)$ is perpendicular to the vector $\overrightarrow{OP}$, it points in a counterclockwise direction, and has magnitude $\omega|\overrightarrow{OP}|$. Figure 30.1 shows some of these velocity vectors for $\omega = 1.2$.   ∎

Definition. *In the situation of Example 1, a vector $\vec{v}(P)$ is attached to every point P in the plane. Such an assignment of a vector to each point in the plane is called a vector field in the plane. The vector $\vec{v}(P)$ is called the field vector at P. We draw the field vector with initial point at P because the field vector varies from point to point. Usually we denote a vector field by $\overrightarrow{F}$, and for the field vector of $\overrightarrow{F}$ at P we write $\overrightarrow{F}(P)$.*

Terminology. The field vectors in Example 1 are velocity vectors. Therefore, such a vector field is called a *velocity field*.

Problem

30.1. The questions below are about the rotating plane of Example 1, with $\omega = 1.2$ (see Figure 30.1 on the page before).

a) Draw the field vector $\vec{v}(U)$ at the point U.

b) Can you find a point W in the plane so that $\vec{v}(W)$ is of length 5 cm, and so that $\vec{v}(W)$ is perpendicular to $\vec{v}(Q)$ at the same time? If yes, draw such a point and its field vector. If no, explain.

c) Now we make another kind of assignment: To a point M in the plane we assign $|\vec{v}(M)|$ where $\vec{v}(M)$ is the field vector of our vector field. Does this assignment define a vector field? If yes, draw the field vector at the point T. If no, explain.

Example 2. The constant vector field. Take a nonzero vector $\vec{a}$. We define a vector field in the plane as follows: For any point P we take $\overrightarrow{F}(P) = \vec{a}$, as shown below. As in Example 1, a vector $\overrightarrow{F}(P)$ is attached to every point P in the plane.

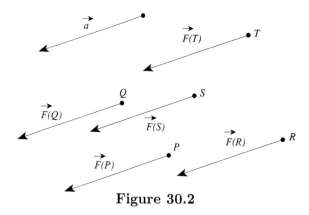

Figure 30.2

Thus, in Example 2 we have again defined a vector field in the plane. Here, the field vector is always $\vec{a}$. We write $\overrightarrow{F}(P) = \vec{a}$ or $\overrightarrow{F} = \vec{a}$ for this vector field.   ∎

Terminology. All the field vectors in Example 2 are the same. Therefore, such a vector field is called a *constant vector field* or simply a *constant field*.

Problem _____

30.2. Can you think of a situation that could be described by a constant vector field? In your situation, what would the constant vector represent? What would its length mean?

_____

Example 3. Temperature distribution in the plane. D is a thin, flat plate of some material. The temperature of D varies from point to point. At the point P the temperature is $T(P)$. We assume that the temperature pattern of D does not vary over time. In other words, the values of the function $T(P)$ depend only on the point P, and not on time.

Go to a point Q of the plate. If you move away from Q in different directions, the temperature will change in different ways. We describe the pattern of temperature changes in different directions by the directional derivatives of the function $T(P)$ at Q. *Remember* from Section 18:

> The directional derivatives at Q are determined by the vector $\operatorname{grad} T|_Q$, the gradient at Q of the function $T(P)$. The vector $\operatorname{grad} T|_Q$ points in the direction of the highest directional derivative at Q, and its magnitude is that highest directional derivative.

Therefore, $\operatorname{grad} T|_Q$ points in the direction in which the temperature rises the fastest. Now we define a vector field $\vec{F}(Q)$ on the plate D as follows:

$$\vec{F}(Q) = \operatorname{grad} T|_Q$$

Figure 30.3 illustrates the situation at three points K, L, and M. The broken arrows indicate directions in which the temperature does not rise as fast as in the direction of $\operatorname{grad} T$. ∎

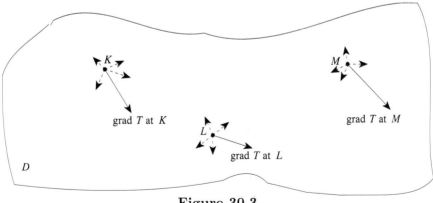

Figure 30.3

Terminology. The field vectors in Example 3 are gradients. Therefore, such a vector field is called a *gradient field*.

Problems _____

30.3. We consider the points P above the horizontal line g, as shown in Figure 30.4 on the following page. For these points we define a function $e(P)$ by $e(P) = 3h$, where h is the distance of P from g, as shown in Figure 30.3. Take 1 cm

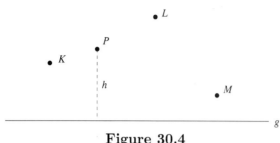

Figure 30.4

as unit. The vector field $\overrightarrow{F}(P)$ is defined by $\overrightarrow{F}(P) = \text{grad } e|_P$. Draw the field vectors at the points K, L, M shown. *Hint:* First, figure out in which direction the function $e(P)$ increases most. Second, take a point Q so that $\overrightarrow{PQ}$ is in the direction of the gradient, and find the average rate of change of e between P and Q:

$$\frac{e(Q) - e(P)}{|\overrightarrow{PQ}|}.$$

Finally, let Q tend to P to find the directional derivative. This gives you the magnitude of the gradient at P.

30.4. This problem is similar to 30.3 in that directional derivatives and gradients come up again. We consider a fixed point P_0 in the plane, as shown below. Then we define a function $r(P)$ for points in the plane as follows: $r(P)$ is the distance of P from P_0, that is, $r(P) = |\overrightarrow{P_0 P}|$. Often we write only r for $r(P)$. As units we take 1 cm.

a) Draw two vectors $\overrightarrow{u}$, $\overrightarrow{v}$ with the following properties:
   - Their initial point is at M.
   - Each is at least 3 cm long.
   - At M, the directional derivative of $r(P)$ is positive in the direction of $\overrightarrow{u}$ and negative in the direction of $\overrightarrow{v}$.

b) At L, draw a vector $\overrightarrow{w}$ of length at least 3 cm and such that the directional derivative of r at L in the direction of $\overrightarrow{w}$ is highest possible.

c) We define a vector field $\overrightarrow{F}$ by $\overrightarrow{F}(Q) = \text{grad } r|_Q$ (the gradient of r evaluated at Q). Draw the field vector at K. *Hint:* To find the magnitude of the field vector, proceed as in Problem 30.3, second hint.

Terminology. The vector field grad r of part c of Problem 30.4 has two properties: (1) At all points Q, the field vector is parallel to the straight line P_0Q. (2) The magnitude of the field vector at a point Q depends only on the distance of Q from the fixed point P_0. Any vector field $\vec{F}$ which satisfies the conditions (1) and (2) is called *radial*. The word "radial" suggests that the field vectors have the direction of radii of a circle centered at P_0.

✍ How do we perform computations with a vector field?

If we want to do computations with vector fields we have to make two choices:

1. We have to choose a coordinate system for the points in the plane.

2. We have to choose basis vectors for the field vectors.

We make the following choices: *As coordinates we take Cartesian xy-coordinates, and as basis vectors we take $\vec{\imath}$ and $\vec{\jmath}$.*

The field vector $\vec{F}$ of a vector field has components (F_1, F_2):

$$\vec{F} = F_1\,\vec{\imath} + F_2\,\vec{\jmath}.$$

The field vector varies from point to point. Therefore, the components must be functions of the coordinates (x, y) of the point:

$$\vec{F} = F_1(x, y)\,\vec{\imath} + F_2(x, y)\,\vec{\jmath}, \quad \text{that is,} \quad \vec{F} = (F_1(x, y), F_2(x, y)).$$

This means that a vector field $\vec{F}$ in the xy-plane is described by two functions $F_1(x, y)$ and $F_2(x, y)$ of two variables (x, y).

Back to Example 1 of the rotating plane. We take the point O as the origin of an xy-system. Let P be a generic point with coordinates (x, y). What are the components $(F_1(x, y), F_2(x, y))$? The field vector $\vec{v}(P)$ is perpendicular to $\overrightarrow{OP} = (x, y)$. Therefore $\vec{v}(P)$ must be of the form $c\,(-y, x)$ for some positive scalar c. The length of $\vec{v}(P)$ must be $\omega|\overrightarrow{OP}|$:

$$|c\,(-y, x)| = c\sqrt{(-y)^2 + x^2} \quad \text{must be equal to} \quad \omega|(x, y)| = \omega\sqrt{x^2 + y^2},$$

and we conclude $c = \omega$. Therefore the components of the vector field of the rotating plane are

$$(F_1(x, y), F_2(x, y)) = (-\omega\,y, \omega\,x),$$

as shown below. Again, we took $\omega = 1.2$.

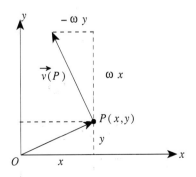

Problem _____

30.5. Both questions are about the vector field of Example 1 with angular velocity $\omega = 1.2$.

    a) Find the components of $\vec{v}(P)$ for $P(3, 7)$.

    b) Find the coordinates of a point Q so that $\vec{v}(Q) = (6, 2.4)$.

Back to Example 2 of the constant vector field. Go back to Figure 30.2 of the constant vector field of Example 2. Draw an xy-system with units of 1 cm. Choose it in such a way that the formulas become as simple as possible.

Problem _____

30.6.   a) Find the components of the constant vector field $\vec{F} = \vec{a}$ of Example 2.

    b) Your colleague worries and asks: "What if you took a different xy-system? Would you not get different components for the same field vector?" What do you say?

Example of a gradient field. The distance of a point P from the origin is a function of the point P in the plane. We write for this function $r(P)$ or only r. We also write $\vec{r}(P)$ or simply $\vec{r}$ for the vector $\overrightarrow{OP}$. Notice that $r = |\vec{r}|$. In Problem 30.7, we consider the gradient field $\vec{F} = \operatorname{grad} r$.

Problem _____

30.7.   a) Write (x, y) for the coordinates of the generic point and work out the formula for $r = |\overrightarrow{OQ}|$.

    b) $\operatorname{grad} r$ has components $(F_1(x, y), F_2(x, y))$. Find the functions $F_1(x, y)$ and $F_2(x, y)$, and rewrite them in terms of r and $\vec{r}$. Then find the magnitude of $\operatorname{grad} r$.

    c) You move from $U(p, q)$ towards the origin along the line

$$(x, y) = (p, q) + t(-p, -q),$$

    as shown in the following figure:

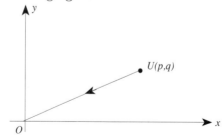

    At each moment of the movement, that is, at each point on the line UO, we look at the field vector. Describe what happens with the direction and magnitude of the field vector as you move from U towards the origin.

Additional Problems

30.8. We continue writing $r(P)$, or only r, for the distance of P from the origin.

a) Now we consider the vector field

$$\vec{G} = \operatorname{grad} r^4.$$

(i) Write out the formulas for the components $G_1(x, y)$ and $G_2(x, y)$ of $\vec{G}$.

(ii) Find the components of $\vec{G}$ at the point $(1, 2)$.

(iii) On page 259, we defined *radial* vector fields. Is $\vec{G}$ a radial field? Give an answer and a justification for the answer.

b) Let $f(t)$ be a function of one variable. We use $f(t)$ to define a vector field $\vec{H}$ as follows:

$$\vec{H}(P) = \operatorname{grad} f(r)$$

(remember that $r = |\overrightarrow{OP}|$). Below are three questions. Write down answers which will convince a colleague.

(i) The three vector fields $\vec{F} = \operatorname{grad} r$, $\vec{G} = \operatorname{grad} r^4$, and $\vec{H}(P) = \operatorname{grad} f(r)$ look related. Is the first, or second, or third a special case of one or more of the others?

(ii) Let M be a point in the plane. Suppose you knew the values $f(t)$, $f'(t)$, and $f''(t)$ for $t = |\overrightarrow{OM}|$. Describe the direction and magnitude of the field vector of $\vec{H}$ at M in terms of this information.

(iii) Is $\vec{H}$ a radial field or not? Or does it depend on the function $f(t)$ whether the field is radial?

c) Now we take for $f(t)$ the specific function $f(t) = e^{t^2}$. Let N be a point at distance 32 units from the origin. Describe the magnitude and direction of $\vec{H}(N)$ in formulas and/or words as well as you can.

30.9. Let A be a fixed point in the plane, and let $\vec{F_0}$ be a fixed vector in the plane, as shown in Figure 30.5 on the following page. We define a vector field $\vec{F}(P)$ by

$$\vec{F}(P) = \frac{|\overrightarrow{AP}|}{2} \vec{F_0}.$$

a) Draw the field vector at the points Q, R, S, and T. Take 1 cm as unit.

b) Draw an xy-system with units 1 cm into Figure 30.5 on the next page. Then $\vec{F}(P)$ can be written in the form

$$\vec{F}(P) = \vec{i}\, F_1(x, y) + \vec{j}\, F_2(x, y) = \Big(F_1(x, y), F_2(x, y)\Big).$$

Find the formulas for the functions F_1 and F_2. *Hint:* Choose the coordinate system such that the components of the fixed vector $\vec{F_0}$ are as simple as possible. Do not worry if the axes are not parallel to the edges of this page.

Figure 30.5

30.10.   a) The vector field $\overrightarrow{F}$ is defined by $F(x, y) = (x, y)$. In the system below left, draw the field vector at the points T, U, V, W as shown.

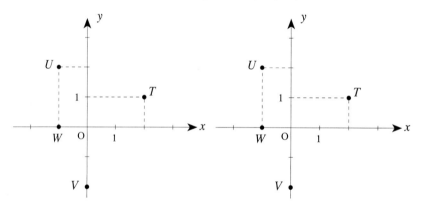

b) The vector field $\overrightarrow{G}$ is defined by $G(x, y) = (-y, x)$. In the system above right, draw the field vector at the points T, U, V, W as shown. They are the same points you worked with in part a.

The vector field $\overrightarrow{F}$ is *radial* (see Problem 30.4 on page 258).

The vector field $\overrightarrow{G}$ has the following property: The field vector at a point S is perpendicular to $\overrightarrow{OS}$. Such a vector field is said to be *rotary*. The word "rotary" suggests that the field vectors have the direction tangent to a circle centered at the origin.

Consider the three vector fields $\overrightarrow{H_1}$, $\overrightarrow{H_2}$, $\overrightarrow{H_3}$ defined as follows:

$$\overrightarrow{H_1} = (237y, -237x), \quad \overrightarrow{H_2} = (237y, 237y), \quad \overrightarrow{H_3} = \frac{1}{\sqrt{x^2 + y^2}}(237x, 237y)$$

For each of the three vector fields decide whether it is radial, or rotary, or neither. Your answer has to consist of three complete sentences.

30.11. $\overrightarrow{F}$ is the vector field defined by $\overrightarrow{F}(P) = \operatorname{grad} xy$.

a) Write out the formulas for the components of $\overrightarrow{F}(P)$.

b) Find $|\overrightarrow{F}(P)|$. Compare with $|\overrightarrow{OP}|$.

c) Draw the field vector at the points T, U, V, W shown in Figure 30.6.

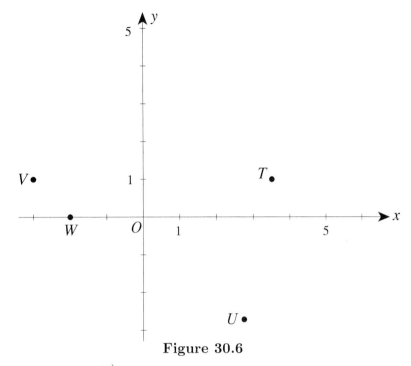

Figure 30.6

30.12. The vector field $\overrightarrow{F}$ in the plane is defined as follows: The field vector at $P(x, y)$ has the same direction as $\overrightarrow{OP}$, and the field vector at P has magnitude $|\overrightarrow{OP}|^2$.

a) Into and xy-system, draw the field vector at $(2, 1)$. Remember: The initial point of the field vector has to be at $(2, 1)$.

b) Write out the formulas for the components $(F_1(x, y), F_2(x, y))$ of the field vector.

Review problems on functions of two variables (level objects, gradient)

30.13. The function $g(x, y)$ is defined by $g(x, y) = -x^2 + y + 4$, and the vector field $\overrightarrow{F}$ in the plane is defined by $\overrightarrow{F} = \operatorname{grad} g$.

a) Find the components of the field vector of $\overrightarrow{F}$ at $P(5, 8)$.

b) At the point $(5, 8)$, the function $g(x, y)$ takes the value $g(5, 8) = -13$. Now we look at *all* points (x, y) for which $g(x, y) = -13$. The collection of all these points is called the level object of level -13 of the function $g(x, y)$ We call it K.

   (i) Is K a line or a curve or a plane or a surface or a solid? Find the equation of K and describe the shape of K in words. *Hint:* Solve the equation of K for y.

   (ii) Is the tangent object of K at $P(5, 8)$ a line or a plane? Find the equation of this tangent object.

   (iii) m is the line which passes through $P(5, 8)$ and which is perpendicular to K. Write down a parametric representation of m.

c) At the point $T(1,1)$ our function $g(x,y)$ has many directional derivatives, one for every direction $\vec{u}$.

    (i) Find the directional derivative of $g(x,y)$ at $T(1,1)$ in the direction of $\vec{u} = (3,4)$.

    (ii) In which direction $\vec{v}$ is the directional derivative at $T(1,1)$ lowest, and what is the value of this lowest directional derivative?

30.14. Go back to Example 3 on page 257 on temperature distributions. Look at Figure 30.3. Through the point L passes a level curve of the temperature function $T(P)$. This level curve has a tangent line at L. Do you have enough information to draw the tangent at L? If yes, draw it; if no, explain.

30.15. Go back to Problem 30.3 and look at Figure 30.4. Draw the level curve of level 4 of the function $e(P)$.

31   Flow Lines of a Vector Field in the Plane

The velocity field of a steady-state flow.   Consider a fluid flowing across the xy-plane in such a way that the flow pattern does not change with time. Such a stable flow pattern is called a *steady-state flow*. For example, the surface of a quietly flowing river presents a picture of a steady-state flow.

Now we drop a speck onto the flowing fluid. The speck will move along a curve. Specks dropped at different locations will move along different curves. The figure below shows curves on which different specks move. These curves are called *flow lines* of the steady-state flow.

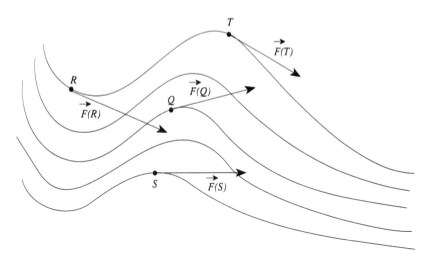

At any position P a floating speck has a velocity vector which we write as $\vec{F}(P)$. This means: A steady-state flow defines a vector field. It is a velocity field because the field vectors represent velocities.

Interpreting a vector field as the velocity field of a steady state flow.   We are given a vector field $\vec{F}$, for example a gradient field or a field defined by formulas. This field may have nothing to do at all with a steady-state flow. *Nevertheless we are free to interpret $\vec{F}$ as the velocity field of a steady-state flow.* When we do this we can relate $\vec{F}$ to experiences in the everyday material world. For example, we can talk about flow lines of $\vec{F}$. Such an interpretation gives us something to hold on to while discussing the mathematics of $\vec{F}$.

Problem

31.1. Each of the three questions asks you to describe something.

   a) $\vec{F}$ is the vector field of the rotating plane of Example 1 of Section 30. Consider $\vec{F}$ as the velocity field of a steady-state flow. Describe the shape of the flow lines of $\vec{F}$.

   b) As in part a, but for the constant vector field of Example 2 of Section 30.

   c) As in part a, but for the vector field of Problem 30.7 of Section 30.

Note:

1. From now on, we will talk about the flow lines of any given vector field. That is, we discuss a vector field as if it were the velocity field of a steady-state flow.

2. The flow lines of a vector field may be curves or they may be straight lines. Remember that we use the word "line" as abbreviation for "straight line." Do not confuse "line" with "flow line."

☞ What are flow lines?

A flow line C of the vector field $\overrightarrow{F}$ in the plane is the trajectory of a speck dropped into the fluid. It represents the path of a moving point whose velocity vector is the field vector of $\overrightarrow{F}$. As time t passes the speck moves. The position of the speck depends on time t. In other words: *The flow line C is a parametrized curve with time t as parameter.* We use position-vector notation and write C as $\vec{r} = \vec{r}(t)$. Then we can summarize:

> *A flow line C of the vector field $\overrightarrow{F}$ is a parametrized curve $\vec{r} = \vec{r}(t)$ with the following property: At every point P of the flow line C the field vector $\overrightarrow{F}(P)$ is the same as the velocity vector $\vec{r}\,'(t)$ of the curve C.* In short: At all points of a flow line, the velocity vector is the field vector.

Flow lines are also called *field lines, stream lines* or *characteristic curves.* In Section 32 we discuss vector fields which represent a force. The flow lines of such force fields are called *lines of force.*

Oriented curves.   A speck moves on a flow line in a specific direction, namely in the direction of the flow. Thus, a flow line is not just a parametrized curve, but a curve on which we have put an arrow to show the direction of the flow. In general, a curve C for which a sense of direction has been chosen is called an *oriented curve.* A chosen direction of a curve C is called an *orientation* of C. By *orienting C* we mean choosing a direction of C. Thus, a flow line of a vector field is an oriented curve, and its orientation is given by the flow.

From now on, most of the parametrized curves we discuss will be oriented. Unless otherwise mentioned, the orientation will be the direction of increasing parameter values. In case the orientation is not the direction of increasing parameter values, we will say so.

✍ How do we find the equations of flow lines?

The given vector field $\overrightarrow{F}$ has components F_1, F_2:

$$\overrightarrow{F} = \Big(F_1(x,y), F_2(x,y) \Big).$$

To find parametric representations $(x,y) = (a(t), b(t))$ of flow lines we apply the "backward search" method:

We assume we already have a flow line $C : (x,y) = (a(t), b(t))$. We find equations which the functions $a(t)$, $b(t)$ must satisfy. Then we turn around and try using the equations to find such functions.

Everything rests on the basic fact "velocity vector = field vector." This is illustrated in Figure 31.1 below. At the point $P(x, y)$ along the flow line C, the position vector

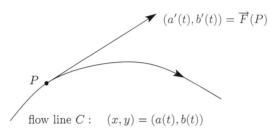

$$(a'(t), b'(t)) = \vec{F}(P)$$

P

flow line C : $\quad (x, y) = (a(t), b(t))$

Figure 31.1

is

$$\vec{r}(t) = (x, y) = (a(t), b(t))$$

so the velocity is

$$\text{velocity} = \vec{r}'(t) = (a'(t), b'(t)).$$

The field vector at $P(x, y)$ is

$$\text{field vector} = \vec{F}(x, y) = \Big(F_1(x, y), F_2(x, y)\Big) = \Big(F_1(a(t), b(t)), F_2(a(t), b(t))\Big).$$

Setting the components of the velocity vector equal to those of the field vector gives

$$\begin{aligned} a'(t) &= F_1(a(t), b(t)) \\ b'(t) &= F_2(a(t), b(t)). \end{aligned}$$

Therefore, the functions $a(t)$, $b(t)$ are solutions of the following problem: Find functions $x(t), y(t)$ which satisfy the two differential equations

$$\begin{aligned} x'(t) &= F_1(x(t), y(t)) \\ y'(t) &= F_2(x(t), y(t)). \end{aligned}$$

Such a set of differential equations is called a *system of ordinary differential equations* or *a system of ODEs*. Thus, to find the equations of the flow lines we must solve this system of ODEs. In this system,

- F_1, F_2 are the components of the given vector field $\vec{F}$, and
- $x(t), y(t)$ are the unknown functions which have to be found, and which will be the components of a flow line $\vec{r} = \vec{r}(t)$.

Notation. If we write down a system of ODE's in order to find solutions, we write $x(t), y(t)$ for the unknown functions. If we already have the functions we use the notation $a(t), b(t)$.

As a rule, finding formulas for solutions to systems of ODE's is difficult. In this course, you will be asked to find flow lines in the following special situations: Either the system of ODE's can be solved with tools from elementary calculus, or the flow lines can be found by geometric reasoning.

Additional Problems

31.2. We consider the vector field $\vec{F} = (x, 0)$.

    a) Draw the field vector at the points P, Q, ... shown in Figure 31.2 below.

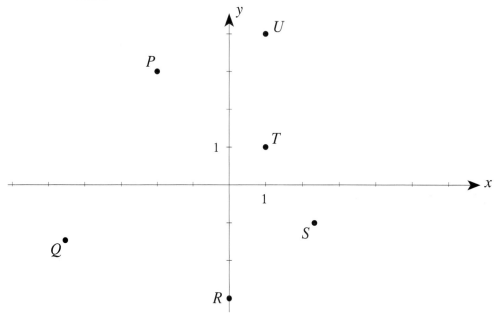

Figure 31.2

    b) To judge from part a, what will be the shape of the flow lines? What will be their orientation?

    c) Write down the system of differential equations you have to solve to find flow lines. Use $x(t), y(t)$ as names for the unknown functions. You will solve the system in part d.

    d) Solve the system of differential equations by solving each equation separately. *Hints:* The best way to write a function $f(t)$ that satisfies $f'(t) = f(t)$ is $f(t) = c_1 e^t$ where c_1 is an arbitrary constant.

    e) To nail down one particular flow line, you must state "at the time $t = \dots$, your speck is at the point $P(\dots)$."

      (i) Find the equation of the flow line of the particle which at time $t = 0$ is at the point $T(1, 1)$ (this is one of the points in Figure 31.2 above).

     (ii) Find the equation of the flow line of the particle which at time $t = 2$ is at the point $Q(-4.5, -1.5)$ (another point shown in Figure 31.2).

   (iii) A colleague says: "I don't get it. The flow lines have the shape of straight lines. But the equations do not look like $5 - 2t$ or any other linear function. What is going on here?" Write down your response.

31.3. Go back to the gradient field

$$\vec{F}(P) = \text{grad}\, |\overrightarrow{OP}|$$

of Problem 30.7 on page 260.

a) Write out the system of differential equations you would have to solve in order to find the equation of a flow line. *Do not try to solve the system.*

b) The system you found in part a looks very complicated. But we already know:

The flow lines are rays (half-lines) issuing from the origin.

The field vector at P is a unit vector in the direction of $\overrightarrow{OP}$.

These facts allow you to write down the equation of any flow line. Write down the equation of the flow line which at time $t = 0$ passes through the point $(7, 48)$. Do not simplify numbers (do not evaluate the square roots, etc.).

31.4. We consider the function

$$f(x, y) = 4x - 3y + 8$$

and the vector field

$$\vec{F}(x, y) = \text{grad}\, f(x, y).$$

a) Into an xy-system, draw $\vec{F}(2, 3)$ (the field vector at $(2, 3)$).

b) The flow line

$$C : (x, y) = (a(t), b(t))$$

passes through $(2, 3)$ at time $t = 0$. Describe the *shape* of C.

c) We consider our vector field $\vec{F}(x, y)$ as the velocity field of a steady-state flow in the plane. At time $t = 0$ we drop a speck at our point $(2, 3)$. The speck starts moving. We write $\vec{v}$ for its velocity vector at $(2, 3)$. Draw $\vec{v}$.

d) d) Your friend asks: Why is it that $\vec{v} = \vec{F}(2, 3)$? Write down your response.

31.5. We consider a vector field $\vec{F}$ in the plane. It is defined by

$$\vec{F} = (x + y, -3x - y).$$

a) Find the components of the field vector at $(x, y) = (3, 2)$.

b) The curve

$$C : (x, y) = (a(t), b(t))$$

is a flow line of $\vec{F}$. It describes the movement of a particle which at time $t = 3$ sec passes through the point $Q(4, 6)$.

(i) Do we have enough information to find $(a(3), b(3))$ without solving the system of differential equations? If yes, find $(a(3), b(3))$. If no, explain.

(ii) Do we have enough information to find $(a'(3), b'(3))$ without solving the system of differential equations? If yes, do it. If no, explain.

c) K is a different curve. It is given by $(x, y) = (p(t), q(t))$ where $p(t)$ and $q(t)$ are defined as follows:

$$
\begin{aligned}
p(t) &= \cos(\sqrt{2}\,t) + \sin(\sqrt{2}\,t) \\
q(t) &= (\sqrt{2} - 1)\cos(\sqrt{2}\,t) - (\sqrt{2} + 1)\sin(\sqrt{2}\,t).
\end{aligned}
$$

Can you decide whether K is a flow line of our vector field $\overrightarrow{F}$? No matter what your answer is, justify your answer.

31.6. The vector field $\overrightarrow{F}$ is given by

$$\overrightarrow{F}(x, y) = (3xy, \sqrt{x^3 + y}).$$

Write down the system of differential equations you would have to solve to find flow lines. *Do not try to solve the system.*

31.7. Let $\overrightarrow{F}$ be a vector field in the plane. We do not know the components $(F_1(x, y), F_2(x, y))$ of $\overrightarrow{F}$, but we are given information on a particular flow line C as follows: C is the flow line which for $t = 0$ passes through $(5, 3)$, and C is given by

$$(x, y) = (t + 5, t^2 + 10t + 3).$$

Do we know enough to find the field vector at $(5, 3)$? If yes, find the field vector; if no, explain.

31.8. We have the following information on a vector field $\overrightarrow{F}$ in the plane:

- $\overrightarrow{F} = \operatorname{grad} h(x, y)$ where $h(x, y)$ is a function of two variables.
- The curve C shown below is a flow line passing through $Q(1, 3)$, and the vector $\vec{v}(Q)$ shown is the velocity vector of the flow line at Q.

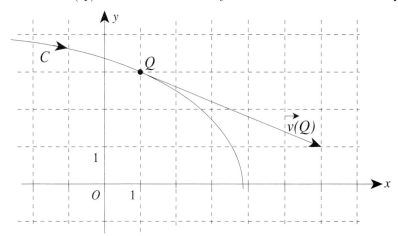

In questions a), b) write "impossible" if the given information is insufficient to find the answer.

a) Find the components of $\operatorname{grad} h(x, y)$ at $Q(1, 3)$.

b) Through Q passes a level curve of the function $h(x, y)$. We call this level curve K. Write down a parametric representation of the tangent line at Q of the level curve K.

31.9. We consider the vector field $\overrightarrow{F}$ given by

$$\overrightarrow{F}(x,y) = (y,y).$$

a) Into an xy-system, draw the field vector at the following points:

$$(-1,1), \quad (1,1), \quad (3,1), \quad (5,1), \quad (0,2), \quad (2,2), \quad (4,2).$$

b) At time $t = 0$ you drop a speck at $(3,1)$. Describe how this speck will move—in which direction? Does it move at constant speed, or increasing speed, or decreasing speed?

c) Find the formulas $x = x(t)$ and $y = y(t)$ for the flow line which at time $t = 0$ passes through $(3,1)$.

32   Vector Fields in Space

Vector fields in space work very much like vector fields in the plane. In computations, there will be a third component. Therefore, to find the equation of flow lines, we have to solve systems of three differential equations.

Example 1.   Space rotating around the x-axis. We let all of space rotate around the x-axis with constant angular velocity ω ($\frac{\omega}{2\pi}$ revolutions per second). We do it in such a way that the rotation is counterclockwise if we look at the yz-plane in the direction of $-\vec{\imath}$. In such a rotation a point P moves on a circle. The center of this circle is on the x-axis. The velocity vector of P is tangent to the circle at P. We write $\vec{v}(P)$ for the velocity vector at P. This is illustrated in the following:

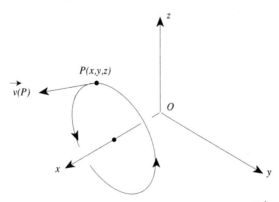

We use the velocity vectors $\vec{v}(P)$ to construct a vector field $\overrightarrow{F}$: We define $\overrightarrow{F}(P)$ to be $\vec{v}(P)$. The field $\overrightarrow{F}$ is a velocity field. Its xyz-components are (F_1, F_2, F_3). Each component is a function of the coordinates (x, y, z) of the point P.   ∎

Problem _____

  32.1. $\overrightarrow{F}$ is the velocity field of Example 1 (space rotating around the x-axis). Find the formulas for the components $F_1(x, y, z)$, $F_2(x, y, z)$, $F_3(x, y, z)$. *Hints:* If you look at the yz-plane in the direction of $-\vec{\imath}$, things look like the yz-plane rotating around the origin, and you can use the work on the rotating plane done in Section 30 (see the paragraph "Back to Example 1 of the rotating plane" on page 259). Note that the points in space move in planes parallel to the yz-plane.

_____

Example 2. The gravitational field. We place a solid of mass 1 at a point P in space. A force $\overrightarrow{F}(P)$, viz. gravity, pulls the solid towards the earth's surface. That is, gravity defines a vector field $\overrightarrow{F}(P)$. It is called the *gravitational field*. What are the formulas for the gravitational field? There are two cases to consider.

  1. We restrict our attention to points near a portion of the earth's surface that is small enough that it looks like a plane. Then $\overrightarrow{F}(P)$ points towards the plane of the earth at a right angle, and its magnitude is a physical constant g, the *acceleration due to gravity*.[1] That is, in this case the gravitational field may be treated as a constant vector field.

_____

    [1]If force is measured in newtons and mass in kilograms, then the value of g is approximately 9.8 m/sec². Since we told you that the gravitational field represents a force, you may be wondering

2. We take into account that the earth is round, and we consider all possible points in space. Then the gravitational field is not constant. Its formula is discussed in Problem 34.3 on page 286.

∎

Terminology. The field vector of the gravitational field is a vector representing a force. Therefore, such a vector field is called a *force field*. Note that a force field has nothing to do with a steady-state flow. Nevertheless, we can consider the flow lines of a force field. The flow lines of a force field are often called *lines of force*.

Problem _____

32.2. $\overrightarrow{F}$ is the constant gravitational field of Example 2 just discussed. Find the parametric representation of the line of force (the flow line) which at time $t = 0$ sec passes through the point $Q(142.9, 378.5, 926.4)$.

Example 3. Gradient fields in space. This field is of the form

$$\overrightarrow{F} = \text{grad } f(x, y, z)$$

where $f(x, y, z)$ is a function of three variables. This is again an example of a gradient field. ∎

Problem _____

32.3. $\overrightarrow{F}$ is the vector field $\overrightarrow{F} = \text{grad}(xyz)$ of Example 3.

   a) Find the components of the field vector at $P(1, 2, 3)$.

   b) Through $P(1, 2, 3)$ passes a level surface of the function $w = xyz$. Find the equation of the tangent plane at P of this level surface. Then write down the components of a vector $\overrightarrow{b}$ which is perpendicular to the tangent plane.

   c) This is a question about parallel vectors. *Remember:* "$\overrightarrow{u}$ parallel to $\overrightarrow{v}$" means "$\overrightarrow{u} = c\,\overrightarrow{v}$ for a nonzero scalar c." Note that c may be positive or negative.

   If the vectors you found in a) and b) are parallel, explain why this is the case. If the two vectors are not parallel, check your work because they should be parallel. Then give the explanation.

Example 4. Radial fields in space. Radial fields in space are defined as radial fields in the plane. As fixed point P_0 we take the origin O. Then the vector field $\overrightarrow{F}$ in space is called *radial* if

1. at all points Q, the field vector is parallel to the straight line OQ, and

2. the magnitude of the field vector at a point Q depends only on the distance of Q from the origin O.

_____

why its magnitude is given in *acceleration* units. The reason is that the gravitational field gives the force that would act on a *unit* mass, and Newton's Second Law tells us that force equals mass times acceleration. Strictly speaking, the gravitational field gives force *per unit mass*.

Condition (1) means that the field vector at Q must be of the form

$$\overrightarrow{F} = a(Q)\,\frac{\vec{r}}{r}$$

where $\vec{r} = \overrightarrow{OQ}$ and $r = |\overrightarrow{OQ}|$, and $a(Q)$ is a scalar. Condition(2) means that $a(Q)$ must be a function $f(r)$ of r alone. Therefore we must have

$$\overrightarrow{F} = f(r)\,\frac{\vec{r}}{r}.$$

$\blacksquare$

Problem

32.4. Consider the vector field $\overrightarrow{F}$ given by $\overrightarrow{F} = (x, y, z)$. Is it or is it not it a radial field? If it is, find the function $f(r)$. If it is not, explain why.

The various kinds of vector fields of Sections 30–32. We worked with velocity fields, constant fields, gradient fields, and force fields. *Have these fields anything to do with each other?* We have come across two kinds of vector fields:

1. Some of the vector fields arise in physics such as velocity fields and force fields. Let us call such fields *physical vector fields*.

2. Other fields are given in a purely mathematical description, that is, without any reference to physics. Let us call such fields *generic vector fields*

What about constant fields and gradient fields? The words "constant" in constant field and "gradient" in gradient field tell us something about the *mathematical properties* of the field. They do not tell us whether the field is a physical field or a generic one. Some physical vector fields are gradient fields, others are not. Some force fields are constant, others are not. Similarly, some generic vector fields are gradient fields, others are not.

Problems

32.5. We consider a vector field $\overrightarrow{F}$. To ask "Is $\overrightarrow{F}$ a gradient field?" means "Is there a function g such that $\operatorname{grad} g = \overrightarrow{F}$?" To say of a vector field "Yes, it is a gradient field" means finding a function g whose partial derivatives are equal to the components of the field in question.

　　a) Is the gravitational field of case 1 of Example 2 on page 272 a gradient field or not?

　　b) Is the constant field of Example 2 on page 256 a gradient field or not?

　　c) Give an example of a gradient field which is not constant.

32.6. Let $P(x, y, z)$ be a generic point in space. We consider the vector field

$$\overrightarrow{F}(P) = \overrightarrow{OP}.$$

Is it constant or not? Is it a gradient field or not?

From now on, when we say "vector field" we mean a *generic* vector field unless otherwise stated.

Additional Problems

32.7. Figure 32.1 shows the vector $\overrightarrow{AB}$ and its shadow $\overrightarrow{A_0 B_0}$. We define a vector field in space by $\overrightarrow{F} = \overrightarrow{AB}$. It is a constant vector field.

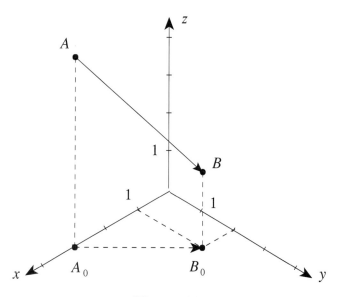

Figure 32.1

a) Find the components $F_1(x, y, x), F_2(x, y, z), F_3(x, y, z)$ of $\overrightarrow{F}$.

b) Describe the shape of the flow lines of $\overrightarrow{F}$.

c) Write down the system of differential equations you would have to solve to find flow lines. Remember to write for the unknown functions $x(t)$, $y(t)$, $z(t)$, and not just x, y, z.

d) We consider $\overrightarrow{F}$ as the velocity field of a steady-state flow in space. At time $t = 3$, we drop a speck into the flow at the point $Q(2, 9, 16)$. Find its position at time $t = 371$.

e) Is there a function $g(x, y, z)$ such that its gradient is our vector field $\overrightarrow{F}$? If yes, find such a function; if no, explain.

32.8. We define a function $f(x, y, z)$ as follows: $f(x, y, z)$ is the distance of the point $P(x, y, z)$ from the z-axis. Now we consider the vector field $\overrightarrow{F} = \operatorname{grad} f(x, y, z)$.

a) Write down the formulas for the components $\overrightarrow{F}$.

b) Consider the point $A(3, 4, 7)$. What can you say about the flow line passing through A? Read all six statements and choose the true statement(s).

- The flow line is a circle in the plane $z = 7$, with center at $(0, 0, 7)$.
- The flow line is a half-line (a ray) parallel to the vector $\vec{a} = (3, 4, 7)$.
- The flow line is a half-line (a ray) perpendicular to the vector $\vec{a} = (3, 4, 7)$.

- The flow line is a half-line (a ray) parallel to the vector $\vec{b} = (3, 4, 0)$.
- The flow line is a half-line (a ray) perpendicular to the vector $\vec{b} = (3, 4, 0)$.
- The flow line is a half-line (a ray) parallel to the z-axis.

32.9. We have the following information on a vector field $\overrightarrow{F}$ in space:

- Its flow lines are circles parallel to the yz-plane with their centers on the x-axis.
- The magnitude of the field vector at P is inversely proportional to the distance of P from the x-axis.

Find the components of $\overrightarrow{F}$.

32.10. $\overrightarrow{F}$ is the vector field $\overrightarrow{F}(P) = (x, 1, y)$.

a) Into and xyz-system, draw the field vector at the points $Q(2, 3, 0)$, $R(0, 3, 0)$, and $S(2, 3, 4)$.

b) Write down the system of differential equations you have to solve to find flow lines.

c) Find the equation $(x, y, z) = (a(t), b(t), c(t))$ of the flow line C which at time $t = 0$ passes through $P(1, 1.5, 2)$.

d) Below are shown the graphs of the functions $a(t)$, $b(t)$, $c(t)$ of the flow line C of b), but the labels are missing – which is which? Label the graphs.

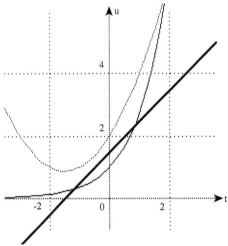

e) Use the graphs of d) to come up with a very rough, but very quick, estimate of the position of the floating particle at time $t = 0.5$.

32.11. The function $f(x, y, z)$ is defined by

$$f(x, y, z) = -x^2 - y^2 + z,$$

and the vector field $\overrightarrow{F}$ is defined by $\overrightarrow{F} = \operatorname{grad} f$.

a) Find the components of the field vector of $\overrightarrow{F}$ at $P(1, 2, 3)$.

b) Into an xyz-system, draw (i) the field vector of $\overrightarrow{F}$ at $P(1, 2, 3)$, and (ii) the shadow of this vector in the xy-plane.

c) The flow line
$$C : (x, y, z) = (a(t), b(t), c(t))$$
of $\vec{F}$ passes through the point $P(1, 2, 3)$ at time $t = 7$. Find the functions $a(t)$, $b(t)$, $c(t)$.

32.12. The vector field $\vec{F}$ is given by $\vec{F} = (x, x, x)$.

    a) Write down the system of differential equations you have to solve in order to find flow lines.

    b) Find the equations for the flow line which at time $t = 3$ passes through $B(1, 5, 9)$.

32.13. The vector field $\vec{F}$ in space is defined as follows: At the point $P(u, v, w)$, the field vector $\vec{F}(P)$ has the same direction as the vector $\overrightarrow{P^*P}$. The point P^* is determined by the point P, as shown below.

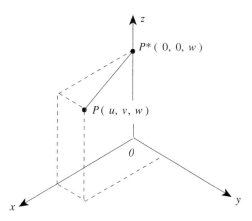

The magnitude of $\vec{F}(P)$ is 1 if P is *not* on the z-axis. If P *is* on the z-axis, the field vector is not defined.

    a) Find the components of the field vector at $Q(1, 4, -5)$.

    b) Describe the shape of the flow line of $\vec{F}$ which at time $t = 0$ passes through $P(u, v, w)$.

    c) The curve $C : (x, y, z) = (a(t), b(t), c(t))$ is the flow line which at time $t = 0$ passes through the point $M(4, 3, 7)$. Write out the formula for the functions $a(t), b(t), c(t)$. *Hint:* What is the shape of the flow line? What is the magnitude of the field vector?

Review problems on functions of three variables (level objects, gradient)

32.14. The function $f(x, y, z)$ is defined by
$$f(x, y, z) = -x^2 - y^2 + z.$$

    a) Find the components of the vector grad f at the point $P(1, 2, 3)$.

    b) At the point $P(1, 2, 3)$, the function $f(x, y, z)$ takes the value -2. Now we look at *all* points (x, y, z) for which $f(x, y, z) = -2$. The collection of all these points is called the level object of level -2 of $f(x, y, z)$. We write K for this level set.

(i) Is K a curve or a surface? Find the equation of K and describe the shape of K in words. *Hint:* Solve the equation of K for z.

(ii) Is the tangent object of K at $P(1, 2, 3)$ a line or a plane? Find the equation of this tangent object.

(iii) m is the line which passes through $P(1, 2, 3)$ and which is perpendicular to K. Write down a parametric representation of m.

c) In which direction $\vec{v}$ is the directional derivative at $T(1, 1, 1)$ highest, and what is the value of this highest directional derivative?

33   The Symbol ∇

This section deals with two operations for vector fields: *divergence* and *curl*. Here we do only the formula work. Their meaning will be discussed later.

Vector fields in space

We consider a vector field $\vec{F} = (F_1, F_2, F_3)$ in space and a function $f(x, y, z)$ of three variables.

The symbolic vector ∇.   The symbol ∇ is pronounced "nabla" or "del." It is an abbreviation for the expression

$$\nabla = \frac{\partial}{\partial x}\,\vec{i} + \frac{\partial}{\partial y}\,\vec{j} + \frac{\partial}{\partial z}\,\vec{k}.$$

The expression looks like a vector, but it's not an ordinary vector because the symbols $\frac{\partial}{\partial x}$, $\frac{\partial}{\partial y}$, $\frac{\partial}{\partial z}$ are not scalars. For this reason, ∇ is also called a *symbolic vector*. Without the vectors $\vec{i}$, $\vec{j}$, and $\vec{k}$, it is written in the form

$$\nabla = \left(\frac{\partial}{\partial x}, \frac{\partial}{\partial y}, \frac{\partial}{\partial z} \right).$$

We are going to work with ∇ as if it were an ordinary vector. First we multiply the vector ∇ with the scalar function $f = f(x, y, z)$. We obtain the vector $\operatorname{grad} f$:

$$\nabla f = \left(\frac{\partial}{\partial x}, \frac{\partial}{\partial y}, \frac{\partial}{\partial z} \right) f = \left(\frac{\partial f}{\partial x}, \frac{\partial f}{\partial y}, \frac{\partial f}{\partial z} \right) = (f_x, f_y, f_z) = \operatorname{grad} f.$$

In other words, ∇f can be used as an abbreviation for $\operatorname{grad} f$.

The divergence of a vector field.   We consider a vector field $\vec{F}$ with components $(F_1(x, y, z), F_2(x, y, z), F_3(x, y, z))$. We take the dot product $\nabla \cdot \vec{F}$:

$$\nabla \cdot \vec{F} = \left(\frac{\partial}{\partial x}, \frac{\partial}{\partial y}, \frac{\partial}{\partial z} \right) \cdot \vec{F} = \frac{\partial F_1}{\partial x} + \frac{\partial F_2}{\partial y} + \frac{\partial F_3}{\partial z}.$$

$\nabla \cdot \vec{F}$ is a *scalar* function called the *divergence* of $\vec{F}$. Usually it is written $\operatorname{div} \vec{F}$.

The curl of a vector field.   We continue with a vector field $\vec{F} = (F_1, F_2, F_3)$, but now we take the cross product $\nabla \times \vec{F}$:

$$\nabla \times \vec{F} = \left(\frac{\partial F_3}{\partial y} - \frac{\partial F_2}{\partial z}, \frac{\partial F_1}{\partial z} - \frac{\partial F_3}{\partial x}, \frac{\partial F_2}{\partial x} - \frac{\partial F_1}{\partial y} \right).$$

$\nabla \times \vec{F}$ is a *vector field* called *curl* of $\vec{F}$. Usually it is written $\operatorname{curl} \vec{F}$. The formula for the curl can be easily remembered using the determinant formula for the cross product (page 61):

$$\operatorname{curl} \vec{F} = \nabla \times \vec{F} = \begin{vmatrix} \vec{i} & \vec{j} & \vec{k} \\ \frac{\partial}{\partial x} & \frac{\partial}{\partial y} & \frac{\partial}{\partial z} \\ F_1 & F_2 & F_3 \end{vmatrix}.$$

When expanding the determinant, in each term, write the partial derivative sign (from the second row) to the left of the component function of the field (from the third row).

Vector fields in the plane

For vector fields $\overrightarrow{F} = (F_1(x,y), F_2(x,y))$ in the plane and functions $f(x,y)$ of two variables, the symbolic vector ∇ has only two components:

$$\nabla = \frac{\partial}{\partial x}\,\vec{i} + \frac{\partial}{\partial y}\,\vec{j} = \left(\frac{\partial}{\partial x}, \frac{\partial}{\partial y}\right).$$

As for vector fields and functions in space, ∇f is the gradient of $f(x,y)$, and the divergence of $\overrightarrow{F}$ is defined as the scalar $\nabla \cdot \overrightarrow{F}$.

The curl of a vector field in the plane. Here is the only place where things are different from vector fields in space.

The curl of $\overrightarrow{F} = (F_1(x,y), F_2(x,y))$ is a scalar. It is defined by

$$\mathrm{curl}\,\overrightarrow{F} = \frac{\partial F_2}{\partial x} - \frac{\partial F_1}{\partial y}.$$

Note that the curl of a vector field in the plane is a scalar.

Many books write the components $\overrightarrow{F}$ as $(P(x,y), Q(x,y))$. When we use subscripts for partial differentiation we end up with a simple expression for the curl in the plane:

$$\mathrm{curl}\,\overrightarrow{F} = Q_x - P_y.$$

Note: The formulas for gradient, divergence, and curl discussed in this sections are for use with Cartesian coordinates. Formulas for calculating these quantities using polar or spherical coordinates can be worked out from the Chain Rule, and are more complicated. Later, in Sections 43 and 44 we will give a geometric interpretation of divergence and curl that doesn't depend on any particular choice of coordinates.

Additional Problems

There may be questions which do not make sense, for example because vectors and scalars are confused. In that case, write "bad question" as your answer. Be sure to distinguish "scalar times a vector" (which will be a vector) from "vector dot a vector" (which will be a scalar).

33.1. We consider the function $g(x,y,z) = x^3 + y + z^2$ and the vector field $\overrightarrow{F} = (2xz, \sin y, e^y)$.

Evaluate the following expressions. That is, if the answer is a vector field, find the components. If the answer is a scalar find the value of the scalar.

a) $\mathrm{grad}\,g$

b) $\mathrm{div}\,g$

c) $\mathrm{div}\,\overrightarrow{F}$

d) $\nabla \times \overrightarrow{F}$

e) $\nabla \cdot (\nabla g)$

f) $\mathrm{curl}(\nabla g)$

g) $\mathrm{div}(\nabla \times \overrightarrow{F})$

33.2. Now $g(x, y, z)$ is a generic function and

$$\overrightarrow{G} = (H(x, y, z), K(x, y, z), M(x, y, z))$$

a generic vector field. Write out the expressions below in terms of the partial derivatives $g_x, \ldots$ and $H_x, \ldots$.
  a) $\nabla \cdot \nabla g$ (often written as $\nabla^2 g$)
  b) $\nabla \times (\nabla g)$
  c) $\nabla \cdot (\nabla \times \overrightarrow{G})$
  d) $\nabla (\nabla \cdot \overrightarrow{G})$
  e) $\nabla \times (\nabla \cdot \overrightarrow{G})$

The curl of a vector field in the plane

33.3. If you start with a generic function $g(x, y)$ of two variables, you can first calculate $\operatorname{grad} g$ to get a vector field, and then calculate the curl $\operatorname{curl} \operatorname{grad} g$ to get a scalar function. In fact, you'll get 0. Write one or more sentences to explain why this is true. *Hint:* Write out the formula for $\operatorname{curl} \operatorname{grad} g$ in terms of the second-order partial derivatives of g.

33.4. We are given three functions of one variable: $u(t)$, $v(t)$, and $w(t)$. We do not know their formulas. Nevertheless, we use these functions to define new functions of three variables and vector fields in space. Your answers will be formulas which contain the derivatives u', v', and w'. To show what we mean we have worked out the answer of a). You have to say whether your answer is a vector or a scalar.
  a) We define a function $F(x, y, z)$ by

$$F(x, y, z) = u(x)v(y)w(z).$$

  Find ∇F.
  Answer:

$$\nabla F = (u'(x)v(y)w(z), u(x)v'(y)w(z), u(x)v(y)w'(z)).$$

  It is a vector.
  b) The function G is defined by

$$G(x, y, z) = u(xyz).$$

  Find $\operatorname{grad} G$.
  c) The vector field $\overrightarrow{H}$ is defined by

$$H(x, y, z) = (u(x)v(x), w(z), u(z)).$$

  Find $\operatorname{div} \overrightarrow{H}$.
  d) The vector field $\overrightarrow{K}(x, y, z)$ defined by

$$\overrightarrow{K}(x, y, z) = \left(w(x), v\left(\frac{y}{z}\right), u(z)\right).$$

  Find $\operatorname{curl} \overrightarrow{K}$.

Chapter 5

Line Integrals

34 Force Fields and Work

In this section, we deal with the physical notions of mass, force, distance, and work. We will usually use standard metric units for these physical quantities: Mass will be measured in kilograms, force in newtons, distance in meters, and work in newton-meters (also called *joules*). At times we will suppress explicit references to units.

The everyday notion of work

A heavy concrete block rests on a rough surface. We want to move it. We can do this only if we apply a force F.

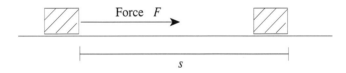

We move the block a distance s. For this, we have to do a certain amount of work.

Problem

34.1. Which of the following expressions would be a reasonable measure of the work we performed in moving the block?

$$F + s, \quad F - s, \quad Fs, \quad -Fs, \quad \frac{F}{s}, \quad \frac{s}{F}$$

Why? Pick one you did not choose and explain why you rejected it.

Motion under the influence of gravity

Rectilinear movement of a particle in a constant force field

Problem 34.1 suggests that the product

$$\text{(force in the direction of displacement)} \times \text{(displacement)}$$

is a reasonable measure for work. We want to make that precise. To start out, we look at the force field of gravity and consider it as a constant vector field, as we did in Example 2 of Section 32. That is, we let the xy-plane represent the ground so that the z-axis points upward. Then the gravity field is given by $\vec{G} = (0, 0, -g)$ where g is a constant. Here g is the magnitude of the gravitational force (in newtons) on a 1 kg mass, which we may treat as constant if our altitude doesn't change much.[1] This is shown below. The horizontal line is the xy-plane seen on edge.

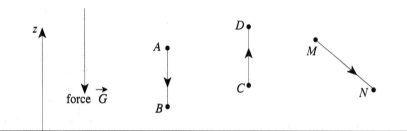

We are going to let a particle of mass 1 kg move consecutively along three line segments AB, CD, and MN. The line segments are the paths for the particle. The line segments AB and CD are parallel to the z-axis.

1. We put the particle at A and let go. Gravity pushes it to B. We describe this by saying "the vector field $\vec{G}$ does work in moving the particle from A to B." The work W done by the field is given by

$$W = \text{(force in the direction of displacement)(displacement)} = |\vec{G}||\overrightarrow{AB}|.$$

   Notice that the units for the two factors on the right are newtons and meters respectively, so the units for the work W are newton-meters. Because $\vec{G}$ and $\overrightarrow{AB}$ have the same direction, $|\vec{G}||\overrightarrow{AB}|$ is the dot product of the two vectors. Result:

$$\text{work } W = \vec{G} \cdot \overrightarrow{AB}.$$

2. We put the particle at C and shoot it straight up toward D. What is the work done by the by the gravitational field as the particle moves from C to D? The force in the direction of the displacement is $(-1)|\vec{G}|$, and we obtain $W = (-1)|\vec{G}||\overrightarrow{CD}|$. The work done by the field is negative. Note that $\vec{G}$ and $\overrightarrow{CD}$ have opposite directions. Therefore, $(-1)|\vec{G}||\overrightarrow{CD}|$ is the dot product of the two vectors. Result:

$$\text{work } W = \vec{G} \cdot \overrightarrow{CD}.$$

3. We put the particle at M and let it move towards N, say by letting it slide down an inclined plane. Now only the part of $\vec{G}$ in the direction of $\overrightarrow{MN}$ acts on the particle. This part is represented by the line segment MQ shown in Figure 34.1 on the next page. The length of MQ is the scalar projection of $\vec{G}$ onto $\overrightarrow{MN}$. Therefore, the work W done by $\vec{G}$ is $W = (\text{scalar projection of } \vec{G} \text{ onto } \overrightarrow{MN})\,|\overrightarrow{MN}|$. That is,

$$W = \left(\vec{G} \cdot \frac{\overrightarrow{MN}}{|\overrightarrow{MN}|} \right) |\overrightarrow{MN}| = \vec{G} \cdot \overrightarrow{MN}.$$

---

[1] At sea level, $g \approx 9.8$ nt.

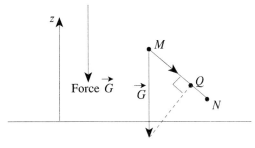

Figure 34.1

For a straight path pointing upward, e.g. for $\overline{NM}$, we would again find that the work is the dot product of the force and the vector.

We summarize (1)–(3) for a vector field $\vec{F}$:

Summary: *A particle moves in a* constant *force field $\vec{F}$ along a line segment $\overrightarrow{PQ}$. Then the work done by $\vec{F}$ is given by $\vec{F} \cdot \overrightarrow{PQ}$. This simple computation of work is possible only in the special situation of "constant force with movement along a straight line."*

Problem _____

34.2. This problem is about a vector field in the plane. It is a ruler-and-pencil problem. We are given a constant force field $\vec{F} = \vec{a}$, as shown below. The picture is at 1/100 scale, so that one cm in the picture corresponds to one meter. The force vector $\vec{a}$ is scaled so that each centimeter of length corresponds to one newton of force.

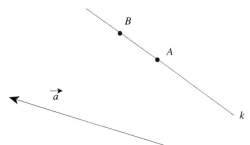

   a) Find the work done by the force $\vec{F}$ in moving a particle from A to B.

   b) Find two points R and S so that the work done by $\vec{F}$ in moving a particle from R to S along the line RS equals the negative of the work you found in a).

   c) Draw a line p which is perpendicular to the field vector $\vec{a}$, and which passes through A. On that line p, draw the point D which is 1.5 cm away from A, and which is to the left of A. How much work is done by $\vec{F}$ in moving a particle from A to D?

   d) Find and draw two points U, T so that the work done by $\vec{F}$ in moving a particle from U to T equals 13.5 nt-m. Is there more than one solution? If yes, draw a second pair of points V, W.

Newton's Law of Gravitation; movement of a particle along a curve in a non-constant force field

In Section 32 and at the beginning of the present section we discussed the constant gravity field $\overrightarrow{G} = (0, 0, -g)$. It is an approximation of what happens in reality. The precise description of gravity is given by Newton's Law of Gravitation. We now describe Newton's Law.

We consider a particle of mass 1 at distance r from the center of the earth.

Newton's Law of Gravitation states that when the particle is at the point P, it is attracted to the center of the earth by a force $\overrightarrow{F}(P)$ which has the following properties:

1. $|\overrightarrow{F}(P)|$ is inversely proportional to the square of the distance r;

2. $\overrightarrow{F}(P)$ has the same direction as the vector which points from P to M.

The vector field $\overrightarrow{F}(P)$ is called the *gravitational force field*.

Problem _____

34.3. Take an xyz-system of coordinates whose origin is the center M of the earth. The gravitational field $\overrightarrow{F}$ has components

$$(F_1(x, y, z), F_2(x, y, z), F_3(x, y, z)).$$

Find the formulas for the functions F_1, F_2, F_3. *Hint:* First, find the components of a unit vector in the direction of the field vector. Then multiply it by the magnitude of the field vector. Use the letter k for the factor of proportionality.

The key questions for work

In this section we worked out the work done by a *constant* force in moving a particle along a *straight line* path? How do we define and calculate work when the force is *not* constant, or the path is *curved*?

We consider a force field $\overrightarrow{F}$ which is *not* necessarily constant, and a *curved* path C. The curve C is oriented. Figure 34.2 on the facing page shows C and the field vectors at some points of C. Note that the curve C is *any* oriented curve, and not necessarily a flow line of $\overrightarrow{F}$. Under these circumstances,

- What do we mean by "work done by $\overrightarrow{F}$ in moving a particle from A to B"?

- How do we compute the work?

These are the key questions for work. They are the topic of Section 35.

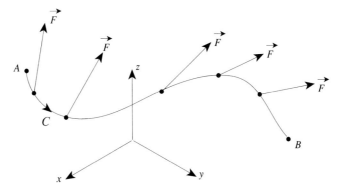

Figure 34.2

Additional Problems

Constant vector fields, movement along a line segment

34.4. $\overrightarrow{F}$ is the constant vector field $\overrightarrow{F} = (4, 1, 5)$.

a) Find the work done by $\overrightarrow{F}$ in moving a point from $A(1, 4, 3)$ to $B(3, 7, 9)$.

b) D is the point $(9, 3, 5)$.

   (i) Find a point K so that the work done by $\overrightarrow{F}$ in moving a point from D to K equals zero.

   (ii) Describe the collection of all points Q such that the work $\overrightarrow{F}$ does in moving a point from D to Q equals zero.

34.5. $\overrightarrow{F}$ is the constant vector field $\overrightarrow{F} = \vec{a}$, and K, L, M are three points in space. We write $W(K, L)$ for the work $\overrightarrow{F}$ does in moving a point from K to L.

a) Express $W(K, L)$ in terms of the points K, L and the vectors one can form with them.

b) We consider two ways to get from K to M: The straight line paths KM and KLM, as illustrated below.

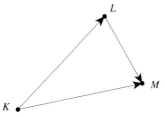

We want to compare the work $\overrightarrow{F}$ does in moving a point along these two paths from K to M. Which of the symbols $<, \le, =, \ge, >$ could go in the box below?

$$W(K, M) \;\boxed{}\; W(K, L) + W(L, M).$$

If you do not have enough information to decide, say so.

c) Explain why your answer to b) is true. *Hint:* Use the fact $W = \overrightarrow{F} \cdot \overrightarrow{AB}$.

Newton's Law of Gravitation

The problems below refer to the gravitational field of the earth which is described in the subsection beginning on page 34.

34.6. h is the line shown below. There are four points on h. If A, B are any two

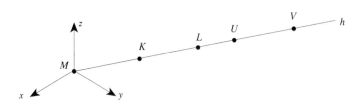

Figure 34.3

points on the line h, we write $W(A, B)$ for the work done by the gravitational force by moving a point from A to B. Note that $W(A, B) = -W(B, A)$.

   a) Is the work $W(L, K)$ positive, or zero, or negative, or do we need more information to decide?

   b) Draw a point N on h so that $W(K, N) < W(K, L)$. Remember that $-43 < -11$.

   c) The two points U and V on h (shown above) are as far apart as K and L. Is $W(V, U)$ greater than $W(L, K)$? Or less? Or are they the same? Or is it impossible to decide without using formulas?

34.7. S is the spherical shell given by $x^2 + y^2 + z^2 = 625$. The points of S are attracted to the origin by the gravitational force. C is a quarter of a great circle on the sphere S. It starts at $K(0, 25, 0)$ on the equator and ends at the north pole $N(0, 0, 25)$. A point P moves from K to N along the path C. Would you expect that the work done by the gravitational force is positive, or zero, or negative, or would we have to use formulas and computations to decide? *Hint:* Take a point Q of the curve C. What is the angle between the field vector at Q and the tangent to C at Q?

35 Work and Line Integrals

Goal of this section. We work out the answers to the *Key Questions for work* at the end of Section 34. The Key Questions read as follows: *We are given a non-constant vector field* $\vec{F}$ *and an oriented curve* C, *as shown in Figure 34.2 on page 287 which is reproduced below. How do we define and describe the work done by F in moving a point along C, and how do we compute the work?*

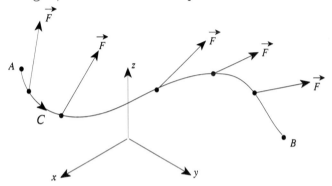

To work out the answers requires mathematics, and we switch from physics to mathematics. Thus, we consider a generic vector field $\vec{F}$ and an oriented curve C, as shown above. There is a mathematical quantity which corresponds to the physical quantity of work. It is called the *line integral* of $\vec{F}$ over C, as shown in the translation table below:

Language of physics:	**Language of mathematics:**
force field $\vec{F}$	generic vector field $\vec{F}$
work done by $\vec{F}$ along C	line integral of $\vec{F}$ over C

In short: A line integral for a generic vector field is what work is for a force field.

Problems

35.1. Let $\vec{H}$ be the constant vector field defined by $\vec{H} = (3, 8, -5)$. Find the line integral of $\vec{H}$ along the straight line segment from $A(1, 7, 2)$ to $B(3, 5, 4)$.

35.2. Go back to Problem 34.6 on the facing page on Newton's gravitational force field $\vec{F}$. Look at the points K and V of Figure 34.3. Is the line integral of $\vec{F}$ over the oriented line segment KV positive? Zero? Negative? Or is it impossible to say anything without making the exact computations?

The ideas behind the line integral

1. For a *constant* field $\vec{F}$ integrated over the *straight line segment* from A to B, the value of the line integral is simply the product of the component of $\vec{F}$ along $\overrightarrow{AB}$ with the distance from A to B:

$$\text{line integral} = \vec{F} \cdot \overrightarrow{AB}.$$

Notice that this is exactly the same formula as the one for the work done by a constant force moving a point along a straight line.

2. For a *non-constant* field over a *curved* path, the integral is built up from the simple case above by approximating the curve by a sequence of short line

segments, and approximating the field by a constant field along each of these segments.

As always we have two questions: *What is the line integral?* and *How do we compute the line integral?* We do the work for vector fields and curves in the plane. For vector fields and curves in space, the process is the same. Thus, until further notice, $\overrightarrow{F}$ is a vector field in the plane, and C is an oriented curve in the plane.

☞ How does one describe and define the line integral?

As we have done previously with double and triple integrals in Chapter 3, we *define* the line integral of a vector field over and oriented curve by specifying a recipe for *approximating* it. The idea is to imagine the field as a force, and to estimate the "work" done by the field in moving a particle along the curve. The approximation is carried out by approximating the curve by a sequence of line segments so that the field is nearly constant on each one. The "work" is then estimated by calculating the work done by a constant force in moving a particle along each line segment, and adding the results.

To evaluate a line integral, you must have

- an oriented curve C
- a vector field $\overrightarrow{F}$ defined on C.

The *line integral* of $\overrightarrow{F}$ over C, which is denoted by the symbol

$$\int_C \overrightarrow{F} \cdot d\vec{r}$$

is the *number* which is estimated by the following procedure.

1. Partition the curve C into small pieces, using division points $P_0, P_1, P_2, \ldots$. Here P_0 is the initial point of the oriented curve C, and the points are arranged in the order they are encountered as the curve is traversed. This is illustrated in Figure 35.1.

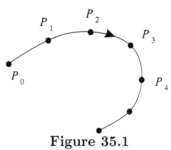

Figure 35.1

2. For each division point P_j, sample the vector field value at P_j, obtaining the vector $\overrightarrow{F}(P_j)$, as shown in Figure 35.2 on the next page.

3. Calculate the vector $\Delta\vec{r}_j$ going from P_j to P_{j+1}, as shown in Figure 35.2 on the facing page.

4. Calculate $\overrightarrow{F}(P_j) \cdot \Delta\vec{r}_j$ for each j, and add up the results. The sum approximates the line integral:

$$\int_C \overrightarrow{F} \cdot d\vec{r} \approx \sum_j \overrightarrow{F}(P_j) \cdot \Delta\vec{r}_j. \tag{35.1}$$

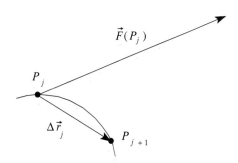

Figure 35.2

This means that the integral on the left can be approximated to any desired accuracy by choosing the division points P_j sufficiently close together.

The approximating sum on the right of (35.1) is called a *Riemann sum* for the line integral. Notice that a typical term $\vec{F}(P_j) \cdot \Delta \vec{r}_j$ in the sum in item 4 is just the work done by the *constant* force $\vec{F}(P_j)$ in moving a particle from P_j to P_{j+1} along a straight line.

Note:

1. **Terminology, notation.** In "line integral," the word "line" is meant to refer to all kinds of curves, and not only to straight lines. There are other notations for the line integral. They will be discussed later.

2. **Orientation of** C. For the line integral $\int_C \vec{F} \cdot d\vec{r}$, we must have an oriented curve C. If C is an oriented curve with initial point A and endpoint B, we write C^* for the same curve with the opposite orientation. The initial point of C^* is B, and its endpoint is A. Then we have

$$\int_C \vec{F} \cdot d\vec{r} = -\int_{C^*} \vec{F} \cdot d\vec{r}.$$

This fact is discussed in Problem 35.3.

Problem _____

35.3. Write down an explanation for the fact that reversing the orientation of the curve C changes the line integral into its negative. *Hint:* Look at a typical Riemann sum and see what happens when you reverse the orientation of the curve.

Basic rules for line integrals

There are three of them:

(i) Let $\vec{F}$ and $\vec{G}$ be two vector fields along the oriented curve C. Then

$$\int_C [\vec{F} + \vec{G}] \cdot d\vec{r} = \int_C \vec{F} \cdot d\vec{r} + \int_C \vec{G} \cdot d\vec{r}.$$

(ii) Let k be a constant. Then

$$\int_C k\vec{F} \cdot d\vec{r} = k \int_C \vec{F} \cdot d\vec{r}.$$

(iii) Let Q be a point of the oriented curve C. We write D for the portion of the curve between A and Q, and E for the part between Q and B, as shown in Figure 35.3. Then we have

$$\int_C \overrightarrow{F} \cdot d\vec{r} = \int_D \overrightarrow{F} \cdot d\vec{r} + \int_E \overrightarrow{F} \cdot d\vec{r}.$$

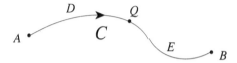

Figure 35.3

Problems

35.4. Your friend says: "Rules (i) and (iii) are the same. They say that you can add line integrals." What do you say?

35.5. Are there corresponding rules for ordinary integrals (the integrals of elementary calculus)? What are they? Write them down. Provide the drawing for (iii).

35.6. Are there corresponding rules for double integrals? What are they? Write them down. Provide the drawing for (iii).

Terminology: Closed curves and circulation

A curve C is called *closed* if its initial point and endpoint are the same, as shown in the following figure:

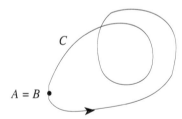

The line integral over such a closed curve C is usually written $\oint_C \overrightarrow{F} \cdot d\vec{r}$. In this notation, it is understood that one integrates once around C from A to A. The line integral of $\overrightarrow{F}$ around a closed curve is often called the *circulation* of $\overrightarrow{F}$ around (or about) C.

✎ How does one evaluate line integrals?

For a line integral we need (i) a vector field $\overrightarrow{F}$ and (ii) an oriented curve C:

(i) The field $\overrightarrow{F}$ is given by its component functions: $\overrightarrow{F} = (F_1(x,y), F_2(x,y))$.

(ii) The curve C is given in a parametric representation $(x,y) = (a(t), b(t))$. The initial point of C corresponds to the parameter value t_A, and the endpoint to t_B. We assume that the orientation of C is the one of increasing parameter values; if it is not, we go ahead with the computation and multiply the result of the computation by -1. We want the point $(a(t), b(t))$ always to move forward on C in the same direction. To make sure that the point never stops and starts backing up, we assume that $(a'(t), b'(t)) \neq (0,0)$ for all values of t between t_A and t_B.

We will use the following abbreviations:

$P(t)$: The point on C corresponding to the parameter value t;

$\vec{r}$ or $\vec{r}(t)$: the vector $\overrightarrow{OP(t)} = (a(t), b(t))$.

To evaluate line integrals we use Riemann sums. The typical term of a Riemann sum for line integrals is of the form

$$\vec{F}(P) \cdot \Delta \vec{r}.$$

In this expression, P stands for $P(t)$, and $\Delta \vec{r}$ is the vector from $P(t)$ to $P(t + \Delta t)$.

Problem _____

35.7. The figure below illustrates the typical term $\vec{F}(P) \cdot \Delta \vec{r}$ of a Riemann sum. Some items are labeled, others are not. Identify and label the following: $\vec{r}(t)$, $\vec{r}(t + \Delta t)$, $\vec{r}'(t) \Delta t$, $\Delta \vec{r}$.

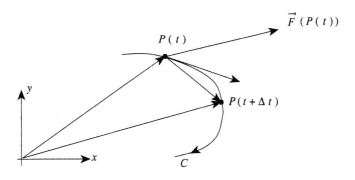

Figure 35.4

The Riemann sum which approximates the integral has the form

$$\sum \vec{F} \cdot \Delta \vec{r} = \sum \vec{F} \cdot \frac{\Delta \vec{r}}{\Delta t} \, \Delta t.$$

Here the $\sum$ means sum all of the terms $\vec{F} \cdot \Delta \vec{r}$. Now we take the increment Δt to be small, by taking a fine subdivision of the curve. Then $\Delta \vec{r}/\Delta t$ is a good approximation to $\vec{r}'(t) = d\vec{r}/dt$, so we get

$$\sum \vec{F} \cdot \Delta \vec{r} \approx \sum \vec{F}(P(t)) \cdot \vec{r}'(t) \, \Delta t,$$

and moreover, the approximation can be made as good as we wish by choosing fine enough subdivisions. This means that the two sides must home in on the *same* value as we take finer and finer partitions of the curve. *However*, notice that the right side of the above approximation formula is a Riemann sum for an integral of for an ordinary function of one variable (t), so taking the above formula to its limiting case gives

$$\int_C \overrightarrow{F} \cdot d\vec{r} = \int_{t_A}^{t_B} \overrightarrow{F} \cdot \vec{r}\,'(t)\, dt = \int_{t_A}^{t_B} \Big[F_1(a(t), b(t))\, a'(t) + F_2(a(t), b(t))\, b'(t) \Big] dt$$

Problem

35.8. C is the directed line segment AB shown below, and $\overrightarrow{F}$ is the vector field
defined by $\overrightarrow{F}(P) = x\,\overrightarrow{OP}$.

    a) Write down a parametric representation of the line defined by the points
       A and B.

    b) Find the components (F_1, F_2) of $\overrightarrow{F}$.

    c) Evaluate $\int_C \overrightarrow{F} \cdot d\vec{r}$.

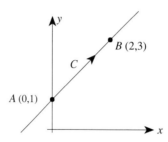

Summary: *The steps to evaluate a line integral are*

Step 1: *You find a parametric representation of the curve C and determine the
parameter values for the endpoints.*

Step 2: *You write down the components of the vector field; they are functions of
(x, y).*

Step 3: *You write down $\overrightarrow{F} \cdot \vec{r}\,'(t)$ as a function of t and integrate with respect to t
from $t = t_A$ to $t = t_B$.*

At times you will be asked to *set up a line integral*. By this we mean doing all the
steps until the line integral is in the form of an ordinary integral of a function of one
variable. That is, all the substitutions have been made, and the dot product has
been worked out. From this point on, anybody who has had single variable calculus
would be able to finish the evaluation.

Notation. The components of $\vec{r}$ are (x, y). We can write $d\vec{r} = (dx, dy)$, where
$dx = x'(t)\, dt$ and $dy = y'(t)\, dt$. This gives rise to another way of writing line
integrals:

$$\int_C \overrightarrow{F} \cdot d\vec{r} = \int_C F_1\, dx + F_2\, dy.$$

Problem

35.9. $\overrightarrow{F} = (xy, x)$

    a) Set up $\int_C F_1 dx + F_2 dy$ where C is the half-circle from $A(1,0)$ to $B(-1,0)$
       shown below.

    b) Set up $\int_K F_1 dx + F_2 dy$ where K is the line segment from $A(1,0)$ to
       $B(-1,0)$.

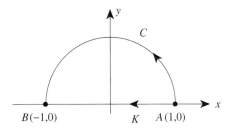

Line integrals for curves and vector fields in space

The formulas remain the same except for a third coordinate for points and a third component for vectors. The evaluation procedure summarized in (6) applies as written. In the notation of (5), there is a third term $F_3 dz$.

Problems

In Problems 35.10 and 35.11 we consider two curves C and K. Both start at the origin and end at $B(4, 2, 16)$. C is a straight line segment, and K is given by $(x, y, z) = (t^2, t, 2t^3)$.

35.10. $\overrightarrow{F}$ is the vector field in space given by $\overrightarrow{F} = (x, xy, z)$.

    a) Find $\int_C F_1 dx + F_2 dy + F_3 dz$.

    b) Find $\int_K F_1 dx + F_2 dy + F_3 dz$.

35.11. $\overrightarrow{G}$ is the vector field in space given by $\overrightarrow{G} = (x^2, z, y)$.

    a) Find $\int_C G_1 dx + G_2 dy + G_3 dz$.

    b) Find $\int_K G_1 dx + G_2 dy + G_3 dz$.

Different curves with the same initial points and endpoints

Problems 35.10 and 35.11 tell you this: Take two different curves which connect the same points A and B. The line integrals of the same vector field over these two curves may or may not be the same. We will revisit this issue in the next section.

The tangential component of a vector field along an oriented curve

In this subsection, we will give an alternate geometric interpretation of line integrals in terms of arc length. Recall (see Section 2) that the *element of arc length* along a parametrized curve is given by

$$ds = \sqrt{\left(\frac{dx}{dt}\right)^2 + \left(\frac{dy}{dt}\right)^2}\, dt.$$

This means that the arc length of a parametrized curve is obtained by integrating the above expression from the starting parameter value t_A to the ending parameter value t_B:

$$\text{arc length} = \int ds = \int_{t_A}^{t_B} \sqrt{\left(\frac{dx}{dt}\right)^2 + \left(\frac{dy}{dt}\right)^2}\, dt.$$

Problem _____

35.12.   a) Write down an integral that measures the arc length of the ellipse $x^2/a^2 + y^2/b^2 = 1$. Do not attempt to evaluate the integral. *Hint:* The ellipse can be parametrized by $x = a \cos t$ and $y = a \sin t$.

b) Using numerical integration with a calculator or computer, calculate the arc length of an ellipse with major and minor axes 4 and 2 units respectively. Your answer should be accurate to 4 decimal places.

_____

In order to interpret a line integral in terms of arc length, notice that the arc length element can be expressed in the form

$$ds = |(x'(t), y'(t))| \, dt = |\vec{r}'(t)| \, dt.$$

Now rewrite the evaluation formula for line integrals as follows:

$$\int_C \overrightarrow{F} \cdot d\vec{r} = \int_{t_A}^{t_B} \overrightarrow{F} \cdot \vec{r}'(t) \, dt = \int_{t_A}^{t_B} \overrightarrow{F} \cdot \frac{\vec{r}'(t)}{|\vec{r}'(t)|} \, |\vec{r}'(t)| \, dt = \int_C \overrightarrow{F} \cdot \overrightarrow{T} \, ds.$$

In this formula, $\overrightarrow{T}(P)$ is the unit vector tangent to C at the point P, pointing in the direction given by the orientation of C. The integrand $\overrightarrow{F} \cdot \overrightarrow{T}$ in the last integral is simply the scalar component of the field vector $\overrightarrow{F}$ along the direction $\overrightarrow{T}$ tangent to the curve. This scalar quantity is called the *tangential component* of the field $\overrightarrow{F}$ along the oriented curve C, and is often written as $F_{\tan}$:

$$F_{\tan} = \overrightarrow{F} \cdot \overrightarrow{T} = \text{tangential component of } \overrightarrow{F}.$$

Notice that $F_{\tan}$ is a scalar quantity that is only defined along the curve C. We can now write

$$\int_C \overrightarrow{F} \cdot d\vec{r} = \int_C F_{\tan} \, ds.$$

This means, in plain English, that the line integral of the field $\overrightarrow{F}$ along the curve C is the integral of the tangential component of $\overrightarrow{F}$ with respect to arc length along C.

Note: Even though a parametrization of C played a role in *deriving* the above formula, the final result is expressed in purely geometric language. This means that, as expected, you get the same result for the line integral, regardless of which parametrization you use, provided that the parametrization is consistent with the orientation of the curve.

Problems _____

35.13. Your friend says: "Something must be wrong with the formula for representing line integrals in terms of arc length and tangential components. I know that changing the orientation of the curve C will change the sign of the line integral $\int_C \overrightarrow{F} \cdot d\vec{r}$, but changing the orientation can't change the sign of arc length, since arc length is always positive!" What do you say?

35.14. Go back to the field $\overrightarrow{F}$ of Example 1 of Section 30. The field describes the velocity of a rigid counterclockwise rotation of the plane about the origin with angular velocity ω.

a) Let A be *any* point in the plane, and let C_1 be the line segment going from the origin to the point A. Calculate the tangential component $F_{\tan}$ of $\overrightarrow{F}$ at any point along C_1. *Hint:* No calculations are needed. Draw a picture.

b) Use your answer to part a to evaluate $\int_{C_1} \overrightarrow{F} \cdot d\vec{r}$.

c) Let C_2 be the counterclockwise circle of radius a, centered at the origin. Find the tangential component of $\overrightarrow{F}$ at each point of C_2. Your answer will have an a (radius) and an ω (angular velocity) in it. *Hint:* Draw a picture.

d) Use your answer to part c to evaluate $\int_{C_2} \overrightarrow{F} \cdot d\vec{r}$.

Estimating a line integral by means of the tangential component. For a given a vector field $\overrightarrow{F}$ and an oriented curve C, it *may* turn out that $F_{\tan}$, the tangential component of $\overrightarrow{F}$ along C, is *constant*. In that case, write F_0 for the constant value of $F_{\tan}$. We can pull out a constant from underneath the integral:

$$\int_C F_{\tan}\, ds = \int_C F_0 \, ds = F_0 \int_C ds.$$

Now the integral $\int_C ds$ is just the arc length of C, which we will abbreviate as $L(C)$.

Result: *IF we know that F_{tan} has a constant value F_0 along C, we can write down the value of the line integral:*

$$\int_C \overrightarrow{F} \cdot d\vec{r} = F_0\, L(C).$$

Similarly, we may be able to see that $F_{\tan}$ is less than or equal to a constant T_0 all along C, that is, $F_{\tan} \le T_0$. This means that the same relation holds for the integrals:

$$\int_C F_{\tan}\, ds \le \int_C T_0 \, ds = T_0 \int_C ds = T_0\, L(C).$$

We omit the two integrals in the middle and have an estimate of the value of the line integral:

$$\int_C \overrightarrow{F} \cdot d\vec{r} \;\le\; T_0\, L(C).$$

The same reasoning shows that if $F_{\tan}$ is *greater than or equal* to a constant t_0 along C, then

$$t_0 L(C) \le \int_C \overrightarrow{F} \cdot d\vec{r}.$$

What is the use of such an estimate? There are situations where the evaluation would be extremely cumbersome or impossible, and where an estimate will do. The estimate may be rough, but a rough estimate is still better than no estimate. For example, we may be able to find the absolute maximum and/or minimum of $F_{\tan}$ along C. These would work as the constants T_0 and t_0.

Problems _____

35.15. The vector field $\overrightarrow{F}$ in space is given by $\overrightarrow{F} = (-y, x, 0)$. In a) and b) we consider the line integrals of the same vector field over two different curves.

Hint for both parts a and b: Take a point P and visualize what the field vector at P looks like. If you see what its direction is, the computations will be short and easy.

a) The curve C is the line segment AB with initial point $A(0, 0, 18)$ and endpoint $B(4, 6, 18)$.

   (i) M is the point on the line segment AB which is 3.9373049287 units away from $A(0, 0, 18)$. Find the angle between $\overrightarrow{AB}$ and the field vector of $\overrightarrow{F}$ at M.

   (ii) Find $F_{\tan}$ along AB.

   (iii) Find $\int_C \overrightarrow{F} \cdot d\vec{r}$.

b) K is the circle given by $(x, y, z) = (12 \cos t, 12 \sin t, 3)$, with $0 \leq t \leq 2\pi$.

   (i) Q is the point on K corresponding to $t = 1.5982236001$. Find the angle between the field vector at Q and the tangent vector of K at Q.

   (ii) Find $F_{\tan}$ along K.

   (iii) Find $\oint_K \overrightarrow{F} \cdot d\vec{r}$.

35.16. $\overrightarrow{F} = -\frac{k}{r^3}(x, y, z)$ is the gravitational field of Problem 34.3. Remember that $r = \sqrt{x^2 + y^2 + z^2}$ and that k is a positive constant. For our curve C we take the line segment from $A(1, 4, 2)$ to $B(3, 12, 6)$. Note that the line defined by A and B passes through the origin. It may help to look again at Figure 34.3 on page 288, or to draw a similar figure.

a) A point P moves from A to B. The tangential component of $\overrightarrow{F}$ at P will be (read all statements before choosing one): *All the time positive — all the time zero — all the time negative — partly positive, partly negative — impossible to predict without making the computation.*

b) The tangential component of $\overrightarrow{F}$ along C has a maximum on C. Call this maximum $\max F_{\tan}$. Find it. *Hint:* Remember that the line defined by A and B passes through the origin. What can you say about the direction of the field vector at the points of C? If you think before starting to compute, the computations will be short and easy.

c) In b) you found the maximum $\max F_{\tan}$ of $F_{\tan}$ along C. It means that $F_{\tan} \leq \max F_{\tan}$. What can you conclude about $\int_C \overrightarrow{F} \cdot d\vec{r}$?

d) Find the minimum $\min F_{\tan}$ of $F_{\tan}$ along C.

e) $\min F_{\tan}$ is the minimum of $F_{\tan}$ along C. Therefore, $F_{\tan} \geq \min F_{\tan}$. What does this mean for $\int_C \overrightarrow{F} \cdot d\vec{r}$? *Hint:* See c), but work with "greater than or equal" instead of "less than or equal."

f) Find constants e and E such that $e \leq \int_C \overrightarrow{F} \cdot d\vec{r} \leq E$. *Hint:* Combine your answers to c) and e). The constants e and E will contain the positive constant k.

Work and kinetic energy

If you push hard enough on an object, it will move. *How* it moves is related to how hard you push and how much the object resists. The precise relation between force and motion is given by Newton's Second Law.

If $\vec{F}$ is the total force acting on a particle of mass m, then the acceleration $\vec{a}$ of the particle is related to the force by

$$\vec{F} = m\vec{a}.$$

We will use Newtons Second Law to "calculate" the work done by a force in moving a particle along a path. We assume that $\vec{F}$ represents the total force acting on the particle, and, as usual, we let $\vec{r}(t)$ denote the position of the particle at time t. We will calculate the work done by the force $\vec{F}$ from an initial time t_0 to a later time t_1. Letting C denote the path of the particle,

$$\text{work} = \int_C \vec{F} \cdot d\vec{r}.$$

Using Newton's Second Law, we can replace $\vec{F}$ by $m\vec{a} = m\vec{r}''$ to get

$$\text{work} = \int_C m\vec{a} \cdot d\vec{r} = m \int_{t_0}^{t_1} \vec{r}''(t) \cdot \vec{r}'(t)\, dt.$$

Now recall that $\vec{r}'$ is the particle velocity $\vec{v}$, so the integrand can be expressed as $\vec{v}' \cdot \vec{v}$. This in turn, by the Product Rule, is the derivative of $\frac{1}{2}\vec{v} \cdot \vec{v} = \frac{1}{2}|\vec{v}|^2$. Letting $v = |\vec{v}|$ denote the speed of the particle, we now have

$$\text{work} = \frac{1}{2}\, m \int_{t_0}^{t_1} \frac{d}{dt}(v(t)^2)\, dt$$

and the Fundamental Theorem of Calculus gives

$$\text{work} = \frac{1}{2}\, m\, v(t_1)^2 - \frac{1}{2}\, m\, v(t_0)^2.$$

In other words, *the work done equals the increase in the quantity* $\frac{1}{2}mv^2$. This quantity is the *kinetic energy* of the particle. To summarize what we have discovered:

> *The work done by the total force acting on a particle equals the increase in kinetic energy of the particle.*

Problem

35.17. The force of gravity on an object of mass 1 kg is directed downward with magnitude 9.8 nt.

a) An object of mass 1 kg is dropped from rest at a height of 1 meter. Calculate

   (i) The work done by the gravitational field in moving the object from its initial position to ground level.

   (ii) Calculate the speed the object attains by the time it reaches the ground.

b) An object of mass 1 kg slides without friction down an inclined plane, tilted at an angle of 30 degrees from horizontal. It starts from rest at a height of 1 meter above ground level. How fast will it be traveling by the time it reaches the ground? *Hint:* The force acting on the object is a sum of the gravitational force $\vec{G}$ and the force $\vec{N}$ exerted by the inclined plane. Since $\vec{N}$ is normal to the direction of motion, it does no work.

c) Repeat part b) with the plane replaced by a slide in the shape of a parabolic arc. Assume that the axis of the parabola is vertical, that the vertex is at ground level, and that the object starts out 1 meter above ground level. *Hint:* In an appropriately chosen xy-system, the parabola has the equation $y = cx^2$ (c a constant). Re-read the hint to part b), and then calculate the work done *by gravity alone*.

Additional Problems

Tangential component

35.18. We consider the function $f(x, y, z) = 2x + y + 4z - 3$ and the curve C given by $(x, y, z) = (t + 1, t^2 + 2, t^5)$. As orientation of C we take the one of increasing values of t.

a) Q is the point of C which corresponds to $t = 1$. Find the field vector of $\vec{F}$ at Q.

b) Find the tangential component of $\vec{F}$ along C at Q.

c) We write P for the *generic* point of C, that is, the point with coordinates $(t + 1, t^2 + 2, t^5)$. Find the tangential component of $\vec{F}$ at P. Your answer will be a function $F_{tan}(t)$ of t.

35.19. The xy-plane rotates around the origin counterclockwise with constant angular velocity ω. We define a vector field $\vec{F}$ as follows: $\vec{F}$ assigns to a point $P(x, y)$ its velocity vector $\vec{v} = (-\omega y, \omega x)$, as shown in the following figure:

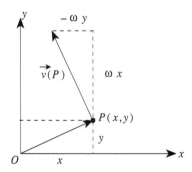

C is the flow line of $\vec{F}$ which at time $t = 0$ passes through $R(5, 0)$. C is given by $(x, y) = (5 \cos \omega t, 5 \sin \omega t)$.

a) Find the tangential component of $\vec{F}$ along C at $R(5, 0)$.

b) The generic point of C is the point $P(5\cos\omega t, 5\sin\omega t)$. Find the tangential component of $\overrightarrow{F}$ at P.

35.20. We consider the gravitational field $\overrightarrow{F}$ in space: $\overrightarrow{F}$ assigns to the point P a vector which has the direction of $-\overrightarrow{OP}$ and whose magnitude equals $k|\overrightarrow{OP}|^{-2}$. In this formula, k is a positive constant. We also consider two oriented curves C and D. C is the x-axis oriented positively, and D is parallel to the z-axis through the point $(a, b, 0)$, but oriented downwards. The curves C and D are shown below.

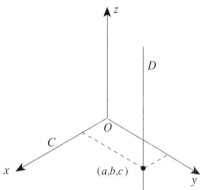

In the following questions, "there is none" may be the correct answer.

a) Find points R_1, R_2, R_3 on C (the x-axis) such that at R_1 the tangential component of $\overrightarrow{F}$ is positive, at R_2 it is zero, and at R_3 it is negative.

b) Find points S_1, S_2, S_3 on D such that at S_1 the tangential component of $\overrightarrow{F}$ is positive, at S_2 it is zero, and at S_3 it is negative.

Estimating a line integral

35.21. We consider a vector field $\overrightarrow{F}$ in the plane. It is defined as follows: The field vector at P is in the same direction as $\overrightarrow{OP}$, and its magnitude is always equal to 4. We also consider an oriented half-circle C of radius 3, with center at $(2, 3)$, as shown in Figure 35.5 on the following page.

a) Use ruler and pencil to find the maximum of the tangential component of $\overrightarrow{F}$ along C. We call this maximum $\max f_{\tan}$. Then find the minimum and call it $\min f_{\tan}$.

b) Go back to the the subsection "Estimating a line integral by means of the tangential component" on page 297. Apply the reasoning done there to our vector field $\overrightarrow{F}$, and use $\max f_{\tan}$, $\min f_{\tan}$ found in a) to estimate the line integral of $\overrightarrow{F}$ over C, that is to find numbers e and E such that

$$e \leq \int_C \overrightarrow{F} \cdot d\vec{r} \leq E.$$

Evaluation

35.22. C is the part of the circle $x^2 + y^2 = 1$ in the first quadrant of the xy-plane (i.e., $x \geq 0, y \geq 0$). We orient C counterclockwise. $\overrightarrow{F}$ is the vector field in the plane defined by $\overrightarrow{F}(x, y) = (x + y, x - y)$.

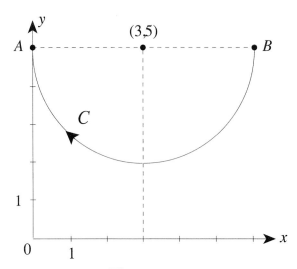

Figure 35.5

a) Set up $\int_C \overrightarrow{F} \cdot d\vec{r}$.

b) Evaluate $\int_C \overrightarrow{F} \cdot d\vec{r}$. *Hint:* $\int \sin^2 x \, dx = (\frac{1}{2})x - (\frac{1}{4})\sin 2x$; $\int \cos^2 x \, dx = (\frac{1}{2})x + (\frac{1}{4})\sin 2x$; $\int \sin x \cos x \, dx = (\frac{1}{2})\sin^2 x$.

35.23. The vector field $\overrightarrow{F}$ in space given by $\overrightarrow{F} = (y, x, z)$. Figure 35.6 shows two paths from the origin to $Q(1, 2, 3)$.

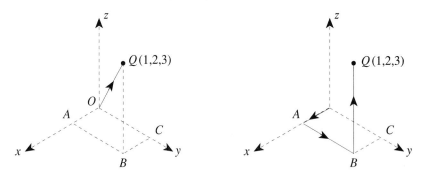

Figure 35.6

a) Find $\int_K \overrightarrow{F} \cdot d\vec{r}$ where K is the path shown at left in Figure 35.6. K is the line segment OQ.

b) Find $\int_K \overrightarrow{F} \cdot d\vec{r}$ where K is the path shown at right in Figure 35.6. K consists of three segments along the axes.

35.24. The vector field $\overrightarrow{F}$ in space given by $\overrightarrow{F} = (y, 2x, z)$ (watch out – it is different from the vector field of Problem 35.23). Figure 35.7 shows two paths from the origin to $Q(1, 2, 3)$.

a) Find $\int_K \overrightarrow{F} \cdot d\vec{r}$ where K is the path shown at left in Figure 35.7. K consists of the two line segments OB and BQ.

b) Find $\int_K \overrightarrow{F} \cdot d\vec{r}$ where K is the path shown at right in Figure 35.7. The

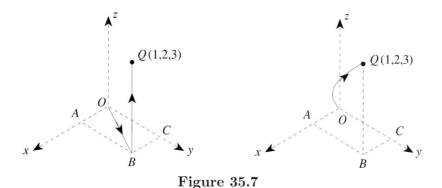

Figure 35.7

curve K is given by $(x, y, z) = (t, 2t^3, 3t^2)$.

35.25. We are working in the xy-plane.

Our curve C is the interval $a \le x \le b$ on the x-axis. We orient C in the positive x-direction. Our vector field $\vec{F}$ is given by $\vec{F}(x, y) = (f(x), 0)$. In this formula, $f(x)$ is a a function of one variable. We do not know the formula for $f(x)$, and we will not need it.

a) Write down a parametric representation of C such that t, the parameter, equals x. What are the parameter values which correspond to the initial point and endpoint of C?

b) Set up the integral $\int_C \vec{F} \cdot d\vec{r}$.

c) Look at your answer to b). Ask yourself: Is the line integral a generalization or special case of the ordinary integral of elementary calculus?

(i) Write down a complete sentence in which the terms "line integral" and "ordinary integral" are linked by the term "special case."

(ii) Write down a complete sentence in which the terms "line integral" and "ordinary integral" are linked by the term "generalization."

35.26. We consider a function $f(x, y)$ and the corresponding gradient field $\vec{F} = \text{grad} f(x, y)$. Two curves are shown below. The curve KL is part of a level curve of the function $f(x, y)$, and MN is part of a flow line of the vector field $\text{grad} f(x, y)$. *Remember:* The flow lines of the vector field $\text{grad} f$ are normal (i.e., perpendicular) to the level curves of the function $f(x, y)$. Therefore, the two curves intersect at a right angle at Q.

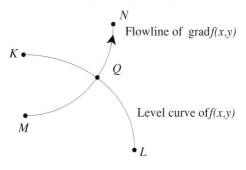

For the following questions on line integrals, work with the tangential component.

a) The line integral of $\operatorname{grad} f(x, y)$ over the level curve from K to L is (choose one):

Positive   *zero*   *negative*   *cannot say without knowing more*

b) The line integral of $\operatorname{grad} f(x, y)$ over the flow line from M to N is (choose one):

Positive   *zero*   *negative*   *cannot say without knowing more*

36   Conservative Vector Fields

This section and Section 37 are about gradient fields.

Line integrals of a constant vector field

We consider a constant vector field $\vec{F}$ and a curve C in the plane:

- The vector field $\vec{F}$ is given by $\vec{F} = (m, n)$, where m and n are constants;
- the curve $C : (x, y) = (a(t), b(t))$ has initial point $A(p, q)$ corresponding to $t = t_A$ and endpoint $B(u, v)$ with $t = t_B$, as shown below.

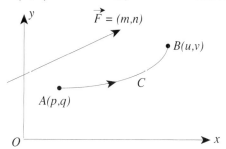

Problem

36.1. a) Set up $\int_C \vec{F} \cdot d\vec{r}$. Your answer will contain constants $m, n, \ldots$ and the functions $a(t), b(t)$ whose formulas are not given.

   b) Evaluate the integral as far as you can. *Hints:* Keep in mind that $(a(t_A), b(t_A)) = (p, q)$, and similarly $(a(t_B), b(t_B)) = (u, v)$. – Remember from elementary calculus: $\int_a^b f'(x)dx = f(b) - f(a)$.

Discussion of Problem 36.1

1. The expression of the line integral contains only the components (m, n) of the vector field and the coordinates (p, q), (u, v) of the initial point and endpoint of C. *The functions $a(t), b(t)$ which describe C between A and B do not figure in the expression.* It means: No matter which curve from A to B we take, the line integral is the same. We say: The line integrals of a constant vector field are *path-independent.* For example, the line integrals of a constant vector field over the curves C_1, C_2, C_3 below have the same value.

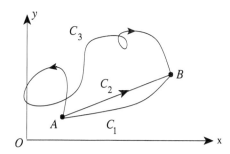

2. In Problem 36.1 we considered a constant vector field in the plane. Everything works the same way for a constant vector field in space:

The line integrals of a constant vector field in space are path-independent.

From now on we consider vector fields in space except when otherwise noted.

The key question for line integrals

We saw that a *constant* vector field has path-independent line integrals. Are there vector fields which are not constant, and which have path-independent line integrals? This is the key question for line integrals:

Which vector fields have path-independent line integrals?

Everything we do between here and the end of the present Section 36 deals with the key question for line integrals.

Line integrals of gradient fields

We begin by looking at line integrals of a gradient field. Thus, we consider:

- A gradient field $\vec{F} = \operatorname{grad} \Phi(x, y, z)$. Here $\Phi(x, y, z)$ (pronounced "fi") is a function of three variables. Thus, the components of our vector field are (Φ_x, Φ_y, Φ_z), and each component is again a function of three variables.
- A curve C given by $(x, y, z) = (a(t), b(t), c(t))$, with initial point A for $t = t_A$ and endpoint B for $t = t_B$.

We want to evaluate $\int_C \operatorname{grad} \Phi(x, y, z) \cdot d\vec{r}$ as far as this is possible for a generic function Φ and a generic curve C. We have

$$\int_C \operatorname{grad} \Phi \cdot d\vec{r} = \int_{t_A}^{t_B} \operatorname{grad} \Phi(a(t), b(t), c(t)) \cdot (a'(t), b'(t), c'(t)) dt.$$

On the right side we have to integrate the function

$$\operatorname{grad} \Phi(a(t), b(t), c(t)) \cdot (a'(t), b'(t), c'(t)).$$

By the Chain Rule (see Section 17) this function is the same as

$$\frac{d\Phi(a(t), b(t), c(t))}{dt}.$$

Thus, the function to integrate is a derivative! Therefore, we apply the Fundamental Theorem of Calculus and obtain

$$
\begin{aligned}
\int_C \operatorname{grad} \Phi \cdot d\vec{r} &= \int_{t_A}^{t_B} \frac{d\Phi(a(t), b(t), c(t))}{dt} dt \\
&= \Phi(a(t_B), b(t_B), c(t_B)) - \Phi(a(t_A), b(t_A), c(t_A)) \\
&= \Phi(B) - \Phi(A).
\end{aligned}
$$

The difference $\Phi(B) - \Phi(A)$ depends only on the value of Φ at the initial point and endpoint of the curve C, and not on the points in between. The line integral is path-independent. This gives a version of the Fundamental Theorem of Calculus for line integrals.

> **First Fundamental Theorem for Line Integrals.** *A   gradient   field*
> $\vec{F} = \text{grad}\,\Phi(x,y,z)$ *has path-independent line integrals. The line integral*
> *over a curve C is given by*
>
> $$\int_C \text{grad}\,\Phi \cdot d\vec{r} = \Phi(B) - \Phi(A).$$
>
> *In this expression, A is the initial point of C, and B is the endpoint of C.*

Problems

36.2. Go back to the constant vector field $\vec{F} = (m,n)$ of Problem 36.1.

    a) Is $\vec{F}$ a gradient field? That is, can you find a function $\Phi(x,y)$ of two variables such that the partial derivative with respect to x equals m and the partial derivative with respect to y equals n? Remember that m and n are constants.

    b) Take the $\Phi(x,y)$ you found in a) and use it to evaluate the integrals by means of the First Fundamental Theorem. Compare with your answer to Problem 36.1.

36.3. The vector field $\vec{F}$ in the plane is given by $\vec{F} = \text{grad}(x^2 y + y^3 + 5)$, and C is the curve $(x,y) = (t^3, t^4)$ from the origin to the point $(1,1)$. Evaluate $\int_C \vec{F} \cdot d\vec{r}$.

36.4. $\vec{F}$ is the constant vector field in space given by $\vec{F} = (3, 9, -1)$.

    a) Is $\vec{F}$ a gradient field? If yes, find a function $\Phi(x,y,z)$ such that $\text{grad}\,\Phi = \vec{F}$.

    b) C is the curve given by $(x,y,z) = (t^3, t^4, t)$ from the origin to the point $(8, 16, 2)$. Evaluate $\int_C \vec{F} \cdot d\vec{r}$.

36.5. C is the line segment OS and K is the curve given by $(x,y) = (t\sqrt{a}, t^2)$, $0 \le t \le \sqrt{a}$, as shown below.

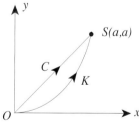

$\vec{F} = (F_1(x,y), F_2(x,y))$ is a vector field. We do not know the formulae for F_1 and F_2, but we do know that $\int_C \vec{F} \cdot d\vec{r} \ne \int_K \vec{F} \cdot d\vec{r}$. What can we say about the functions F_1 and F_2? Read all statements and choose one: *F_1 and F_2 are both constant—F_1 and F_2 are not both constant—we cannot make any prediction of this sort.* Explain your answer.

Terminology. Gradient fields have been used in different areas of the physical sciences for a long time. Therefore a special terminology has developed. A gradient field is a vector field $\vec{F}$ of the form $\vec{F} = \text{grad}\,\Phi(x,y,z)$. The function $\Phi(x,y,z)$ is called a *potential function* or just a *potential* of the vector field.

A common name for a gradient field is *conservative* vector field. The word "conservative" is used because gradient fields come up in discussing conservation of potential and kinetic energy (see page 311).

Note:

1. A vector field $\overrightarrow{F}$ may already be *given* as a gradient $\operatorname{grad}\Phi(x, y, z)$. Then $\overrightarrow{F}$ is conservative, and $\Phi(x, y, z)$ is a potential of $\overrightarrow{F}$.

2. A line integral of a conservative field is path-independent.

3. The two functions $\Phi(x, y, z)$ and $\Psi(x, y, z) = \Phi(x, y, z) + \text{constant}$ have the same gradient (Ψ: "capital psi"). It means: If $\Phi(x, y, z)$ is a potential of the conservative field $\overrightarrow{F}$, then so is $\Psi(x, y, z)$ On the other hand, if Φ and Ξ are two potentials of the same field $\overrightarrow{F}$, then $\operatorname{grad}\Phi = \operatorname{grad}\Xi$ (Ξ: "capital xi"). This is only possible if $\Phi = \Xi + k$ where k is a constant. Thus: A conservative field has many different potentials. Any two of them differ only by a constant.

Vector fields with path-independent line integrals

We have seen that a conservative vector field has path-independent line integrals. Now we ask the question the other way around:

> $\overrightarrow{F}$ *is a vector field with path-independent line integrals. Is it conservative?*

We consider a vector field $\overrightarrow{F}$ which has path-independent line integrals. The question just asked means:

> $\overrightarrow{F}$ *is a vector field with path-independent line integrals. Can we find a potential for $\overrightarrow{F}$?*

Now we work out the answer. Thus, we are given a vector field $\overrightarrow{F} = (F_1, F_2, F_3)$. We know that it has path-independent line integrals. We are going to construct a potential for $\overrightarrow{F}$. That is, the answer to our last two questions is "yes." There are two steps.

Step 1: Definition of a function $g(x, y, z)$

We pick a point P_0 and keep it fixed all the way through. Now take any curve C which connects the fixed point P_0 with the generic point $P(x, y, z)$, as shown below.

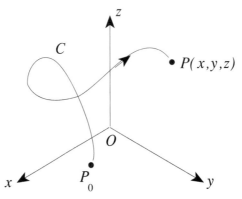

Since our field has path-independent line integrals, the value of $\int_C \overrightarrow{F} \cdot d\vec{r}$ does not depend on the the particular choice of a path C from P_0 to P.

We let $\int_{P_0}^{P} \overrightarrow{F} \cdot d\vec{r}$ stand for the common value of all line integrals over paths from P_0 to P. In other words, to evaluate $\int_{P_0}^{P} \overrightarrow{F} \cdot d\vec{r}$, you choose any path C going from P_0 to P and calculate:

$$\int_{P_0}^{P} \overrightarrow{F} \cdot d\vec{r} = \int_{C} \overrightarrow{F} \cdot d\vec{r}.$$

We now *define* a function $g(x, y, z)$ by

$$g(x, y, z) = \int_{P_0}^{P} \overrightarrow{F} \cdot d\vec{r}.$$

This definition makes sense only because our vector field $\overrightarrow{F}$ has path-independent line integrals. If it did not have path-independent line integrals, we could not define the function $g(x, y, z)$ the way we did. The function $g(x, y, z)$ is our candidate for a potential. We still have to check whether it really is a potential of $\overrightarrow{F}$. This is done in step 2.

Step 2: Computation of $\operatorname{grad} g$

We want to find the partial derivative g_x at the generic point $P(x, y, z)$. For this, we consider the point $P'(x + h, y, z)$ and the line segment K from P to P', as shown on the left of Figure 36.1 below. By definition, the

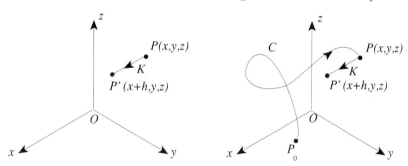

Figure 36.1

partial derivative g_x is

$$g_x = \lim_{h \to 0} \frac{g(x + h, y, z) - g(x, y, z)}{h}.$$

Now we transform the right side. It takes some doing.

We begin by remembering the definition of the function $g(x, y, z)$ as a line integral:

$$g(x, y, z) = \int_{C} \overrightarrow{F} \cdot d\vec{r}$$

where C is a curve going from P_0 to P. We can get $g(P')$ by integrating along *any* curve going from P_0 to P'. We will use the one obtained by combining P and K as shown in Figure 36.1. At the same time we break up the integral over $C + K$ into integrals over C and K:

$$g(x + h, y, z) = \int_{C+K} \overrightarrow{F} \cdot d\vec{r} = \int_{C} \overrightarrow{F} \cdot d\vec{r} + \int_{K} \overrightarrow{F} \cdot d\vec{r}.$$

The integral over C is $g(x, y, z)$, so we obtain

$$g(x + h, y, z) = g(x, y, z) + \int_K \overrightarrow{F} \cdot d\vec{r},$$

and rearranging gives and dividing by h gives

$$\frac{g(x + h, y, z) - g(x, y, z)}{h} = \frac{1}{h} \int_K \overrightarrow{F} \cdot d\vec{r}. \tag{36.1}$$

We can now interpret the integral on the right side as the integral of the tangential component of $\overrightarrow{F}$ along K, taken with respect to arc length along K. We make two observations:

1. Since K is parallel to the x-axis, the tangential component of $\overrightarrow{F}$ along K is F_1. In other words,

$$\int_K \overrightarrow{F} \cdot d\vec{r} = \int_K F_1 \, ds. \tag{36.2}$$

2. If h is small, then the value of F_1 along K never gets far from (x, y, z) (the initial point of K). In other words, along K,

$$F_1 \approx F_1(x, y, z). \tag{36.3}$$

By this we mean that we can ensure that the values of F_1 along K are as close as we wish to $F_1(x, y, z)$ by making sure that h is small enough.

Combining (36.1), (36.2), and (36.3) now gives

$$\frac{g(x + h, y, z) - g(x, y, z)}{h} \approx \frac{1}{h} \int_K F_1(x, y, z) \, ds$$

$$= F_1(x, y, z) \frac{\text{length of } K}{h} = F_1(x, y, z).$$

Since the approximation becomes exact as $h \to 0$, we get

$$\frac{\partial g}{\partial x}(x, y, z) = F_1(x, y, z).$$

Thus, at the limit, $g(x, y, z)$ (the function we constructed) equals the first component $F_1(x, y, z)$ of $\overrightarrow{F}$ (the given vector field with path-independent line integrals). For the second and third component we reason the same way, replacing the line segment K by line segments parallel to the y and z axis respectively. We have shown that the gradient of the function $g(x, y, z)$ equals the given vector field $\overrightarrow{F}$. That is, the function $g(x, y, z)$ is a potential of $\overrightarrow{F}$.

We combine Steps 1 and 2 and obtain the

Second Fundamental Theorem for Line Integrals. *A vector field* $\overrightarrow{F}$ *with path-independent line integrals is conservative. Moreover, a potential for* $\overrightarrow{F}$ *is given by*

$$\Phi(x, y, z) = \int_{P_0}^{P(x,y,z)} \overrightarrow{F} \cdot d\vec{r}.$$

Note that all the reasoning for the Second Fundamental Theorem worked only because we assumed that the given field $\overrightarrow{F}$ had path-independent line integrals.

The answer to the key question on line integrals. The key question is *Which vector fields have path-independent line integrals?* The answer follows from the two Fundamental Theorems above.

> *A conservative vector field has path-independent line integrals.*
> *Conversely, a vector field with path-independent line integrals is conservative.*

In other words: The properties "conservative" and "having path-independent line integrals" are equivalent in the following sense: If a field has one of the two properties, it also has the other.

Circulation. The property "path-independence" can be formulated in terms of integrals around closed curves:

- If $\overrightarrow{F}$ has path-independent line integrals then $\oint_C \overrightarrow{F} \cdot d\vec{r} = 0$ for *every* closed curve C.
- If $\oint_C \overrightarrow{F} \cdot d\vec{r} = 0$ for *all* closed curves C then $\overrightarrow{F}$ has path-independent line integrals.

These facts are worked out in Problem 36.9. Since we now know that the conservative fields are exactly those that have path-independent line integrals, we can conclude

> *A conservative vector field has zero circulation around every closed path.*
> *Conversely, a field with zero circulation around every close path is conservative.*

Problem ⎯⎯⎯⎯⎯⎯⎯⎯⎯⎯⎯⎯⎯⎯⎯⎯⎯⎯⎯⎯⎯⎯⎯⎯⎯⎯⎯⎯⎯

36.6. Go back to the field $\overrightarrow{F}$ of Example 1 of Section 30, which describes the velocity of a rigid counterclockwise rotation of the plane about the origin with angular velocity ω. Is this field conservative? Explain your answer. *Hint:* Try to find a closed curve around which the circulation is not 0. You may want to revisit Problem 35.14 of Section 30

Conservation of energy

Consider a force field $\overrightarrow{F}$ that acts on a particle, moving it along a path C from an initial point P_0 to a terminal point P_1. In Section 35 (page 299), we saw that the work done by $\overrightarrow{F}$ equals the increase in kinetic energy for the particle:

$$\text{Work} = \text{increase in kinetic energy}.$$

We now add the additional assumption that the field $\overrightarrow{F}$ is conservative: $\overrightarrow{F} = \operatorname{grad} \Phi$ for some potential Φ. By the First Fundamental Theorem for Line Integrals, we get another representation of the work:

$$\text{Work} = \Phi(P_1) - \Phi(P_0).$$

The *potential energy* of the moving particle when it is at the point P is defined by

$$\text{Potential energy at } P = -\Phi(P)$$

so we get

Work = decrease in potential energy.

Thus the *increase* in kinetic energy and the *decrease* in potential energy are *both* equal to the work, so we get

Increase in kinetic energy = decrease in potential energy.

In other words, the *total* energy (potential plus kinetic) remains constant. This fact is sometimes called the *Law of Conservation of Energy.* The reason that "energy" is often conserved in physical systems is that many of the forces of nature are represented by conservative fields.

Problems _____

36.7. Near the surface of the earth, its gravitational field may be treated as a constant field, with magnitude 9.8 nt/kg.

    a) Choose a coordinate system, and find a formula for the gravitational potential.

    b) You toss an object straight up with initial velocity of 10 m/sec. Use part a to predict how high it will go. *Hint:* You could, of course, solve this by elementary calculus. Instead, use conservation of energy, and the fact that the kinetic energy is zero when it reaches its highest point.

    c) Toss the same object with an initial velocity of 10 m/sec, but this time at an angle of 45 degrees with the ground. How fast will it be traveling when it returns to ground level? Justify your answer.

36.8. A planet of mass m orbits about a star of mass M, following an elliptical path. If $\vec{r}$ is the position vector of the planet relative to the star, then the gravitational force acting on the planet is

$$\vec{F} = \frac{-GMm\vec{r}}{|\vec{r}|^3}$$

where G is the universal gravitation constant. The associated potential is

$$\Phi = \frac{GmM}{|\vec{r}|}.$$

    a) As the planet moves closer to the star, does its potential energy increase or decrease? Explain.

    b) As the planet moves closer to the star, does its speed increase, or decrease? Explain.

Summary of Section 36

- A *conservative* vector field $\vec{F}$ is one that can be expressed as a gradient:

$$\vec{F} = \text{grad } \Phi.$$

The scalar function Φ is called a *potential* for the vector field $\vec{F}$.

- The line integral of a conservative field $\vec{F} = \operatorname{grad} \Phi$ over any curve C from the point A to the point B equals the change in the potential function:

$$\int_C \vec{F} \cdot d\vec{r} = \int_C (\operatorname{grad} \Phi) \cdot d\vec{r} = \Phi(B) - \Phi(A).$$

- A given vector field $\vec{F}$ has any *one* of the following three properties, then it also has the other two:

  1. $\vec{F}$ is conservative.
  2. Line integrals of $\vec{F}$ are path-independent.
  3. The circulation of $\vec{F}$ around *every* closed curve is 0.

Application to physics: For motion under the influence of a conservative force, the sum of kinetic and potential energy remains constant.

Additional Problems

36.9. a) Explain why the following statement is true, and make a sketch to illustrate your explanation: "If the line integral of $\vec{F}$ is path-independent, then $\oint_C \vec{f} \cdot d\vec{r} = 0$ for all closed curves C." *Hint:* Make a sketch showing the axes and a closed curve C. Pick two points A and B on your curve. On the curve C, you can get from A to B in two ways. Look at the line integrals over these two curves. What is the relation between the value of the two line integrals and the value of the line integral over the closed curve C?

b) Explain why the following statement is true, and make a drawing to illustrate your explanation: "If the line integral of $\vec{F}$ is zero for all closed curves C, then the line integral is path-independent." *Hint:* Adjust the hints for a). Take two curves C_1 and C_2 from A to B. C_1 followed by the opposite of C_2 is a closed curve. What do you know about closed curves? How are the integrals over C_1 and C_2 related to the integral over the closed curve "C_1 followed by the opposite to C_2"?

c) Your friend says: "This is ridiculous. Part a) and b) are the same. What is the use of doing the same thing twice in a row?" Is your friend right or wrong? Write down a response.

36.10. C is the curve $(x, y, z) = (t, 2t^3, 3t^2)$. It connects the origin with $Q(1, 2, 3)$. Evaluate $\int_C \vec{F} \cdot d\vec{r}$ where $\vec{F}$ is given by $\vec{F} = \operatorname{grad} e^{x+y+z+1}$.

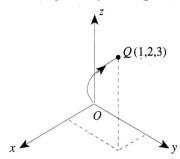

36.11. $\overrightarrow{F}$ is the vector field in the plane defined by $\overrightarrow{F} = (xy, 0)$.

  a) Evaluate the circulation of $\overrightarrow{F}$ along the circle C shown below left. *Hint:* For $\int \sin^2 t \cos t\, dt$, use the substitution $u = \sin t$.

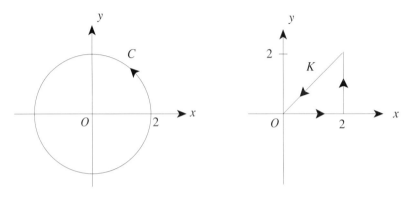

  b) Person A on your right says: "The computation just done shows that $\overrightarrow{F}$ is conservative." Person B on your left disagrees. Who is right? Write down an answer which would make sense to both persons.

  c) Evaluate the circulation of $\overrightarrow{F}$ along the closed polygonal path K shown above right.

  d) Go back to the answer you gave in b). Do you you want to leave it as is? Change it? Add to it?

36.12. Figure 36.2 shows two closed curves. On the left we have a circle C, and on the right we have a triangular path K which consists of the segments RS, ST, and TR.

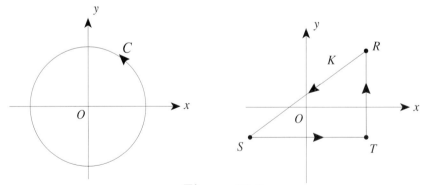

Figure 36.2

  a) For a certain vector field $\overrightarrow{F}$ we find that $\oint_C \overrightarrow{F} \cdot d\vec{r} = 0$. Consider the following statement:

$$\int_{RS} \overrightarrow{F} \cdot d\vec{r} + \int_{ST} \overrightarrow{F} \cdot d\vec{r} = \int_{RT} \overrightarrow{F} \cdot d\vec{r}.$$

Is the statement true, or false, or do we need more information on $\overrightarrow{F}$ in order to decide?

b) Now we consider the vector field

$$\overrightarrow{G} = \operatorname{grad}\left(\frac{xy}{\sqrt{x^2 + y^2}}\right).$$

Consider the following statement:

$$\oint_C \overrightarrow{G} \cdot d\vec{r} = \int_{RS} \overrightarrow{G} \cdot d\vec{r} + \int_{ST} \overrightarrow{G} \cdot d\vec{r} + \int_{TR} \overrightarrow{G} \cdot d\vec{r}.$$

Note that the last integral is taken in the direction from T to R. – Is the statement true, or false, or do we need more information on $\overrightarrow{G}$ in order to decide?

c) For a certain vector field $\overrightarrow{H}$ we find that

$$\oint_C \overrightarrow{H} \cdot d\vec{r} \neq 0.$$

Does $\overrightarrow{H}$ have a potential? Choose one:

        yes   *no*   *impossible to decide*

36.13. We consider the vector field $\overrightarrow{F}$ in the plane given by $\overrightarrow{F} = (e^x \sin y, e^x \cos y)$.

a) By $OR + RS$ we mean the path "from the origin to R and then on to S," as shown in the following figure:

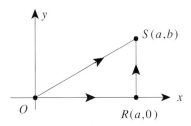

(i) Find $\int_{OR+RS} \overrightarrow{F} \cdot d\vec{r}$.

(ii) Your answer to (i) contains a and b. That is, it is a function of (a, b). Usually, independent variables are called x and y. Rewrite your answer to (i) in terms of (x, y) instead of (a, b). Your result is a function which we call $f(x, y)$.

(iii) Find grad $f(x, y)$ where $f(x, y)$ is the function you came up with in (ii).

(iv) For each of the sentences 1–4, true or false:

    1. $\overrightarrow{F}$ is a potential for $f(x, y)$.

    2. $\overrightarrow{F}$ is conservative.

    3. The function $f(x, y)$ is conservative.

    4. $f(x, y)$ is a potential for $\overrightarrow{F}$.

b) We continue with the vector field $\overrightarrow{F} = (e^x \sin y, e^x \cos y)$.

(i) Set up the line integral of $\overrightarrow{F}$ over the line segment OS. By "set up the line integral" we mean the following: Write the line integral as an ordinary integral of one variable, but do not carry out the integration. The ordinary integral must show the limits of integration. In other words: When you set up a line integral you end up with an integral of elementary calculus.

(ii) What will you obtain when you evaluate the integral you just set up? You can write down the value of the integral on the spot. *Hint:* Go back to your work in a).

36.14. Below is shown part of the level curve $f(x, y) = 6$ of the function $f(x, y) = xy + 4$.

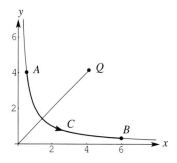

a) The curve C is part of the level curve. Its initial point is $A(1/2, 4)$, and its end point is $B(6, 1/3)$. Evaluate $\int_C \operatorname{grad} f \cdot d\vec{r}$.

b) The point Q lies on the line $x = y$. Find the coordinates of Q such that $\int_{OQ} \operatorname{grad} f \cdot d\vec{r} = 17$.

c) R is the point with coordinates $(3, 21)$. Describe the collection of all points P for which the line integral of $\operatorname{grad} f$ over the line segment RP equals 22.

36.15. A plot of a 2 dimensional vector field is shown below. Say whether the field is conservative, and justify your answer. *Hint:* Can you find a closed curve C so that the circulation around C is not zero?

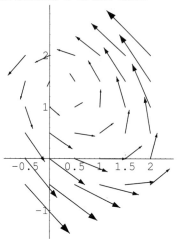

Review exercises on the Chain Rule

Some functions are given by a formula, others only by their name. In the latter case, push the computations as far as you can. For example, consider the function $q(t)$. We do not know its formula. If you are asked to find the derivative of $[q(t)]^2$, your answer should be $2\,q(t)\,q'(t)$.

36.16. Find both partial derivatives of $f(x,y) = \sqrt{x^2 + y^3}$.

36.17. $p(t)$ is a function of one variable, and we do not know its formula. Find both partial derivatives of $h(x,y) = p(e^{xy})$.

36.18. The function $b(t)$ is defined by $b(t) = \sin(t^2 + t)$, and $g(x,y)$ is a function of two variables, the formula of which we do not know. Find

$$\frac{\partial b(g(x,y))}{\partial y}.$$

36.19. With

$$\varphi(x,y,z) = \frac{1}{(x^2 + y^2 + z^2)^{3/2}}$$

write out the components of $\nabla\varphi$.

37   The Curl-Test

The key question for conservative vector fields

In many situations in science and engineering the question arises whether a given vector field is conservative. If the vector field is already given as a gradient field, the question is answered. If the vector field is given by its components, we need tools to decide whether it is conservative, and that is the topic of the present section. Thus, the key questions are as follows:

- *Given a vector field, how can we decide whether it is conservative?*
- *Once we know a vector field is conservative, how do we find a potential?*

We discuss a test, called the *curl-test*, which tells us whether a field is conservative. Like any test, it has limitations and drawbacks. It does not work for all vector fields, and vector fields in the plane and in space are handled differently.

   We do not discuss why the curl-test works. Such a discussion requires more advanced mathematics.

Vector fields in the plane

We consider a gradient field $\overrightarrow{F} = \operatorname{grad} \Phi(x, y)$ in the plane. That is, we already know that the field is conservative, and we already have a potential. To say "$\overrightarrow{F} = (F_1, F_2) = \operatorname{grad} \Phi(x, y)$" means in components

$$F_1 = \Phi_x \text{ and } F_2 = \Phi_y. \tag{37.1}$$

Now we look at the second partial derivatives of $\Phi(x, y)$. Remember from Section 15: As a rule[1] the order in which we take partial derivatives does not matter:

$$(\Phi_x)_y = (\Phi_y)_x,$$

In view of the relation (37.1) above, this becomes

$$(F_1)_y = (F_2)_x. \tag{37.2}$$

Now recall (see Section 33) that the curl of the vector field $\overrightarrow{F}$ is defined by

$$\operatorname{curl} \overrightarrow{F} = (F_2)_x - (F_2)_y$$

so (37.2) can be expressed in by the vector equation

$$\operatorname{curl} \overrightarrow{F} = \overrightarrow{0}.$$

We conclude: *If the vector field in the plane is conservative, then its curl is zero. In other words: If the curl is different from zero, then the the field is not conservative.*

Problems _____

37.1. A colleague asks you: "I am confused. Which of the following statements are true: *If the curl is zero, then the field is conservative. If the field is conservative, then its curl is zero. If the curl is different from zero, then the field may or may not be conservative. If the curl is different from zero, then the field is not conservative.*" What do you say?

---

[1]The technical condition that ensures equality of the mixed partial derivatives is continuity of the functions Φ, Φ_x, Φ_y, Φ_{xy}, and Φ_{yx}.

37.2. Let $\vec{u}$, $\vec{v}$, and $\vec{w}$ be vector fields in the plane given by

$$\vec{u} = (x^2 y, xy^2), \quad \vec{v} = (y + 3, x - 1), \quad \vec{w} = \left(\frac{x}{\sqrt{x^2 + y^2}}, \frac{y}{\sqrt{x^2 + y^2}} \right).$$

For each field, choose one of the following:
It is conservative—it is not conservative—we cannot decide.

Now, the question is: *If the curl of a vector field in the plane vanishes, is the field conservative?* The curl-test gives a partial answer, as follows.

The curl-test for vector fields in the plane

We are given a vector field $\vec{F} = (F_1, F_2)$ in the plane. Calculate

$$\operatorname{curl} \vec{F} = (F_2)_x - (F_1)_y.$$

1. *If* $\operatorname{curl} \vec{F}$ is different from zero at even one point, then $\vec{F}$ is *not conservative*.
2. If $\operatorname{curl} \vec{F}$ equals zero everywhere, look at the four partial derivatives of the component functions:

$$(F_1)_x, (F_1)_y, (F_2)_x, (F_2)_y.$$

Inspect the formulas for these derivatives and ask: Are they defined for all points in the plane, or are there points where they are not defined?[2]
There are two cases:

Case A Each of the four partial derivatives of the component functions is defined for all points of the plane. Then $\vec{F}$ is *conservative*.

Case B At least one the four partial derivatives has a point where it is not defined. Then $\vec{F}$ may or may not be conservative. The *test is inconclusive*.

Note that the possible outcomes of the curl-test are "yes, conservative"; "no, not conservative"; and "we do not know."

Problem _____

37.3. Apply the curl-test to the vector fields of Problem 37.2.

Vector fields in space

We reason as we did at the beginning of the section for vector fields in the plane.

Problem _____

37.4. We consider the field $\vec{F} = \operatorname{grad} \Phi(x, y, z)$; it is conservative. Compute its curl.

As we did for vector fields in the plane, we conclude
If the vector field $\vec{F}$ in space is conservative, then its curl must be the zero vector. That is: If the curl is not the zero vector, the field is not conservative.

---

[2]Strictly speaking, we mean "defined as continuous functions on the plane."

Problem _____

37.5. Let $\vec{u}$, $\vec{v}$, and $\vec{w}$ be three vector fields in space. They are given by

$$\vec{u} = (yz, xz, xy), \quad \vec{v} = (x, x, x), \quad \vec{w} = (3, -1, 7).$$

For each field, choose one of the following:
It is conservative—it is not conservative—we cannot say anything.

_____

The curl-test for vector fields in space.

The curl-test for vector fields in space looks like the one for the plane, except that part 2 is different. We are given a vector field $\vec{F} = (F_1, F_2, F_3)$ in space. Write out the formulas for curl $\vec{F}$.

1. If curl $\vec{F}$ is different from the zero vector then $\vec{F}$ is _not conservative._
2. If curl $\vec{F} = \vec{0}$, look at the partial derivatives of the component functions:

$$(F_1)_x, \quad (F_1)_y, \quad (F_1)_z,$$
$$(F_2)_x, \quad (F_2)_y, \quad (F_2)_z,$$
$$(F_3)_x, \quad (F_3)_y, \quad (F_3)_z.$$

Inspect the formulas for these derivatives and ask: Are they defined for all points in space, or are there points where they are not defined?[3] There are two cases:

Case A Each partial derivative is not defined for at most a _finite_ number of points. Then $\vec{F}$ is _conservative._

Case B One or more of the nine partial derivatives are not defined for _infinitely_ many points in space. Then $\vec{F}$ may or may not be conservative. The _test is inconclusive._

The difference between the test in the plane and in space

Both versions of the curl-test have two parts 1 and 2. Part 1 reads "If the curl is different from zero, then the field is not conservative." Part 1 is the same for both versions. In part 2 we have to look at the partial derivatives of the component functions, and here things are different in the plane and in space.

- **$\vec{F}$ is a vector field in the plane, and its curl is zero.** If just one of the four partial derivatives of the components has one point where it is not defined, the test is inconclusive. It takes _one point_ where a partial derivative is not defined to make the test _inconclusive._

- **$\vec{F}$ is a vector field in space, and its curl is zero.** Each of the nine partial derivatives may have a finite number of points where it is not defined, and the vector field is still conservative. It takes _infinitely many points_ where a partial derivative is not defined to make the test _inconclusive._

_____

[3]As with the curl test in the plane, we mean here "defined as continuous functions."

Problems ⎯⎯⎯⎯⎯⎯⎯⎯⎯⎯⎯⎯⎯⎯⎯⎯⎯⎯⎯⎯⎯⎯⎯⎯⎯⎯⎯⎯⎯⎯⎯⎯

37.6. Apply the curl-test to the three vector fields of Problem 37.5. Which of them are conservative? Which not? For which is the test inconclusive?

You may find Problems 37.7–37.9 to be difficult. They are about the difference between the curl-tests for the plane and for space. You are given two vector fields in space and one in the plane. In each case, the curl of the field is zero. (You may take our word for this. You do not have to compute the curl yourself.) Your mission is apply the curl-test to each field. In some cases, you may have to report that the curl-test is inconclusive.

37.7. $\overrightarrow{F} = -\dfrac{k}{(x^2 + y^2 + z^2)^{3/2}}(x, y, z)$ is the gravitational field of a point mass. It is a vector field in space.

37.8. $\overrightarrow{F} = \left(\dfrac{x}{x^2 + y^2}, \dfrac{y}{x^2 + y^2}\right)$ is a vector field in the plane.

37.9. $\overrightarrow{F} = \left(\dfrac{x}{x^2 + y^2}, \dfrac{y}{x^2 + y^2}, 0\right)$ is a vector field in space. It is obtained from the field of 37.8 by adding 0 as the third component.

How to find a potential for a conservative field

We are given a vector field $\overrightarrow{F}$, and we found out that it is conservative. *How do we find a potential?* There are several methods. They work the same way in space and in the plane. No matter which method you choose, you *have to check* whether the function Φ you found is a potential of $\overrightarrow{F}$, that is, whether grad $\Phi(x, y, z)$ is the given field $\overrightarrow{F}$.

Note that a given conservative field will have many potentials, since you can add a constant to one potential to produce another, but *any two potentials differ by a constant.* Thus finding potentials is much like finding antiderivatives in elementary calculus: Once you find one, you get all the others by adding constants.

Guess-and-check method

If the formulas of the components F_1, F_2, F_3 are simple, you can often guess a potential *and then check it.* For example, a constant vector field (a, b, c) has a potential $ax + by + cz$; $(2x, 2y, 2z)$ has a potential $x^2 + y^2 + z^2$, etc.

Problems ⎯⎯⎯⎯⎯⎯⎯⎯⎯⎯⎯⎯⎯⎯⎯⎯⎯⎯⎯⎯⎯⎯⎯⎯⎯⎯⎯⎯⎯⎯⎯⎯

37.10. Let $\vec{u}$, $\vec{v}$, and $\vec{w}$ be the vector fields in the plane given by

$$\vec{u} = (12x^2 - 6, 5y^4 + 2), \quad \vec{v} = (y, x), \quad \vec{w} = (y, -x)$$

Find a potential for each of them by guessing.

37.11. Let $\vec{u}$, $\vec{v}$, and $\vec{w}$ be the vector fields from Problem 37.5. They are given by $\vec{u} = (yz, xz, xy)$, $\vec{v} = (x, x, x)$, $\vec{w} = (3, -1, 7)$. Find a potential for those that are conservative.

Compute-and-check method

The field

$$\vec{F} = (1 + 4y + 5z, 2 + 4x, 3 + 5x)$$

is conservative. We want to find a potential $\Phi(x, y, z)$. We know that

$$\Phi_x = 1 + 4y + 5z.$$

We consider y and z as constants and integrate both sides with respect to x:

$$\Phi = x + 4xy + 5xz + g(y, z). \tag{37.3}$$

In this equation, $g(y, z)$ is the undetermined constant of the integration with respect to x. It may contain y and z. To determine $g(y, z)$ we use the fact $\Phi_y = 2 + 4x$. Therefore differentiating (37.3) with respect to y gives

$$2 + 4x = \frac{\partial \Phi}{\partial y} = \frac{\partial(x + 4xy + 5xz + g(y, z))}{\partial y} = 4x + \frac{\partial g(y, z)}{\partial y}.$$

We conclude

$$\frac{\partial g(x, y)}{\partial y} = 2$$

and integrating with respect to y gives

$$g(y, z) = 2y + h(z)$$

where $h(z)$ is still unknown. Therefore we can write

$$\Phi(x, y, z) = x + 4xy + 5xz + 2y + h(z). \tag{37.4}$$

To determine $h(z)$ we use the fact $\Phi_z = 3 + 5x$, and we find $h'(z) = 3$. Therefore $h(z) = 3z + k$ where k is a constant which contains none of x, y, z. Finally, inserting this into (37.4) gives the result:

$$\Phi(x, y, z) = x + 4xy + 5xz + 2y + 3z + k.$$

We check: grad $\Phi = (1 + 4y + 5z, 2 + 4x, 3 + 5x)$, as it should be.

Problems _____

37.12. $\vec{F} = (1 + 2y + 4xy, 2 + 2x + 2x^2)$ is a vector field in the plane. It is conservative. Use the compute-and-check method to find a potential.

37.13. $\vec{u}$ is the vector field given by $\vec{u} = (yz + 1, xz + 2, xy + 3)$. Use the compute-and-check method to find a potential.

Integrate-and-check method

This is the method discussed in Section 36. $\vec{F}$ is a conservative vector field. Choose a fixed point P_0. Then pick a curve C which connects P_0 with the generic point $P(x, y, z)$. Define a function $\Phi(x, y, z)$ by

$$\Phi(x, y, z) = \int_C \vec{F} \cdot d\vec{r},$$

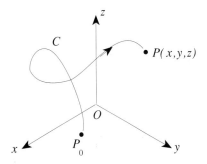

Figure 37.1

as illustrated in the Figure 37.1. We saw in Section 36 that the function $\Phi(x, y, z)$ is a potential function of $\overrightarrow{F}$. The idea is to choose a point P_0 and to pick a curve C such that the line integral is easy to compute. For example, the fixed point P_0 may be the origin, and C may be a straight line from the origin to P. Or C may consist of three line segments parallel to the coordinate axes, as shown on the right of Figure 35.6 on page 302.

Note:

1. We have to integrate over the curve C. Therefore, C may not pass through any point for which the vector field is not defined.

2. Check whether the function you found IS a potential of the field.

Problems _____

37.14. The field $\overrightarrow{F} = (z, 1, x)$ is conservative. Use the integrate-and-check method to find a potential of $\overrightarrow{F}$, with the origin as fixed point P_0, and with a straight line as curve C.

Caution: The endpoint of C is (x, y, z). If you integrate over C, there will be (x, y, z) representing a point which moves on the line C from the origin to the endpoint (x, y, z). To avoid confusion between the moving point and the endpoint, use (a, b, c) as temporary name for the endpoint of C. You will find the potential as a function $\Phi(a, b, c)$. Now you can replace a, b, c in the formula for Φ with x, y, z, and you get a function $\Phi(x, y, z)$ as desired.

37.15. The field $\overrightarrow{F} = (ye^x, e^x)$ is conservative. Pick $(1, 1)$ as fixed point P_0, and as curve C take a line from $(1, 1)$ to $P(x, y)$. Set up, but do not evaluate, the line integral which defines a potential function $\Phi(x, y)$. See the caution of Problem 37.14, and adapt it to 37.15.

Additional Problems

37.16. $\overrightarrow{F}$ is a conservative vector field in the plane.

   a) Are the line integrals of $\overrightarrow{F}$ path-independent? Choose one: "yes," "no," or "impossible to decide based on the given information."

   b) Are the partial derivatives of the components of $\overrightarrow{F}$ defined at all points of the plane? Choose one: "yes," "no," or "impossible to decide."

c) K is the curve that goes from the origin to A, then once around the circle C, then back to the origin, as shown in Figure 37.2. The line integral of $\overrightarrow{F}$ over K is (choose one): positive, zero, negative, or impossible to decide.

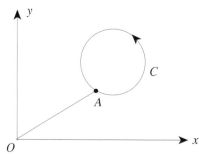

Figure 37.2

37.17. The vector field $\overrightarrow{F}$ is defined by $\overrightarrow{F} = (y, x, 2z)$. Is it conservative? If yes, find a potential $\phi(x, y, z)$ such that $\phi(1, 1, 1) = 378$. If it is not conservative, or if you cannot decide, explain.

37.18. The vector field $\overrightarrow{G}$ is defined by $\overrightarrow{G} = \operatorname{grad} f$, $\quad f(x, y, z) = xyz^2$. Is it conservative? If yes, find a potential $\phi(x, y, z)$ such that $\phi(1, 1, 1) = 2239$. If it is not conservative, or if you cannot decide, explain.

37.19. The vector field $\overrightarrow{H}$ is defined by $\overrightarrow{H} = (y, x, 2z + x)$. Is it conservative? If yes, find a potential $\phi(x, y, z)$ such that $\phi(1, 1, 1) = 67$. If it is not conservative, or if you cannot decide, explain.

37.20. $\overrightarrow{F}$ is the vector field in space given by $\overrightarrow{F} = \left(\dfrac{-y}{x^2 + y^2}, \dfrac{x}{x^2 + y^2}, -4 \right)$. Its curl is zero.

a) Explain why the curl-test is inconclusive.

b) C is the circle given by $x^2 + y^2 = 1$, $z = 12$. We orient C as follows: A point P moves on C. It starts out at $(1, 0, 12)$. Then it goes through $(0, 1, 12)$, then through $(-1, 0, 12)$, etc. Evaluate $\oint_C \overrightarrow{F} \cdot d\vec{r}$.

c) Is $\overrightarrow{F}$ conservative? Is it not conservative? Or is it impossible to decide? Explain your answer.

37.21. The vector field $\overrightarrow{F}$ in the plane is given by $\overrightarrow{F} = \left(\dfrac{x}{\sqrt{x^2 + y^2}}, \dfrac{y}{\sqrt{x^2 + y^2}} \right)$.

a) Apply the curl-test to $\overrightarrow{F}$.

b) Find the two partial derivatives of the function $h(x, y) = \sqrt{x^2 + y^2}$.

c) Is $\overrightarrow{F}$ conservative? Or is it not? Or do we have to do more work to decide?

37.22. $\overrightarrow{F}$ is the vector field in space given by $\overrightarrow{F} = (x^{-2}, z, y)$.

a) Explain why the curl-test is inconclusive.

b) The components of $\overrightarrow{F}$ are not complicated. Find a potential and check it.

c) Is $\overrightarrow{F}$ conservative?

Chapter 6

Surface Integrals

38 Parametrized Planes

Parametric representations of planes

You are familiar with the equation of a plane of the form $ax + by + cz = d$. Here we discuss a different way of representing planes.

Problem ───

38.1. The three points $A(2, 5, 4)$, $B(2, 0, 2)$ and $C(0, 2, 5)$ shown in Figure 38.1 on the following page determine a plane α.

   a) Draw the point Q whose position vector $\overrightarrow{OQ}$ is given by

$$\overrightarrow{OQ} = \overrightarrow{OA} + 2\,\overrightarrow{AB} + (1.4)\,\overrightarrow{AC}.$$

   b) Is Q a point of the plane α? Explain your answer.

   c) Let u and v stand for two real numbers. Then consider the point R whose position vector $\overrightarrow{OR}$ is given by

$$\overrightarrow{OR} = \overrightarrow{OA} + u\,\overrightarrow{AB} + v\,\overrightarrow{AC}.$$

   It follows from your reasoning in part b that R must lie in the plane α, no matter what u and v are. What are the values (u, v) that correspond to the point B? To A? To C?

   d) The point S shown in Figure 38.1 on the next page lies in α. Find numbers (u, v) so that

$$\overrightarrow{OS} = \overrightarrow{OA} + u\,\overrightarrow{AB} + v\,\overrightarrow{AC}.$$

───

What you just did with the plane α and the points A, B, C works for *any* three points K, L, and M that lie in a plane β but that do not lie on a line.

Result: *We are given a plane β. Pick three points K, L, and M that lie in β but that do not lie on a line. Then the position vector $\overrightarrow{OP}$ of any point P in β can be written in the form*

$$\overrightarrow{OP} = \overrightarrow{OK} + u\,\overrightarrow{KL} + v\,\overrightarrow{KM}$$

325

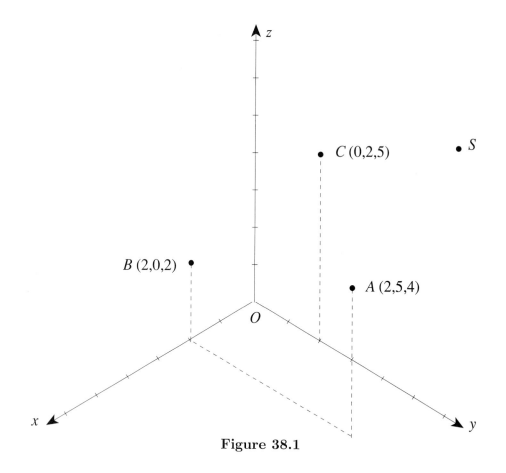

Figure 38.1

where (u, v) is a pair of real numbers. Conversely, given a pair (u, v) of real numbers, the point Q defined by

$$\overrightarrow{OQ} = \overrightarrow{OK} + u\,\overrightarrow{KL} + v\,\overrightarrow{KM}$$

lies in the plane β. In other words: Once we have chosen three points K, L, and M in β, we can describe the points of β by pairs (u, v) of real numbers. The procedure of describing the points P of a plane in the form

$$\overrightarrow{OP} = \overrightarrow{OK} + u\,\overrightarrow{KL} + v\,\overrightarrow{KM}$$

is called a parametric representation of β. The real numbers (u, v) are called the parameters.

> *Throughout this section, we will always write α for the plane of Problem 38.1, and we will always write β for a generic plane.*

In Problem 38.2 we continue working with the plane α.

Problem _____

38.2. Write out the parametric representation $\overrightarrow{OQ} = \overrightarrow{OA} + u\,\overrightarrow{AB} + v\,\overrightarrow{AC}$ of the plane α from Problem 38.1 in components. By "write out in components" we mean: Write down in formulas the fact "first component of the left side equals first component of the right side," "second ...," etc.

How to work with a parametric representation of a plane

Using parameters as coordinates. Given a plane β, we choose three points K, L, M to establish a parametric representation of β. Then the position vector of the generic point $P(x, y, z)$ of β is written in the form

$$\overrightarrow{OP} = \overrightarrow{OK} + u\,\overrightarrow{KL} + v\,\overrightarrow{KM}.$$

Once we have chosen such a parametric representation of β, a point P of β has two sets of coordinates, namely:

- The coordinates (x, y, z) as a point in space;
- The pair (u, v) which describes the position of P within β by means of the vectors $\overrightarrow{KL}$ and $\overrightarrow{KM}$.

Thus, there are two questions:

1. Given the space coordinates (x, y, z) of P in β, how do we find the corresponding parameters (u, v)?
2. Given the parameters (u, v) of a point P in β, how do we find its space coordinates (x, y, z)?

Problem 38.3 works through these two questions.

Problem _____

38.3. We continue working with the plane α and its parametric representation in Problem 38.1.

   a) The point S shown in Figure 38.1 of Problem 38.1 lies in α. Find its (x, y, z)-coordinates. *Hint:* You have already found its parameters (u, v) in Problem 38.1 d). Now use Problem 38.2.

   b) $T(1, 1, z)$ is a point of α. Find z. *Hint:* Use the first two equations of your answer to Problem 38.2 to find (u, v).

The two ways of representing a plane. A plane is given in the form $ax + by + cz = d$. How do we find a parametric representation? We pick three points in the plane and make sure that they do not lie on a straight line. Then we can write down a parametric representation.

Conversely, a plane is given in a parametric representation. How do we find its equation in the form $ax + by + cz = d$? We pick three pairs (u, v) of parameter values and find the corresponding (x, y, z) values. Then we have three points in the plane and use the usual procedure for "given three points of a plane, find its equation in the form $ax + by + cz = d$."

Problems _____

38.4. Find the equation of α of Problem 38.1 in the form $ax + by + cz = d$.

38.5. We consider the plane given by $x + 3y - 2z = 6$. Write out a parametric representation of the plane. Note that there are infinitely many correct answers.

The parameter plane. Now we look at a parametric representation of the generic plane β from another point of view, as follows.

Consider a uv-plane separate from the plane β. It has a u-axis and a v-axis, as shown below left. We have to distinguish points in the uv-plane from points of β. Therefore, we write P°, Q°, etc. for points in the uv-plane. Below, the xyz-space with the plane β is shown on the right. A point P° in the uv-plane on the left has

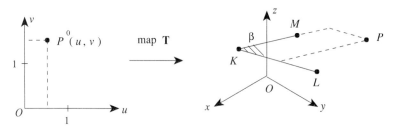

Figure 38.2

coordinates (u, v). The pair (u, v) defines a point P of the plane β on the right. In this way, we associate with any point P° in the uv-plane a point P in the plane β. The plane β lies in xyz-space. Therefore, we associate with P° on the left a point P in xyz-space on the right. Such a procedure which assigns to points in a uv-plane points in xyz-space is called a *map* from the plane into space. We write **T** for this map, as shown in Figure 38.2 above. The map **T** allows us to go back and forth between the uv-plane on the left and the plane β in xyz-space on the right.

Problem

38.6. This is about the map **T** of Figure 38.2 above. Note that the units in the uv-plane on the left are *not* 1 cm.

    a) Find the uv-coordinates of the points K°, L°, and M° which **T** maps into the points K, L, and M.

    b) Draw the point $R^\circ(1.5, -0.5)$ in the uv-plane on the left and the corresponding point R on the right

The uv-plane is covered by a net of lines $u = constant$ and $v = constant$. To each of these lines corresponds a line in the plane β. Thus, the plane β is also covered by a net of lines. We call these lines the *coordinate curves* or *lines* of β. We label these lines as follows: Instead of saying "the line of β corresponding to $v = 7$ we say simply "the coordinate curve (or line) $v = 7$ of β."

Problems

38.7. This is again about the plane α of Problem 38.1.

The figure from Problem 38.7 is reproduced below. Draw the following coordinate curves of α: $u = -0.5$, $u = 0$, $u = 0.5$, $u = 1$, $u = 1.5$, $v = -1$, $v = -0.5$, $v = 0$, $v = 0.5$, $v = 1$, $v = 1.5$.

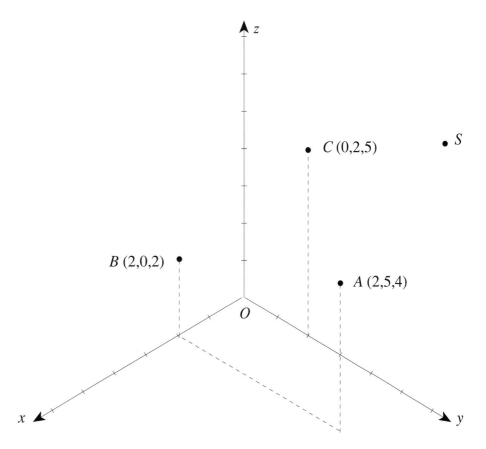

38.8. We continue with the plane α.

a) In the uv-plane the lines $u = 1$ and $v = 0.5$ are perpendicular. Are the coordinate lines $u = 1$ and $v = 0.5$ of α also perpendicular? *Hint:* Keep in mind that a drawing like the one you made in Problem 38.7 is like a photograph: A right angle in reality does not appear necessarily as a right angle on the picture, and what looks like a right angle in the picture might not be a right angle in reality. You have to find the answer by computation.

b) Find points D and E in α which have the following two properties at the same time:

(i) In the parametric representation $\overrightarrow{OP} = \overrightarrow{OA} + r\,\overrightarrow{AD} + s\,\overrightarrow{AE}$ the coordinate curves $r = constant$ and $s = constant$ intersect at right angles.

(ii) $|\overrightarrow{AD}| = |\overrightarrow{AE}| = 1$.

Hint: First, find points G, H so that $\overrightarrow{AG}$ is perpendicular to $\overrightarrow{AH}$. For G take the given B. $\overrightarrow{AH}$ must be parallel to α and perpendicular to $\overrightarrow{AG}$. Therefore, $\overrightarrow{AH} = \overrightarrow{AG} \times \vec{n}$ where $\vec{n}$ is normal to α. Now, condition (1) is satisfied for G and H. To satisfy condition (2), you take the unit vectors in the direction of $\overrightarrow{AG}$ and $\overrightarrow{AH}$.

Comparison between parametric representations of lines and planes. In the figures below, P is the generic point of a line g (on the left) and of a plane β (on the right).

Line g in space:
defined by two points A, B
$$\overrightarrow{OP} = \overrightarrow{OA} + t\,\overrightarrow{AB}$$
parameter t

Plane β in space:
defined by three points K, L, M
$$\overrightarrow{OP} = \overrightarrow{OK} + u\,\overrightarrow{KL} + v\,\overrightarrow{KM}$$
parameters (u, v)

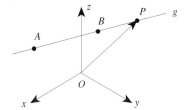
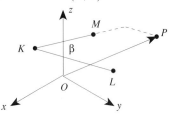

Figure 38.3

Look at Figure 38.3 and compare the formulas for the line and the plane. We notice:

- For the line, we have a *linear* expression in one variable t, and for the plane we have a *linear* expression in two variables u and v. Thus, the two formulas have in common that they are *linear* expressions.

- For a line you need *two* points, and you have *one* parameter t. For a plane you need *three* points, and you have *two* parameters (u, v).

Problems _____

38.9. What is the use of discussing similarities of formulas for parametric representations of lines and planes?

38.10. A parametric representation of a plane can be viewed as a map from the uv-plane into xyz-space. If there is a similarity between parametric representations of lines and planes, is there also a way of viewing a parametric representation of a line as a map $\mathbf{T}$ of some sort? What kind of assignment would this map define?

_____

Additional Problems

38.11. Figure 38.4 on the facing page shows two planes α and β. Remember that the pictures show only the part of the planes in the first octant and that the planes extend behind and below the coordinate planes.

    a) Create a parametric representation $\overrightarrow{OP} = \overrightarrow{OK} + u\,\overrightarrow{KL} + v\,\overrightarrow{KM}$ of α. Your answer will express each coordinate of the point P of α as a function of u and v: $x = \ldots, y = \ldots, z = \ldots$. *Hint:* α is perpendicular to the xz-plane.

    b) Do the same with β on the right. *Hint:* α is perpendicular to the xy-plane.

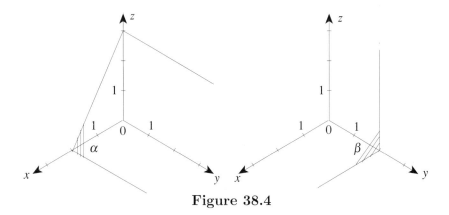

Figure 38.4

38.12. A plane α is given by the parametric representation

$$
\begin{aligned}
x &= 2 + 3u - v \\
y &= 1 - u + 4v \\
z &= 5 + v.
\end{aligned}
$$

a) Can you find a value of m so that $R(2, m, 1)$ lies in the plane α? If yes, find such an m; if no, explain.

b) Can you find a value of p so that $S(2, p, 1)$ does *not* lie in the plane α? If yes, find such a p; if no, explain. Note that R and S have the same x-coordinate and z-coordinate.

38.13. a) We consider the plane α given by $2x - y + 2z = 4$.

(i) Find a parametric representation $(x, y, z) = \ldots$ of α.

(ii) Is that the only parametric representation of α? If no, find another one.

b) We consider the plane β given by

$$
\begin{aligned}
x &= 2 + 3u - v \\
y &= 1 - u + 4v \\
z &= 5 + v.
\end{aligned}
$$

(i) Find for β an equation of the form $AX + By + Cz = D$.

(ii) Is there more than equation of that form for β? If yes, find another one.

38.14. Here we use the following facts about two planes and in space: Either the two planes are the same, or they are parallel and distinct, or they are not parallel. Two planes which are not parallel have a line in common, their line of intersection.

a) The equations $6x + 12y + 30z = 18$ and $8x + 16y + 40z = 32$ represent planes. Are these planes the same, or are they parallel, or are they not parallel?

b) We are given two parametric representations

$$(x, y, z) = (0, 0, 2) + u(1, 0, 0) + v(0, 2, -2)$$
$$(x, y, z) = (3, 2, 0) + u(-3, -2, 2) + v(-3, 0, 0)$$

Do these equations represent two planes which are the same, or are parallel, or are not parallel?

38.15. The following figure shows three points K, L, M, and their shadows (which do not carry any labels). These three points define a plane α.

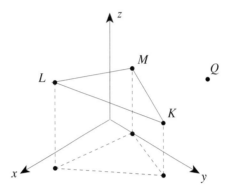

Also shown is a point Q. We know that Q is in α. Therefore, there must be scalars u, v such that

$$\overrightarrow{OQ} = \overrightarrow{OK} + u\,\overrightarrow{KL} + v\,\overrightarrow{KM}.$$

a) Use ruler and pencil to find u, v and to draw the shadow Q^0 of Q.

b) Determine from what you draw whether the coordinates of Q are positive, or zero, or negative. For example, if the x-coordinate and z-coordinates are negative and the y-coordinate positive write write *(negative, positive, negative)* as answer.

39   Parametrized Surfaces

In Section 38 we worked out a recipe for parametric representations of planes. A parametric representation for a plane is a special case of a *parametrized surface*. We begin this section by reviewing some features of parametrized planes, which will lead to the idea of a parametrized surface.

A parametric representation of a plane β is a map $\mathbf{T}$ from the uv-plane to xyz-space.

- *What is the map $\mathbf{T}$, what does it do?*

  It assigns to a point (u, v) in the uv-plane a point (x, y, z) in xyz-space. It works by expressing each of the coordinates x, y, z as functions $a(u, v)$, $b(u, v)$, $c(u, v)$ of u and v:

$$
\begin{aligned}
x &= a(u, v) \\
y &= b(u, v) \\
z &= c(u, v).
\end{aligned}
$$

- *Why is it that the points $(a(u, v), b(u, v), c(u, v))$ in xyz-space make up a plane?*

  The functions $a(u, v)$, $b(u, v)$, $c(u, v)$ are linear. Therefore any line in the uv-plane is transformed into a line in xyz-space. The grid of lines $u = $ constant and $v = $ constant in the xy-plane is transformed into a grid of lines in the plane β, as shown in Figure 39.1. The lines making up the grid of the plane

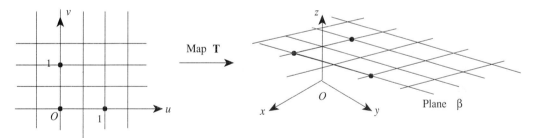

Figure 39.1

are the coordinate lines (or curves) of the parametric representation of β.

- *What if the functions $a(u, v)$, $b(u, v)$, $c(u, v)$ of the map $\mathbf{T}$ are not linear?*

  Then each of the lines $u = $ constant and $v = $ constant in the uv-plane is transformed into a curve in xyz-space. All these curves make up a surface S in xyz-space. The map $\mathbf{T}$ is a *parametric representation* of the surface S, as shown in Figure 39.2 on the next page. The surface S is covered by the web of curves that correspond to the lines $u = $ constant and $v = $ constant in the uv-plane. These curves are called the *coordinate curves* of the parametric representation of the surface S.

Example 1. Right circular cylinder. Let S be the surface of a right circular cylinder. Its axis is the z-axis, and its radius equals h, as shown in Figure 39.3 on the following page. C is the circle of radius h in which the surface S intersects the xy-plane. $P(x, y, z)$ is a point on S. ∎

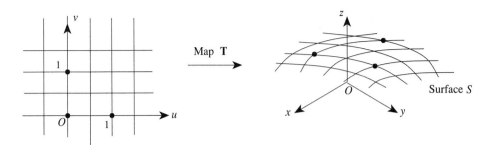

Figure 39.2

Problem _____

39.1. a) Write down a parametric representation

$$x = a(u, v)$$
$$y = b(u, v)$$
$$z = c(u, v).$$

of the cylinder S of Example 1 above, as follows: u is the parameter of a parametric representation of the circle C; the value of u determines the location of the point $P_0(x, y, 0)$ vertically below the point $P(x, y, z)$ of S. Let v be the elevation of $P(x, y, z)$ above the xy-plane (v will be negative if P is below the xy-plane).

b) The coordinate curves $u = $ constant and $v = $ constant pass through the point P, shown in Figure 39.3. Use two different colors to draw these curves into the figure, one color for each.

c) The point $Q(0, -h, m)$ lies on S (remember that h is the radius of the

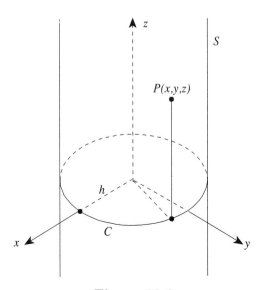

Figure 39.3

cylinder). The coordinate curves $u = k_1$ and $v = k_2$ pass through Q. What are the values of k_1 and k_2? Write down parametric representations for these two coordinate curves. The formulas will contain h, u, m.

d) Cut off from the cylinder S everything above the plane $z = 2$ and everything below $z = -1$. The cut-off cylinder looks like a piece of pipe. We call it T. Now draw a separate uv-system. Draw in this uv-plane the region T^0 which corresponds to the cut-off cylinder T.

Example 2. Surfaces given by $z = f(x, y)$**.** For such a surface, we can write down a parametric representation as follows:

$$
\begin{aligned}
x &= u \\
y &= v \\
z &= f(u, v).
\end{aligned}
$$

▪

Problems _____

39.2. The surface S given by $z = f(x, y) = xy - x - y + 2$ is shown below. Note that the picture is not taken from our usual viewpoint and that, therefore, the units on the three axes appear of different length.

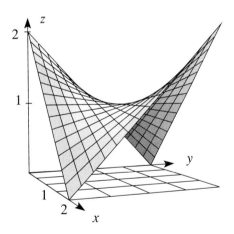

Figure 39.4

a) Write down a parametric representation of S as explained in Example 2. What is the shape of the coordinate curves $u = $ constant and $v = $ constant?

b) In Figure 39.4 above, the surface S is shown with the web of coordinate curves of the parametric representation of a).
The point $Q(\frac{3}{2}, \frac{1}{2}, \frac{3}{4})$ lies on S. Identify Q in the figure above. Also, identify the coordinate curves passing through Q.

c) The surface S is shown above the square $0 \leq x \leq 2$, $0 \leq y \leq 2$. Notice that the square does not look like a square because of our viewpoint.

Draw a separate uv-plane. Show the region S^0 in the uv-plane which corresponds to the part of the surface S shown.

39.3. Consider a generic surface S given by $z = f(x, y)$. We parametrize S as we discussed just before Example 2: $(x, y, z) = (u, v, f(u, v))$. Let $P(k, l, m)$ be a point of S. Write down parametric representations of the two coordinate curves that pass through P. Indicate which of the many symbols stand for the parameters of the coordinate curves.

Example 3. The sphere. Below is shown a sphere S centered at the origin. Its radius equals 4 units. The point $P(x, y, z)$ is on S, and $P_0(x, y, 0)$ is the shadow of P in the xy-plane.

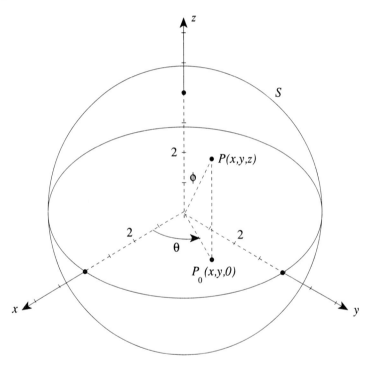

The angle θ is the angle of polar coordinates in the xy-plane. The angle ϕ is the angle between the z-axis and OP, measured from the z-axis on downward. The coordinates of $P(x, y, z)$ on the sphere S and the angles θ, ϕ are related as follows:

$$
\begin{aligned}
x &= 4 \sin \phi \cos \theta \\
y &= 4 \sin \phi \sin \theta \\
z &= 4 \cos \phi.
\end{aligned}
$$

These formulas provide us with a parametric representation of the sphere S with parameters θ and ϕ. You obtain the formulas by looking at the right triangle OP_0P.

■

Problems _____

39.4. Let S be the sphere $x^2 + y^2 + z^2 = 16$, parametrized by θ and ϕ.

   a) Describe in words the shape of the coordinate curves which pass through a point on S such as P shown above.

   b) At a point like P, what angle is formed by the two coordinate curves?

   c) The point P shown has coordinates $(x, y, z) = (1, 2, \sqrt{11})$. Write down parametric representations of the coordinate curves which pass through P.

39.5. By the "upper half-sphere" we mean the part of S which lies in and above the xy-plane. We parametrize the upper half-sphere not by θ and ϕ, but by the method of Example 2:

$$(x, y, z) = (u, v, \sqrt{16 - u^2 - v^2}).$$

   a) Describe in words the shape of the coordinate curves which pass through $P(1, 2, \sqrt{11})$.

   b) Write down parametric representations of the two coordinate curves which pass through $P(1, 2, \sqrt{11})$.

   c) The two coordinate curves at P intersect at an angle θ. Find $\cos \theta$.

Normal vectors, tangent planes. In the figure below, we consider a point P on a generic surface S given by $(x, y, z) = (a(u, v), b(u, v), c(u, v))$. P has coordinates (x, y, z), and the position vector $\overrightarrow{OP}$ is written $\vec{r}$.

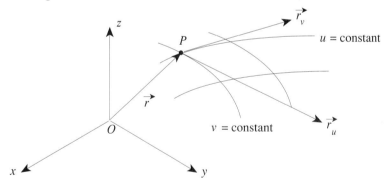

The point P lies at the intersection of the coordinate curves $u = $ constant and $v = $ constant. On the curve $u = $ constant, only v varies. That is, v is the parameter. Therefore, the vector

$$\vec{r}_v = \left(\frac{\partial a(u, v)}{\partial v}, \frac{\partial b(u, v)}{\partial v}, \frac{\partial c(u, v)}{\partial v} \right) = (a_v, b_v, c_v)$$

is tangent to the curve $u = $ constant at the point P. Similarly, the vector $\vec{r}_u = (a_u, b_u, c_u)$ is tangent to the curve $v = $ constant at P. The two vectors $\vec{r}_u$ and $\vec{r}_v$ define the tangent plane of the surface S at the point P.

Result: *Let S be the surface given by $\vec{r} = (x, y, z) = (a(u, v), b(u, v), c(u, v))$, with P the point corresponding to (u, v). We write $\vec{r}_u$ for the vector (a_u, b_u, c_u), and we write $\vec{r}_v$ for the vector (a_v, b_v, c_v). Then*

$$\vec{n} = \vec{r}_u \times \vec{r}_v$$

is normal to the tangent plane of S at the point P. We call $\vec{n} = \vec{r}_u \times \vec{r}_v$ the standard normal of the surface $\vec{r} = (x, y, z) = (a(u, v), b(u, v), c(u, v))$.

For surfaces such as spheres and cylinders, the tangent plane can be found by direct geometric reasoning, and it is not necessary to compute $\vec{n} = \vec{r}_u \times \vec{r}_v$.

Problems

39.6. Let S be the right circular cylinder of Example 1 on page 333, and P the point on S with coordinates $(x, y, z) = (h\frac{\sqrt{3}}{2}, \frac{h}{2}, 4)$. Note that $\cos\frac{\pi}{6} = \frac{\sqrt{3}}{2}$ and $\sin\frac{\pi}{6} = \frac{1}{2}$.

   a) Write down the equation of the tangent plane at P in the form $Ax + By + Cz = D$.

   b) Take the parametric representation for S of Problem 39.1. Use the vectors $\vec{r}_u$ and $\vec{r}_v$ at P to write down a parametric representation of the tangent plane at P.

39.7. Let S be the surface $z = xy - x - y + 2$ of Problem 39.2 on page 335, and Q the point on S with coordinates $(x, y, z) = (\frac{3}{2}, \frac{1}{2}, \frac{3}{4})$.

   a) In Section 16, surfaces like S were discussed as graphs $z = f(x, y)$ of functions of two variables. There you found that a normal vector of the tangent plane of $z = f(x, y)$ is given by $(f_x, f_y, -1)$. Use this to find the equation of the tangent plane at Q in the form $Ax + By + Cz = D$.

   b) Take the parametric representation for S of Problem 39.2. Use the vectors $\vec{r}_u$ and $\vec{r}_v$ at Q to find the components of a normal vector of the tangent plane at Q.

39.8. Let S be the sphere of Example 3 on page 336. We parametrize S with θ and ϕ. The point P on S has coordinates $(x, y, z) = (1, 2, \sqrt{11})$.

   a) Find the components of a normal vector of the sphere S at the point P. *Hint:* What is the angle between $\overrightarrow{OP}$ and the normal of S at P?

   b) In Problem 39.4.c you found parametric representations of the curves $\theta = $ constant and $\phi = $ constant at P. Use tangent vectors to these curves to write down a parametric representation of the tangent plane at P. *Hint:* Use the computations made in Problem 39.4 on page 336.

Note on cylindrical coordinates (r, θ, z) in space. These coordinates are given by

$$
\begin{aligned}
x &= r\cos\theta \\
y &= r\sin\theta \\
z &= z.
\end{aligned}
$$

In these formulas, (r, θ) are the usual polar coordinates in the xy-plane. The coordinate surfaces of cylindrical coordinates are the surfaces $r = $ constant, $\theta = $ constant, and $z = $ constant.

In Problem 39.1, you considered a cylinder S, and you found the following parametric representation for S:

$$x = h\cos u$$

$$y = h \sin u$$
$$z = v.$$

Write r instead of h, θ instead of u, and z instead of v. Then we see: *Our surface S is nothing but the coordinate surface $r = h$ of cylindrical coordinates.*

Note on spherical coordinates (ρ, θ, ϕ) in space. These coordinates are given by

$$x = \rho \sin \phi \cos \theta$$
$$y = \rho \sin \phi \sin \theta$$
$$z = \rho \cos \phi.$$

In these formulas, ρ (pronounced "ro") is the distance of $P(x, y, z)$ from the origin, θ is the angle of polar coordinates in the xy-plane, and ϕ is the angle between the z-axis and OP, measured downward from the z-axis. The coordinate surfaces of spherical coordinates are the surfaces $\rho = $ constant, $\theta = $ constant, and $\phi = $ constant.

In Example 3 and Problem 39.4, you considered a sphere S. You used the following parametric representation for S:

$$x = 4 \sin \phi \cos \theta$$
$$y = 4 \sin \phi \sin \theta$$
$$z = 4 \cos \phi.$$

Write ρ instead of 4. Then we see: *Our sphere S is nothing but the coordinate surface $\rho = 4$ of spherical coordinates.*

Additional Problems

39.9. The surface S is shown below. It is obtained as follows: You take the piece from $C(0, 0, c)$ to $D(0, b, 0)$ of the curve $z = g(y)$ in the yz-plane and move it the distance a in the direction of the positive x-axis. All that is covered by that piece of curve makes up S. The surface looks like a sheet of metal which has been bent. The four corners of the surface are marked with bullets.

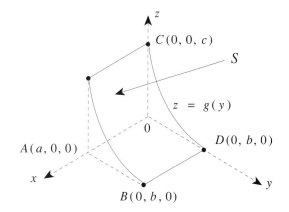

a) Find a parametric representation of S. *Hint:* S is a surface given in the form $z = f(x, y)$. What is $f(x, y)$? Then apply the usual procedure to get a parametric representation $(x, y, z) = (a(u, v), b(u, v), c(u, v))$.

b) Let S^0 be the region in the uv-plane which corresponds to S. Describe S^0 in one or more complete sentences.

c) What is the shape of the coordinate curves of our parametric representation of S? Describe them in complete sentences: "The coordinate curves $u = $ constant are ....."

d) Find the components of the standard normal $\vec{n} = \vec{r}_u \times \vec{r}_v$ of your parametric representation. Does it point towards you or away from you?

e) The general, or generic, point of S has coordinates

$$(x, y, z) = (a(u, v), b(u, v), c(u, v)),$$

with $a(u, v), \ldots$ as in part a. Find the equation of the tangent plane to S at the generic point P in the form $Kx + Ly + Mz = N$. The coefficients K, L, M, N will contain the surface parameters u, v.

39.10. Below is shown the surface S. It is part of a right cylinder of radius b and length h. Its axis is the x-axis. One of its end circles rests on the yz-plane.

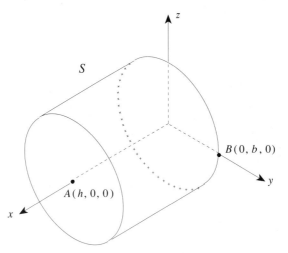

a) Write down a parametric representation of S. *Hint:* Go back to Example 1 on page 333 where a similar cylinder is discussed.

b) Q is the following point of S: Q lies in the xy-plane, and its distance from the yz-plane equals $\frac{h}{3}$. Take the parametric representation that you established in part a. What are the parameter values that correspond to Q?

c) What is the shape of the coordinate curves of your parametric representation of S? Describe them in complete sentences: "The coordinate curves $u = const$ are ....."

d) S^0 is the region in the uv-plane which corresponds to S in your parametric representation. Describe S^0.

e) Find the components of the standard normal $\vec{n} = \vec{r}_u \times \vec{r}_v$ of your parametric representation. Does it point into the cylinder or out of it?

f) Take the point Q of part b. Find a parametric representation of the tangent plane to S at Q. *Hint:* For the parameters of the tangent plane use letters different form u and v to avoid confusion with the parameters of the surface. For example, take s and t.

39.11. Here you will find a parametric representation of a circular cone S which rests on the xy-plane. Its height is h, and its radius is b, as shown in the following figure:

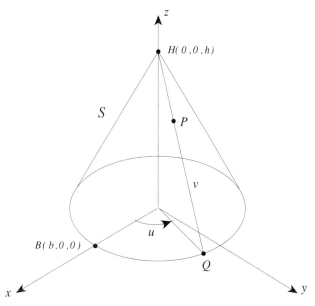

$P(x, y, z)$ is a generic point of S, and Q is the point where the line HP intersects the base circle. As parameter u of S, we take the angle that Q forms with the positive x-axis. As v we take the distance QP. Then we have

$$\overrightarrow{OP} = \overrightarrow{OQ} + \overrightarrow{QP}.$$

a) "To find a parametric representation" means "to express the components of $\overrightarrow{OP}$ in terms of u, v." To do that, we have to express the components of $\overrightarrow{OQ}$ and $\overrightarrow{QP}$ in terms of u, v. We do that in four steps.

(i) Find the components of $\overrightarrow{OQ}$ in terms of u. Note that the z-coordinate must be zero.

(ii) Find the components of the unit vector in the direction of QP in terms of u, v.

(iii) Use (ii) to find the components of $\overrightarrow{QP}$.

(iv) Combine (i)–(iii) to write the components of OP in terms of u, v. Now you have a parametric representation

$$(x, y, z) = (a(u, v), b(u, v), c(u, v))$$

of S.

b) Describe the shape of the curves $u = constant$ and $v = constant$, which pass through P.

c) Describe the region S^0 in the uv-plane which corresponds to S.

d) Go to the point Q of S which corresponds to $(u, v) = (\frac{\pi}{4}, \frac{1}{2}\sqrt{b^2 + h^2})$. It is a point at half-height of the cone. Write down a parametric representation of the tangent plane of S at Q. *Hint:* For the parameters of the tangent plane use letters different form u and v to avoid confusion with the parameters of the surface. For example, take s and t.

40   Velocity Fields and Flux

Flow across a surface, flux.   All through this section, we work with vector fields and surfaces from a physical point of view:

- We consider vector fields as velocity fields of fluids flowing through space. We assume that the flow patterns do not change over time. Such flows are called *steady-state flows.*

- We consider surfaces immersed in a steady-state flow. We assume that the surfaces are porous so that the liquid flows across them without obstruction.

The following figure shows fluid flowing across a surface S. The flow is represented by flow lines that cross the surface.

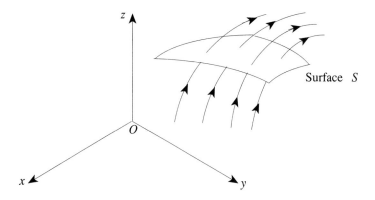

In such a situation we ask:

> *How much fluid passes through the surface per unit time? Or, what is the same: What is the rate of flow across the surface?*

The rate of flow of $\overrightarrow{F}$ across S is called the *flux* of $\vec{F}$ across S, or through S. Our goal is to compute the flux of a given vector field across a given surface. We can do this only if we say exactly what we mean by flux. There are two points to clear up:

1. The flux of a velocity field measures the *volume* that passes through the surface per unit time. There is no simple relation between *volume* per unit time and *mass* per unit time if the mass density of what flows varies from point to point. The latter may happen when we consider the flow of a gas.

2. A fluid can flow across S in two directions. How can we distinguish the two directions of flow? A surface has two sides, and the flow is from one side to the other.[1] How do we keep the two sides apart? We use normal vectors, as follows. At a point P of the surface S, there are normal vectors pointing in two opposite directions, as shown below.

---

[1]There are surfaces that have only one side (e.g., the Moebius strip). We do not consider such one-sided surfaces.

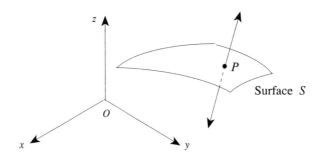

At every point we choose one of the two directions. Choosing a direction of a normal vector is called "choosing a normal vector" or "orienting the surface." Once we have oriented our surface S, we say that the flux is positive if the flow is in the direction of the chosen normal, and negative otherwise.

In short:

> *(1) We measure flux by units of volume per unit of time. (2) We consider only oriented surfaces.*

Problem _____

40.1. Let S be the rectangle in the yz-plane shown in Figure 40.1. We orient S by choosing $\vec{i}$ as normal. In parts a to c, we consider three vector fields $\vec{F}$, $\vec{G}$, $\vec{H}$. For each of them, we ask: *Is the flux across S positive, or zero, or negative, or do we need more information to decide?* Hint: Visualize the field vectors. Per unit of time, is more flowing from back to front or the other way around?

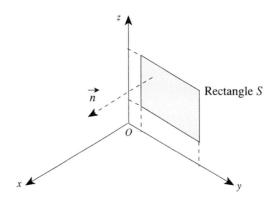

Figure 40.1

a) $\vec{F}$ is the constant vector field with components $(-3, -7, 2)$.

b) $\vec{G}$ is the vector field given by $\vec{F}(P) = 2\,\overrightarrow{OP}$.

c) $\vec{H} = (e^{-x-y-z}, \cos(-x-y-z), \sin(-x-y-z))$. Hint: Write $\vec{H}$ as a sum $(e^{-x-y-z}, 0, 0) + (0, \cos(-x-y-z), \sin(-x-y-z))$. The flux of the sum of two vector fields is the sum of the flux of each vector field.

Flux of a constant vector field across a flat surface. We consider a rectangle R, and we orient R by choosing a normal $\vec{n}$. $\vec{G}$ is a constant vector field. Below left are shown the rectangle R, its normal $\vec{n}$, and several copies of $\vec{G}$ along the edges of the rectangle ("general view of R"). Now we change our point of view so that our eyes are in the plane of R. Then the rectangle is seen on edge. This is shown below right ("view of R on edge").

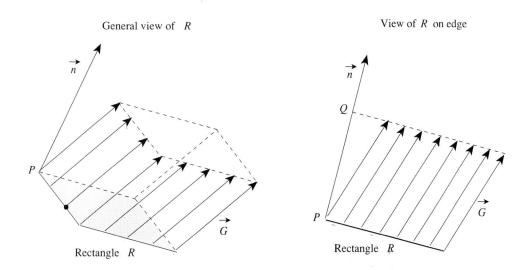

During one unit of time, a particle of the fluid such as the one indicated by a black dot (on the left) travels from the initial point of $\vec{G}$ to its endpoint. The quantity of fluid flowing through the dot in one unit of time can be thought of as a string of length $|\vec{G}|$ of fluid particles. We look at the strings of all points of R. The strings fill up a prism that has R as base and $\vec{G}$ as edge. We conclude: *The volume of fluid flowing across R in one unit of time is given by the volume of the prism that has R as base and $\vec{G}$ as edge.* The volume of a prism is "area of base times height." In our case: The height PQ is the scalar projection of $\vec{G}$ onto the normal $\vec{n}$ (see view of R on edge). What is true for the rectangle R is true for any plane surface S.

To compute the flux of a constant field $\vec{G}$ across a flat surface S we multiply the scalar projection of $\vec{G}$ onto $\vec{n}$ by the area $A(S)$ of S:

$$flux = \left(\vec{G} \cdot \frac{\vec{n}}{|\vec{n}|} \right) A(S).$$

The flux is $\pm$ the volume of the prism which has S as base and $\vec{G}$ as edge.

Note: The boxed formula holds only if the field is constant and the surface is flat.

We look at a special case of the boxed formula. Let $\vec{G}$ be a constant vector field, and the flat surface S a parallelogram spanned by two vectors $\vec{p}$ and $\vec{q}$. As normal we choose $\vec{n} = \vec{p} \times \vec{q}$, as shown below.

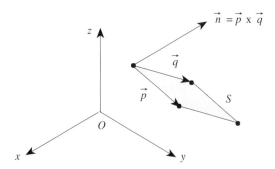

The formula tells us that

$$\text{Flux} = (\text{normal projection of } \overrightarrow{G} \text{ onto } \vec{n})(\text{area of } S).$$

The area of S equals $|\vec{p} \times \vec{q}| = |\vec{n}|$, so we obtain

$$\text{Flux} = \left(\overrightarrow{G} \cdot \frac{\vec{p} \times \vec{q}}{|\vec{p} \times \vec{q}|} \right) |\vec{p} \times \vec{q}| = \overrightarrow{G} \cdot (\vec{p} \times \vec{q}).$$

This easy formula for the flux works only in the special situation "vector field constant, surface a parallelogram, orientation of the parallelogram chosen in a particular way."

Some authors call an expression such as $\overrightarrow{G} \cdot (\vec{p} \times \vec{q})$ a *triple product*, or *triple scalar product*, or *mixed product*, or *box product*. Often, it is written $[\overrightarrow{G}, \vec{p}, \vec{q}]$. Thus, given any vectors $\vec{a}$, $\vec{b}$, $\vec{c}$, the expression $[\vec{a}, \vec{b}, \vec{c}]$ is $\pm$ the volume of the prism with base spanned by $\vec{a}$ and $\vec{b}$, and with edge $\vec{c}$.

Problem

40.2. S is the rectangle of Problem 40.1, oriented by $\vec{\imath}$, as shown in Figure 40.1. The coordinates of the corners of S are $(0, a, b)$, $(0, a + p, b)$, $(0, a + p, b + q)$, $(0, a, b + q)$. $\overrightarrow{F}$ is the constant vector field given by $\overrightarrow{F} = (-2, 3, 7)$. Find the flux of $\overrightarrow{F}$ across S.

Additional Problems

40.3. Let $P(1, 2, \sqrt{11})$ be a point on the sphere $S : x^2 + y^2 + z^2 = 16$. We orient S by the normal which points towards the origin (the "inside" normal).

  a) Write down the components of an inside normal vector of S at P.

  b) At the origin there is a source of a liquid. The liquid flows away from the origin in the following way: At the point Q, the velocity vector of the flow equals $2\overrightarrow{OQ}$. Is the flux across the sphere S positive, or zero, or negative, or do we need more information to decide?

40.4. Let S be the portion $0 \le z \le 6$, $0 \le \theta \le \frac{\pi}{2}$ of the cylinder $(x, y, z) = (3\cos\theta, 3\sin\theta, z)$, as shown in Figure 40.2 on the next page. We orient S by choosing the normal pointing to the outside.

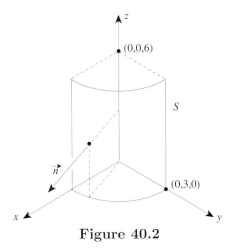

Figure 40.2

a) The point $P(x, y, z)$ lies on the surface S, and $\vec{n}$ is the outside normal of length 1 at the point P. Write down the components of $\vec{n}$.

b) $\overrightarrow{F}$ is the constant vector field $(-1, -1, -1)$. Is the flux of $\overrightarrow{F}$ across S positive, or zero, or negative? What about the flux of $\overrightarrow{G} = (-y, x, 0)$?

40.5. The following figure shows a parallelogram $ABCD$ in space.

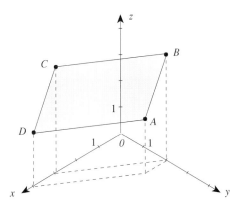

a) Read off the coordinates of the points A, B, C, and D, and find the surface area of the parallelogram. *Remember:* The surface area of a parallelogram is the magnitude of the cross product of the two vectors which define the parallelogram.

b) We orient the parallelogram by picking the normal whose z-component is positive. It is the normal that points toward us. Find the components of such a normal that is of magnitude 1.

c) $\overrightarrow{F}$ is the constant vector field with components $(1, 4, 3)$. Find the flux of $\overrightarrow{F}$ across $ABCD$.

d) Can you find a constant vector field $\overrightarrow{G}$ such that the flux of $\overrightarrow{G}$ across $ABCD$ is zero?

If yes: (i) Find such a vector field $\overrightarrow{G}$. (ii) Is there only one such $\overrightarrow{G}$ or is there more than one? Explain.

If no: Explain.

e) Can you find a constant vector field $\overrightarrow{H}$ such that the flux of $\overrightarrow{H}$ across $ABCD$ equals 8?

If yes: (i) Find such a vector field $\overrightarrow{H}$. (ii) Is there only one such $\overrightarrow{H}$ or is there more than one? Explain.

If no: Explain.

40.6. We consider the constant vector field $\overrightarrow{F} = (2, 5, 0)$. K is the triangular solid shown below. On the faces of K we take the normal which points away from K.

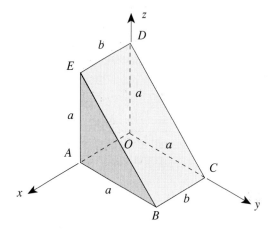

a) Find the flux of $\overrightarrow{F}$ across the bottom face $OABC$.

b) Find the flux of $\overrightarrow{F}$ across the side face ABE.

c) Find the flux of $\overrightarrow{F}$ across the side face OCD.

d) Find the flux of $\overrightarrow{F}$ across the back face $OAED$.

e) Find the flux of $\overrightarrow{F}$ across the slanted face $BCDE$.

f) When you add up the flux across all faces you get the net volume of fluid that enters or leaves the interior of the solid per unit of time. Add up the flux across all faces. Express your result in a sentence about the net amount "Per unit of time, so and so much ...."

41 Flux and Surface Integrals

In Section 40, we worked out the flux of a *constant* vector field across a *flat* surface. Here we will extend the ideas of Section 40 to apply to *non-constant* vector fields and *curved* surfaces. This leads to a new type of integral called a *flux integral* or *surface integral*. When the vector field represents the velocity field of a fluid, the flux integral will measure the volume of fluid that flows across the surface per unit time.

☞ How are flux integrals defined?

Preliminaries

The unit normal vector to an oriented surface. We work with an *oriented* surface S. At each point P of S, there are two unit vectors that are normal to S. We select the one which is consistent with the orientation of S, and call it $\vec{n}(P)$. When we refer to the *unit normal vector* to S at P, we mean $\vec{n}(P)$.

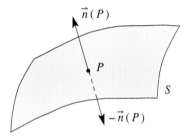

The normal component of a vector field on a surface. Now take a vector field $\overrightarrow{F}$ defined along the oriented surface S. At any point P on the surface S, the *normal component* of $\overrightarrow{F}$ at P is the component of $\overrightarrow{F}(P)$ along the unit normal vector $\vec{n}(P)$. We denote the normal component of $\overrightarrow{F}$ at P by $F_n(P)$. In the picture below, $F_n(P)$ is the length of the leg of of the indicated right triangle that runs along the normal vector $\vec{n}(P)$.

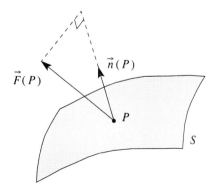

Since $\vec{n}$ is a unit vector, we have the simple formula

$$F_n(P) = \overrightarrow{F}(P) \cdot \vec{n}(P).$$

Definition of flux integrals

As we have done with double integrals, triple integrals, and line integrals, we define flux integrals by giving a recipe for approximating them. To define a flux integral, you must have:

- An oriented surface S;
- A vector field $\overrightarrow{F}$ defined along the surface S.

The flux integral of the vector field $\overrightarrow{F}$ over the surface S is denoted by the symbol $\iint_S \overrightarrow{F} \cdot d\overrightarrow{A}$. It is the number which is approximated by the following procedure.

1. Partition the surface S into small[1] surfaces $S_1, S_2, \ldots$ that fill up S and do not overlap.

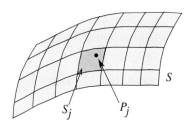

Figure 41.1

2. For each small surface S_j, sample the normal component of $\overrightarrow{F}$ at a point P_j in S_j. The result is a number $F_n(P_j)$.

3. Multiply the sampled values $F_n(P_j)$ by the areas of the small surfaces S_j, and add up the results. The sum approximates the flux integral:

$$\iint_S \overrightarrow{F} \cdot d\overrightarrow{A} \approx \sum F_n(P_j) A(S_j). \tag{41.1}$$

In the preceding formula, $A(S_j)$ denotes the area of the small surface S_j.

The value of the flux integral $\iint_S \overrightarrow{F} \cdot d\overrightarrow{A}$ can be approximated to any precision you wish by choosing a partition with small enough subdivisions S_j.

Problems

41.1. $\overrightarrow{F}$ is the velocity field for a fluid flowing along a pipeline from east to west. The surface S is an imaginary membrane stretched across the pipeline at a right angle.

   a) If S is oriented by a westward pointing normal, is the flux integral $\iint_S \overrightarrow{F} \cdot d\overrightarrow{A}$ positive, negative or zero? Please explain.

   b) Answer the same question when S is oriented by the eastward pointing normal.

41.2. $\overrightarrow{F}$ is the gravitational field of the earth. At any point P outside the earth, the field vector $\overrightarrow{F}(P)$ points directly toward the center of the earth.

   a) The surface S is a portion of a plane that passes through the center of the earth. The surface S does not intersect the earth. Is the flux integral $\iint_S \overrightarrow{F} \cdot d\overrightarrow{A}$ positive, negative, or zero? Please explain.

---

[1] Here "small" means that the maximum distance between any two points in S_j is small. Having small area is not enough.

b) The surface S is a sphere which completely encloses the earth. The center of S need not coincide with the center of the earth. We orient S by the normal that points outward. Is the flux integral $\iint_S \vec{F} \cdot d\vec{A}$ positive, negative, or zero? Please explain.

✍ How do we evaluate flux integrals?

Special case: fields and surfaces with constant normal component

In applications of flux integrals, one often encounters situations where the normal component of the field $\vec{F}$ is constant on the surface S. Two such situations are the following:

- $\vec{F}$ is constant and S is part of a plane (see the subsection on constant fields and flat surfaces on page 345).
- $\vec{F}$ is a radial field and S is a sphere centered at the origin (this is discussed in Example 1 below).

In such cases, the approximating sums (41.1) for the flux integral all have the same value: If $F_n(P) = c$ (a constant) for *every* point P on S, then every approximating sum reduces to c times the area of S. This gives a simple recipe for evaluating $\iint_S \vec{F} \cdot d\vec{A}$ in such cases, provided you can calculate the area of S.

If $F_n(P) = c$ (a constant) for every point P on the oriented surface S, then

$$\iint_S \vec{F} \cdot d\vec{A} = cA(S).$$

Here $A(s)$ denotes the area of the surface S.

Example 1. Look at the field

$$\vec{F}(P) = G \frac{\overrightarrow{OP}}{|\overrightarrow{OP}|^3}$$

where G is some constant. Physically, $\vec{F}$ could represent either the gravitational field associated with a point mass, or the electrostatic field of a point charge. Let us calculate its flux through a sphere S_a, with radius a and center O, and oriented by the outward normal. To simplify, we first observe that, since G is constant, we can factor it out of the surface integral, so we can calculate the flux as if $G = 1$, and then put G back in as a factor at the end, so we follow that strategy, putting $G = 1$ for the time being. Notice that $\vec{F}(P)$ is itself normal to the sphere at P, and points outward, away from the origin, in the same direction as the normal that orients the sphere. That means that the normal component of $\vec{F}$ is just its *magnitude*:

$$F_n = |\vec{F}(P)| = \frac{|\overrightarrow{OP}|}{|\overrightarrow{OP}|^3} = \frac{a}{a^3} = \frac{1}{a^2}.$$

Since the normal component is constant, the flux is

$$\int_{S_a} \vec{F} \cdot d\vec{A} = \frac{1}{a^2} A(S_a) = \frac{1}{a^2} 4\pi a^2 = 4\pi.$$

All this was under the assumption that $G = 1$. To get the answer for general G, we have to multiply by G. We get

$$\int_{S_a} \overrightarrow{F} \cdot d\overrightarrow{A} = 4\pi G.$$

Notice that something unexpected happened here. The answer does not contain the radius of the sphere! That is, the flux is the same for *all* values of a, it does not depend on the radius a! This fact is a special case of a fact in electrostatics called Gauss' Law. The law asserts that the flux of such a field across a closed surface (such as a sphere or the faces of a cube) is proportional to the charge enclosed by the surface. The constant which we called G represents the enclosed charge.   ■

Problems

41.3. S is the filled in triangle ABC, where A, B, C are as indicated in Figure 38.1 on page 326. We orient the triangle by the normal which points away from the origin. The vector field $\overrightarrow{F}$ is given by $\overrightarrow{F} = 2\,\vec{\imath} + \vec{\jmath} - 3\,\vec{k}$.

    a) Find the normal component of $\overrightarrow{F}$ at the following points: (i) B; (ii) the center of gravity of the triangle ABC.

    b) Evaluate the flux integral $\int\int_S \overrightarrow{F} \cdot d\overrightarrow{A}$.

41.4. S is the cylinder defined by $x^2 + y^2 = a^2$ with $0 \leq z \leq 1$, oriented by the normal that points away from the z-axis.

    a) $\overrightarrow{E}$ is the field given by

$$\overrightarrow{E}(x, y, z) = \frac{(x, y, 0)}{x^2 + y^2}.$$

    Physically, $\overrightarrow{E}$ could represent the electrostatic field of an infinitely long charged wire running along the z-axis. Evaluate the flux integral $\int\int_S \overrightarrow{E}\, d\overrightarrow{A}$.

    b) $\overrightarrow{B}$ is the field defined by

$$\overrightarrow{B}(x, y, z) = \frac{(x, -y)}{x^2 + y^2}.$$

    Physically, $\overrightarrow{B}$ could represent the magnetic field produced by an electrical current flowing along the z-axis. Evaluate the flux integral $\int\int_S \overrightarrow{B} \cdot d\overrightarrow{A}$.

Using parametric representations

In this subsection, we work out a method for expressing a flux integral $\int\int_S \overrightarrow{F} \cdot d\overrightarrow{A}$ as an ordinary double integral in the parameter plane. In the language of flows: *The fluid flows across the surface in three-dimensional space. But when it comes down to evaluate the flux, we can work in the two-dimensional parameter plane.* The idea is to use the coordinate curves given by the parametric representation to partition the surface S. The approximating sums (41.1) for the flux integral can then be reinterpreted as approximating sums for an ordinary double integral in the parameter plane.

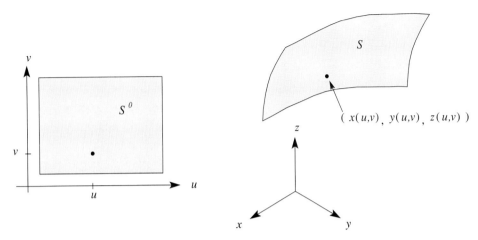

Figure 41.2

The setup. Start with a parametric representation for a surface S in xyz-space. This gives x, y, and z along S as functions of parameters (u, v):

$$x = x(u, v) \qquad y = y(u, v) \qquad z = z(u, v).$$

Here (u, v) varies over a region S^0 in the (u, v)-plane, as shown in Figure 41.2. We call S_0 the *parameter domain*. We let $\vec{r} = \vec{r}(u, v)$ denote the position vector of the point on S with parameter values (u, v):

$$
\begin{aligned}
\vec{r} &= x\vec{i} + y\vec{j} + z\vec{k} \\
&= x(u, v)\vec{i} + y(u, v)\vec{j} + z(u, v)\vec{k}.
\end{aligned}
$$

Partitioning the surface. We overlay the parameter domain S^0 with a rectangular grid and use the parametric representation to transfer the grid to the surface S, as shown in Figure 41.3. The rectangular grid on the parameter domain S^0 partitions S^0 into small rectangles $R_1, R_2, \ldots$. The grid on S partitions S into corresponding

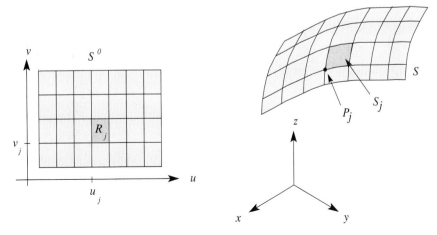

Figure 41.3

small surfaces $S_1, S_2 \ldots$. The small surfaces $S_1, S_2, \ldots$ are approximately filled in parallelograms.

Estimating area of S_j. Here we estimate the flux of the field $\overrightarrow{F}$ across the small surface S_j. We do this by replacing the non-constant field $\overrightarrow{F}$ by a sampled value $\overrightarrow{F}(P_j)$, and replacing the surface S_j by an approximating filled parallelogram.

Let (u_j, v_j) be the lower left corner of the rectangle R_j in the uv-plane, and let P_j be the corresponding point on the surface S. Then the vectors

$$
\begin{aligned}
\vec{r}_u &= \vec{r}_u(u_j, v_j) = \left(\frac{\partial x}{\partial u}(u_j, v_j), \frac{\partial y}{\partial u}(u_j, v_j), \frac{\partial z}{\partial u}(u_j, v_j) \right) \\
\vec{r}_v &= \vec{r}_v(u_j, v_j) = \left(\frac{\partial x}{\partial v}(u_j, v_j), \frac{\partial y}{\partial v}(u_j, v_j), \frac{\partial z}{\partial v}(u_j, v_j) \right)
\end{aligned}
$$

are tangent to the coordinate curves passing through P_j as shown below.

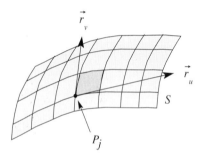

Now let Δu and Δv denote the horizontal and vertical dimensions of the small rectangle R_j. Then the vectors $\vec{r}_u \Delta u$ and $\vec{r}_v \Delta v$ span a parallelogram which approximates the small surface S_j, as shown in Figure 41.4.

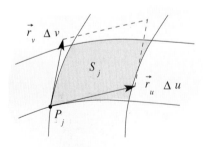

Figure 41.4

Thus, we get

$$
\begin{aligned}
\text{area of } S_j &\approx \text{ area of parallelogram} \\
&= |\vec{r}_u \times \vec{r}_v| \Delta u \, \Delta v \\
&= |\vec{r}_u \times \vec{r}_v| \, (\text{area of } R_j) \qquad (41.2)
\end{aligned}
$$

Sampling the normal component of $\overrightarrow{F}$. In order to find the normal component of $\overrightarrow{F}$ in terms of the parameters u and v, we need a way of representing the unit normal vector to the oriented surface S in terms of the parametrization. We can do so using the vectors $\vec{r}_u$ and $\vec{r}_v$ of the preceding paragraph. Each of these vectors is tangent to the surface S, so their cross product $\vec{r}_u \times \vec{r}_v$ is normal to S.

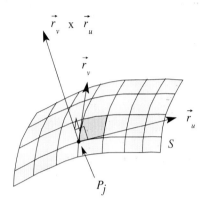

Thus, assuming $\vec{r}_u \times \vec{r}_v$ is not the zero vector, we get

$$\frac{\vec{r}_u \times \vec{r}_v}{|\vec{r}_u \times \vec{r}_v|} = \pm\vec{n}(P_j). \tag{41.3}$$

Terminology. *We say that the parametric representation is consistent with the orientation of S if we get a plus sign on the right of (41.3).*

We proceed under the assumption that the parametric representation of S is consistent with the orientation.[2] In this case, the normal component of $\overrightarrow{F}$ at P_j is

$$F_n(P_j) = \overrightarrow{F}(P_j) \cdot \vec{n}(P_j) = \overrightarrow{F}(P_j) \cdot \frac{\vec{r}_u \times \vec{r}_v}{|\vec{r}_u \times \vec{r}_v|}. \tag{41.4}$$

Estimating the flux across S_j. Now that we have worked out an estimate for the area of S_j and sampled the normal component of $\overrightarrow{F}$, it is easy to estimate the flux across S_j:

$$\text{Flux across } S_j \approx (\text{sampled value of } F_n)(\text{area of } S_j)$$
$$\approx \left(\overrightarrow{F}(P_j) \cdot \frac{\vec{r}_u \times \vec{r}_v}{|\vec{r}_u \times \vec{r}_v|}\right) (|\vec{r}_u \times \vec{r}_v| \, (\text{area of } R_j)).$$

Canceling the term $|\vec{r}_u \times \vec{r}_v|$ gives

$$\text{Flux across } S_j \approx \overrightarrow{F}(P_j) \cdot (\vec{r}_u \times \vec{r}_v)(\text{area of } R_j)$$
$$= \overrightarrow{F}(x(u_j, v_j), y(u_j, v_j), z(u_j, v_j)) \cdot (\vec{r}_u \times \vec{r}_v)(\text{area of } R_j). \tag{41.5}$$

---

[2]Working with a parametrization that is *not* consistent with the orientation will have the effect of reversing the sign of the final result. If you are must work with such a parametrization, just proceed, and remember to change the sign at the end.

Calculating the flux through S**.** Summing the flux estimates (41.5 through the small surfaces S_j gives an estimate for the flux through the entire surface S:

$$\text{Flux through } S \approx \sum_j \overrightarrow{F}(x(u_j, v_j), y(u_j, v_j), z(u_j, v_j)) \cdot (\vec{r}_u \times \vec{r}_v)(\text{area of } R_j). \quad (41.6)$$

Notice that the sum on the right is an approximating sum for the ordinary area integral

$$\iint_{S^0} \overrightarrow{F}(x(u, v), y(u, v), z(u, v)) \cdot (\vec{r}_u \times \vec{r}_v) \, du \, dv. \quad (41.7)$$

Thus, by using a fine enough grid in the parameter domain, the left side of (41.6) can be made as close as we wish to the flux flux integral $\iint_S \overrightarrow{F} \cdot d\overrightarrow{A}$, and the right side can be made as close as we wish to the area integral (41.7). The conclusion:

$$\boxed{\iint_S \overrightarrow{F} \cdot d\overrightarrow{A} = \iint_{S^0} \overrightarrow{F}(x(u, v), y(u, v), z(u, v)) \cdot (\vec{r}_u \times \vec{r}_v) \, du \, dv. \quad (41.8)}$$

The vector area element. You can read the formula (41.8) as follows. To evaluate the flux integral $\iint_S \overrightarrow{F} \cdot d\overrightarrow{A}$, write out the vector quantity $\overrightarrow{F}(x, y, z)$ in terms of the parameters u and v, and replace $d\overrightarrow{A}$ with $\vec{r}_u \times \vec{r}_v \, du \, dv$. Then integrate with respect to u and v. For this reason, we sometimes write

$$d\overrightarrow{A} = \vec{r}_u \times \vec{r}_v \, du \, dv. \quad (41.9)$$

The above expression is called the *vector area element* on the surface.

Example 2. Flux through a half-sphere. Let S be the top half of the unit sphere $x^2 + y^2 + z^2 = 1$, oriented by the normal vector that points away from the origin. $\overrightarrow{F}$ is the vector field defined by $\overrightarrow{F}(x, y, z) = z\vec{k}$. We want to calculate $\iint_S \overrightarrow{F} \cdot d\overrightarrow{A}$.

We first choose a parametric representation for S. There are many possibilities. We will use the angles ϕ and θ of spherical coordinates as parameters. The parametric representation is

$$\begin{aligned} x &= \sin\phi\cos\theta \\ y &= \sin\phi\sin\theta \\ z &= \cos\phi. \end{aligned}$$

To get the top half of the sphere, we use the parameter domain S^0 defined by $0 \le \phi \le \frac{\pi}{2}$ and $0 \le \theta \le 2\pi$. The position vector $\vec{r}$ is

$$\vec{r} = \sin\phi\cos\theta\,\vec{\imath} + \sin\phi\sin\theta\,\vec{\jmath} + \cos\phi\,\vec{k}.$$

Differentiating with respect to ϕ and θ gives

$$\begin{aligned} \vec{r}_\phi &= \cos\phi\cos\theta\,\vec{\imath} + \cos\phi\sin\theta\,\vec{\jmath} - \sin\phi\,\vec{k} \\ \vec{r}_\theta &= -\sin\phi\sin\theta\,\vec{\imath} + \sin\phi\cos\theta\,\vec{\jmath}. \end{aligned}$$

The cross product is

$$\vec{r}_\phi \times \vec{r}_\theta = \sin^2\phi\cos\theta\,\vec{\imath} + \sin^2\phi\sin\theta\,\vec{\jmath} + \sin\phi\cos\phi\,\vec{k}. \quad (41.10)$$

Before proceeding, we check whether our parametrization is consistent with the orientation. It is enough to check that the above cross product points away from the origin at a single point of the surface. We check the point $(x, y, z) = (1, 0, 0)$, which corresponds to the parameter values $(\phi, \theta) = (\pi/2, 0)$. Substituting these values in the above formula gives the vector $\vec{\imath}$, which does indeed point away from the origin at the point $(1, 0, 0)$. Thus, our parametrization *is* consistent with the orientation.

We next express the vector area element $d\vec{A}$ in terms of the parameters (ϕ, θ):

$$
\begin{aligned}
d\vec{A} &= \vec{r}_\phi \times \vec{r}_\theta \, d\phi \, d\theta \\
&= (\sin^2\phi \cos\theta \, \vec{\imath} + \sin^2\phi \sin\theta \, \vec{\jmath} + \sin\phi \cos\phi \, \vec{k}) \, d\phi \, d\theta. \qquad (41.11)
\end{aligned}
$$

Next, we express the field vectors $\vec{F} = z\vec{k}$ in terms of the parameters ϕ and θ:

$$
\vec{F} = z\vec{k} = \cos\phi \, \vec{k}.
$$

Combining this with the parameter expression for $d\vec{A}$ gives the integrand in (41.8):

$$
\begin{aligned}
\vec{F} \cdot d\vec{A} &= (\cos\phi \, \vec{k}) \cdot (\sin^2\phi \cos\theta \, \vec{\imath} + \sin^2\phi \sin\theta \, \vec{\jmath} + \sin\phi \cos\phi \, \vec{k}) \, d\phi \, d\theta \\
&= \sin\phi \cos^2\phi \, d\phi \, d\theta.
\end{aligned}
$$

Finally, integrating over the parameter domain gives the value of the flux integral:

$$
\iint_S \vec{F} \cdot d\vec{A} = \int_0^{2\pi} \int_0^{\pi/2} \sin\phi \cos^2\phi \, d\phi \, d\theta = \frac{2\pi}{3}.
$$

∎

Problem _____

41.5. Verify the formula (41.11) for the vector area element $d\vec{A}$ in Example 2 by working out the formula (41.10) for $\vec{r}_\phi \times \vec{r}_\theta$.

Measuring area

In this subsection we will show how to measure the area of a parametrized surface. We stick with the setup from the discussion of flux integrals over parametrized surfaces (except for the field $\vec{F}$, which plays no role here). We can partition the surface S in the same way, using a rectangular grid in the parameter domain. We already estimated the area of each small surface in the resulting partition. It is given in (41.2). Summing these estimates gives gives an estimate for the area of the entire surface S. But the sum *also* gives an approximating sum for a double integral over the parameter domain, so we obtain an integral formula for the area of the surface S:

$$
\text{area of } S = \iint_{S^0} |\vec{r}_u \times \vec{r}_v| \, du \, dv.
$$

The expression

$$
dA = |\vec{r}_u \times \vec{r}_v| \, du \, dv \qquad (41.12)
$$

is called the *area element* or the *element of surface area* on the surface S.

We now ask: *What is the relation between the vector area element $d\vec{A}$ and the scalar area element dA?* Comparing (41.3), (41.9), and (41.12) reveals the answer:

$$d\vec{A} = \vec{n}\, dA$$

where n is the unit normal vector to the surface. Notice that this tells you that flux integrals can be expressed in terms of either $d\vec{A}$ or dA:

$$\iint_S \vec{F} \cdot d\vec{A} = \iint_S \vec{F} \cdot \vec{n}\, dA = \iint_S F_n\, dA.$$

Problems

41.6. S is a sphere of radius 1.

   a) Work out the formula for the area element on S in terms of the parameters ϕ and θ used in Example 2.

   b) The sphere S is cut into two parts by a horizontal plane b units above the equator. Here $0 \leq b < 1$. Find the area of each piece.

41.7. S is the half-sphere of Example 2, given by $x^2 + y^2 + z^2 = 1$, $z \geq 0$, with the same orientation and parametric representation. T is a rectangular-like patch on S, similar to the patch S_j of Figure 41.3. To the patch T on S there corresponds a rectangle T^0 in the $\phi\theta$-plane. T^0 is a true rectangle given by $0.5 \leq \phi \leq 0.7$, $0.8 \leq \theta \leq 0.9$ (angle measures are in radians). $A(T)$ is the true surface area of T on the half-sphere, and $A_{\text{approx}}(T)$ is the approximation of $A(T)$ obtained by using the tangent vectors $\vec{r}_\phi$ and $\vec{r}_\theta$.

   a) Evaluate $A_{\text{approx}}(T)$. In all your numerical work, use three decimal places only.

   b) $A(T^0)$ is the true surface area of T^0 in the $\phi\theta$-plane. It equals 0.02. Find the constant k such that

$$A(T^0) \cdot k = A_{\text{approx}}(T).$$

   The constant k tells you by how much you have to multiply the area of T^0 in the ϕ, θ-plane in order to get the approximate area of T on S.

   c) What is the relation between $\vec{r}_\phi$, $\vec{r}_\theta$, and k? Write down an answer in complete sentences.

Additional Problems

The normal component of a vector field on a surface

41.8. S is the surface given by $z = 3xy$. As normal of S we pick the one whose third component is negative. $\vec{F}$ is the vector field given by $\vec{F}(x, y, z) = (z, z, z)$. Find the normal component of $\vec{F}$ on S at the point $Q(1, 2, 6)$.

41.9. S is the surface given by $(x, y, z) = (uv, u + v, v)$. As normal of S we take the standard normal $\vec{r}_u \times \vec{r}_v$.

a) The standard normal of S at the generic point corresponding to (u, v) has components (n_1, n_2, n_3). Find these components. They are functions of u and v.

b) $\overrightarrow{F}$ is the vector field given by $\overrightarrow{F}(x, y, z) = (z, z, 0)$. Find the normal component of $\overrightarrow{F}$ on S at the generic point corresponding to (u, v). Your answer will contain u and v.

Estimates of surface integrals

41.10. S is the triangle OKL in the xz-plane, as shown below. As normal, we take $\vec{\jmath}$. The vector field $\overrightarrow{F}$ is defined by $\overrightarrow{F} = (z^2, x^2, y^2)$.

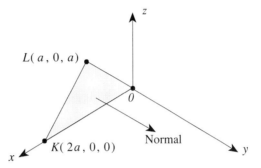

Estimate the flux of $\overrightarrow{F}$ across S. That is, find constants k and K such that

$$m \leq \int\int_S \overrightarrow{F} \cdot d\overrightarrow{A} \leq M$$

Evaluation of surface integrals by reasoning with the normal component

41.11. S is the straight circular cylinder of radius a and height h shown below. We orient it by picking the normal which points away from the z-axis.

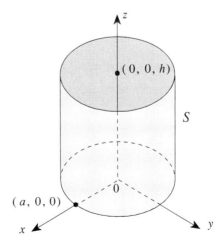

a) $\overrightarrow{F}$ is the vector field defined by $\overrightarrow{F} = (x, y, xyz)$. Find the flux of $\overrightarrow{F}$ across S. *Hint:* Look at F_n.

b) $\overrightarrow{G}$ is the vector field defined by $\overrightarrow{G} = (-y, x, x^2 + y^2 + z^2)$. Find the flux of $\overrightarrow{G}$ across S. *Hint:* Look at F_n.

41.12. The vector field $\overrightarrow{F}$ is defined by $\overrightarrow{F} = 4(x, y, z)$.

    a) R is the rectangle in the xz-plane shown below left. As its normal, we take $\vec{\jmath}$. Find the flux of $\overrightarrow{F}$ across R.

    b) S is the half-sphere $x^2 + y^2 + z^2 = b^2$, $z \geq 0$, with the normal pointing away from the origin, as shown below right. Find the flux of $\overrightarrow{F}$ across S.

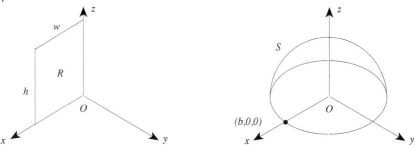

Evaluation of surface integrals using parametric representations

41.13. S is the filled in parallelogram $ABDC$ shown below. We orient S by the normal which points away from the origin.

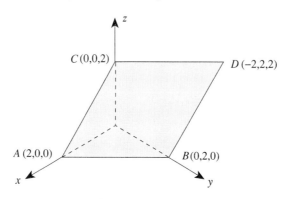

    a) As parametric representation of the plane of S we take

$$\overrightarrow{OP} = \overrightarrow{OA} + u\,\overrightarrow{AB} + v\,\overrightarrow{AC}.$$

    Write out this representation in components.

    b) Find the the flux of $\overrightarrow{F} = (0, x, 0)$ across the parallelogram ABD. *Use the parametric representation of a).*

41.14. The surface S is given by $(x, y, z) = (uv, u^2, u+v)$. We orient S by means of the standard normal. Figure 41.5 on the facing page shows the coordinate curves $u = 2$, $u = 2.3$, $v = 1$, $v = 1.2$. The figure is not to scale, it only explains the situation.

    a) P is the point on S corresponding to the parameter values $(u, v) = (2, 1)$. $\vec{a}$ is a tangent vector to the coordinate curve $u = 2$ at P, and $\vec{b}$ is a tangent vector to the coordinate curve $v = 1$ at P. Find the components of $\vec{a}$ and $\vec{b}$.

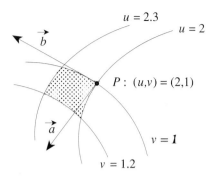

Figure 41.5

b) $\overrightarrow{F}$ is the vector field given by $\overrightarrow{F} = (0, 0, 3x)$. Use the field vector at P and tangent vectors to the coordinate curves at P to give an approximate value of the flux of $\overrightarrow{F}$ across the shaded patch on the surface.

c) To find the exact value of the flux of the field $\overrightarrow{F} = (0, 0, 3x)$ across the shaded patch we have to to evaluate a double integral which looks like $\int\int g(u, v) \, du \, dv$. Set up the integral but do not evaluate it.

41.15. We let all of space rotate around the x-axis with constant angular velocity ω ($\frac{\omega}{2\pi}$ revolutions per second). We do it in such a way that the rotation is counterclockwise if we look at the yz-plane in the direction of $-\vec{\imath}$. In this rotation, every point P in space has a velocity vector $\overrightarrow{v}(P)$. That is, the rotation defines a vector field $\overrightarrow{F}(P)$ defined by $\overrightarrow{F}(P) = \overrightarrow{v}(P)$. The field $\overrightarrow{F}(P)$ is a velocity field.

S is the rectangle in the xy-plane shown below. We use $\vec{k} = (0, 0, 1)$ as normal of S.

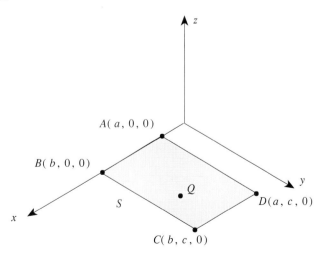

a) We are interested in the flux of $\overrightarrow{F}$ across S. Would you expect the flow to be positive, or zero, or negative, or is it impossible to say anything without making the computation? Explain your answer.

b) $Q(p, q, 0)$ is the generic point of S. Find the normal component of $\overrightarrow{F}$ at Q. Your answer will contain letters.

c) Set up $\iint_S \overrightarrow{F} \cdot d\overrightarrow{A}$ as an ordinary double integral in the xy-plane.

d) Evaluate the flux of $\overrightarrow{F}$ across S. Your answer will contain letters.

Measuring area

41.16. The surface S is the graph of a function $z = f(x, y)$. Work out a formula for the element of surface area dA on S in terms of x, y, and $f(x, y)$. *Hint:* You can parametrize the surface using x and y as parameters.

41.17. Calculate the surface area of the paraboloid given by $z = x^2 + y^2$, $0 \le z \le 1$.

41.18. The surface shown below is given parametrically by

$$\begin{aligned} x &= s \cos t \\ y &= s \sin t \\ z &= t \end{aligned}$$

with $-1 \le s \le 1$ and $0 \le t \le 2\pi$.

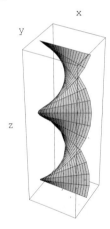

a) Write out an integral that measures the area of this surface.

b) Find the area of the surface by evaluating the integral you found in part a. *Hint:* You may want to consult an integral table or a computer.

Chapter 7

The Theorems of Green, Gauss, and Stokes

42   The Boundary of Solids, Surfaces, and Curves

Five kinds of integrals

By now, you have worked with five integrals in calculus: Integrals of functions $f(x)$, $g(x, y)$, and $h(x, y, z)$ of one, two, and three variables; line integrals, surface integrals. In each case, we integrate a function or vector field over a geometric object such as an interval, or a region, or a solid, or a curve, or a surface, as shown in Table 42.1 on the next page.

Problem

42.1. Is there anything that some or all of the five types of integrals have in common? In the way they are defined? In the way they are computed? In what they mean?

Now we look closely at the geometric objects over which we integrate. These objects are called *domains of integration*. The domains of integration have different "dimensions" in the following sense (the boldface numbers refer to the table of the five integrals):

1, 4 The domain of integration for an integral of $f(x)$ is an interval, and the domain for a line integral is a curve. To locate a point in an interval I on the x-axis requires *one* coordinate, and to locate a point on a curve C requires *one* parameter value t. Therefore, we say: Intervals and curves are *one-dimensional*.

Interval I: $m \le x \le n$ Curve C: $(x, y, x) = (a(t), b(t), c(t))$

1.	The integral $\int_m^n f(x)\,dx$ of a function $f(x)$ of one variable over an interval $I = [m, n]$ on the x-axis:	
2.	The double integral $\int\int_D g(x, y)\,dA$ of a function $g(x, y)$ of two variables over a region D in the xy-plane:	
3.	The triple integral $\int\int\int_K h(x, y, z)\,dV$ of a function $h(x, y, z)$ of three variables over a solid K in xyz-space:	
4.	The line integral $\int_C \vec{F} \cdot d\vec{r}$ of the vector field $\vec{F}$ over an oriented curve C in xyz-space:	
5.	The surface integral $\int\int_S \vec{F} \cdot d\vec{A}$ of the vector field $\vec{F}$ over an oriented surface S in xyz-space:	

Table 42.1: Five kinds of integrals

2, 5 The domain of integration for an integral of $g(x, y)$ is a region in the xy-plane. To locate a point in a region D requires *two* coordinates, and to locate a point on a surface S requires *two* parameter values (u, v). Therefore, we say: Regions and surfaces are *two-dimensional*.

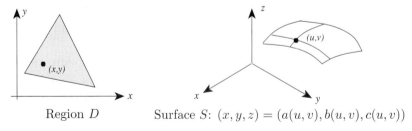

Region D Surface S: $(x, y, z) = (a(u, v), b(u, v), c(u, v))$

3 The domain of integration for an integral of $h(x, y, z)$ is a solid in xyz-space. To locate a point in a solid requires *three* coordinates. Therefore, we say: Solids in space are *three-dimensional*.

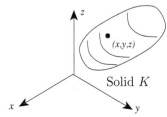

Solid K

Problem _____

42.2.   a) Let U be the set of all points in space that satisfy the conditions $y = e^x$
with $3.9 \leq x \leq 22.7$ and $z = 15.3$. What is the dimension of U?

  b) Let V be the set of all points in the plane that satisfy

$$(x - 22)^2 + (y + 49)^2 = 2.$$

What is the dimension of V?

  c) Let W be the set of all points in space that satisfy

$$(x - 22)^2 + (z + 49)^2 \leq 2$$

and $y = 0$. What is the dimension of W?

_____

The boundary of intervals and curves

The boundary of a finite interval consists of the points where it ends, namely the
endpoints. Similarly, consider a curve from initial point A to endpoint B in the
plane or in space. Its boundary consists of the points A and B. A *closed* curve has
neither an initial point nor an endpoint. That is, the boundary of a closed curve is
empty.

A common notation for "boundary of" is the round delta ∂. Thus, the sentence
"the boundary of the curve C consists of the points A and B is written as

$$\partial C = \{A, B\}.$$

The boundary of intervals and curves is illustrated in the Table 42.2.

Summary: *Intervals and curves are one-dimensional. The boundary of an interval
or a curve either consists of two points or is empty.*

Interval or curve C	Boundary ∂C

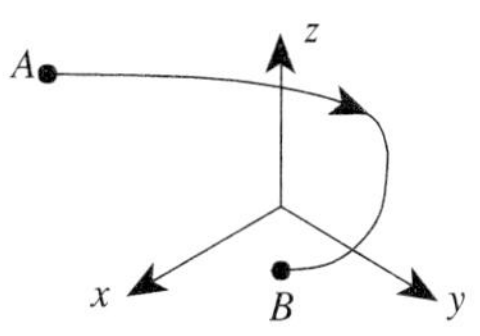	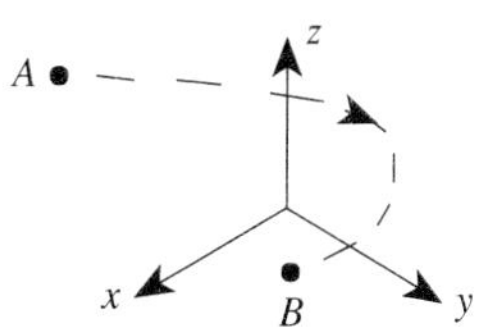
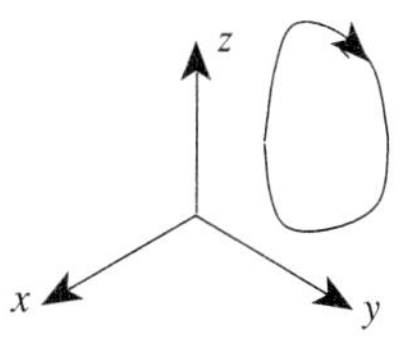	Empty

Table 42.2: Boundaries of curves and integrals

Region D	Boundary curve ∂D
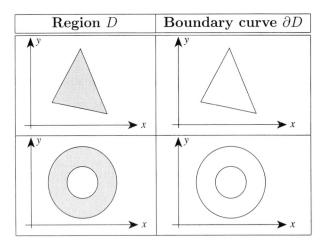	

Table 42.3: Boundaries of plane regions

The boundary of regions and surfaces

We work only with regions and surfaces of finite extent.

The boundary ∂D of a region D in the xy-plane is the curve that contains the region. The curve ∂D is closed. In the region shown in the first row of Table 42.3, the closed curve ∂D has sharp corners because it consists of three line segments joined end to end.

A region in the plane such as the disk $x^2 + y^2 \leq 1$ has a boundary that has no sharp corners.

It is possible for the boundary ∂D of a plane region to consist of *several distinct* closed curves. For example, the region D shown in row two of Table 42.3 is a disk with a smaller disk removed. Its boundary ∂D consists of the two circles.

The boundary ∂S of a surface S in space is the curve which limits the surface. ∂S is either empty or is made up of one or several closed curves. It consists of the points where you can go from one side of the surface to the other. Again, ∂S may have sharp corners, as shown in the first row of Table 42.4 on the next page. As illustrated in the second row of Table 42.4, a surface may have several parts and sharp edges. The surface in row two consists of a rectangle and a triangle joined together along the edge LM. The boundary of this surface consists of the closed curve $KLQMNK$. The surface in the last row consists of all six faces of the rectangular box. *This surface has no boundary points because there is no point where you can get from the inside to the outside.* It is a *closed* surface. The boundary of a closed surface is empty.

Another example of a closed surface is the sphere $x^2 + y^2 + z^2 = 144$.

Summary: *Regions in the plane and surfaces in space are two-dimensional. The boundary of a region in the plane or of a surface either consists of one or more closed curves or is empty.*

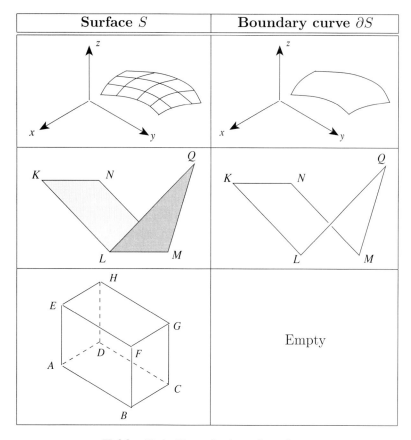

Surface S	Boundary curve ∂S

Table 42.4: Boundaries of surfaces

The boundary of a solid in space

We use the word "solid" for a solid which is of finite extent. For instance, the cylinder $x^2 + z^2 \leq 4$ is *not* of finite extent because it extends all along the entire y-axis.

The boundary of a solid in space consists of one or more closed surfaces which separate the solid from the rest of three-dimensional space. For example, if the solid K is the ball defined by $x^2 + y^2 + z^2 \leq 4$, then the boundary ∂K is the sphere defined by $x^2 + y^2 + z^2 = 4$. If K is the hollowed out ball defined by $1 \leq x^2 + y^2 + z^2 \leq 4$, then its boundary ∂K consists of the two spheres $x^2 + y^2 + z^2 = 1$ (the inner boundary) and $x^2 + y^2 + z^2 = 4$ (the outer boundary). It is possible for the boundary of a solid to consist of many closed surfaces. Think of a Swiss cheese.

The boundary of a solid may or may not have sharp edges. For example, the boundary of a solid box is a closed surface which has sharp edges.

Summary: *Solids in space are three-dimensional. The boundary of a solid consists of one or more closed surfaces which separate the solid from the rest of space.*

Problem _____

42.3. Let K be the solid pyramid shown in Figure 42.1 on the following page. Its corners are the points Q, R, U, V, its edges are the line segments QR,

QU, QV, RU, RV, UV, and its faces are the filled-in triangles QRU, QRV, QUV, RUV. We provide several copies of K so that you can draw on them.

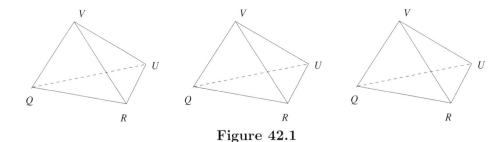

Figure 42.1

a) Describe ∂K in words in the following form: First, state what the boundary is (two points, or a closed curve, etc.). Then identify the boundary by means of the corners, edges, etc.

b) C is the curve that consists of the edges QR, RV, VU joined end to end. Describe ∂C in two parts, as in part a.

c) S is the surface that consists of the faces QRU and RUV joined along the edge RU. Describe ∂S in two parts, as in part a.

d) Can you use corners, edges, and faces to form a surface S_1 such that ∂S_1 consists of the edges QR, RV, VQ joined end to end? If yes, do it; if no, explain.

e) Can you use corners, edges, and faces to form another surface S_2 that has the same boundary as S_1 of part d? In other words: Can you find two different surfaces S_1 and S_2, which have the same boundary consisting of the edges QR, RV, VQ joined end to end? If yes, find such a surface S_2; if no, explain.

Additional Problems

42.4. K is the solid cube $0 \leq x \leq 1, 0 \leq y \leq 1, 0 \leq z \leq 1$. a) Describe ∂K in words. b) Describe $\partial(\partial K)$ in words.

42.5. Below is shown a half-sphere of radius r which rests on the xy-plane.

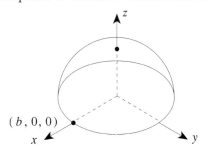

The following objects can be seen in this picture:

$$C : x^2 + y^2 = b^2, \ z = 0$$
$$D : x^2 + y^2 \le b^2, \ z = 0$$
$$S : x^2 + y^2 + z^2 = b^2, \ z \ge 0$$
$$K : x^2 + y^2 + z^2 \le b^2, \ z \ge 0$$

a) Describe each of C, D, S, K in one or more complete sentences.

b) From now on, X stands for any of the four objects C, D, S, K. What is the dimension of X? (four questions)

c) The symbol $\emptyset$ denotes the empty set. Find all X such that $\partial X = \emptyset$.

d) Find all X so that ∂X is a closed curve.

e) Are there pairs X, Y so that $\partial X = \partial Y$? If yes, give all of them. If no, explain.

f) The plane $z = \frac{b}{2}$ intersects K. We write L for the figure of intersection. What is the dimension of L? Describe ∂L by formulas.

g) Find all X so that ∂X consists of two separate points.

h) Find all X so that ∂X is of dimension 2.

42.6. K is the triangular solid shown below. It is the solid over which you integrated in Problem 27.9 on page 234.

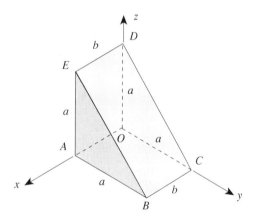

K has five faces: The solid rectangles $OABC$, $OADE$, $BCDE$, and the triangles OCD and ABE. K has nine edges: OA, AB, BC, CO, BE, CD, OA, DE, and AE.

a) Find the dimension of the following:

(i) ABE; (ii) $OADE$; (iii) $BE + AE$.

By "+" we mean that we consider the two edges BE and AE as one new object.

b) Find the boundary of the following:

(i) $OABC + OADE$;   (ii) $OC + CD + DO$;
(iii) $OABC + OADE + BCOE$.

Hint: Remember one way of describing the boundary of a surface is as the collection of all points where you can go from one side of the surface to the other.

42.7. K is the solid circular cone shown in Figure 42.2.

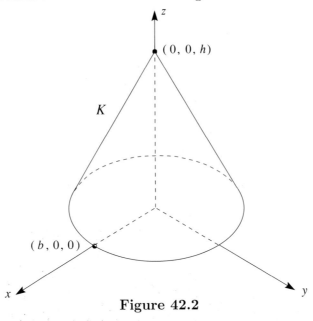

Figure 42.2

a) Describe ∂K in words, without using any equation(s).

b) We take the boundary of ∂K and remove the solid circle $x^2 + y^2 \le b^2$, $z = 0$ in the xy-plane. We write S for the surface we obtain in this way. Describe ∂S in words, without using any equation(s).

43   Gauss' Divergence Theorem

In this section, we come to one of the central results of the subject: the Divergence Theorem of Gauss. Gauss' Theorem relates the flux of a vector field through a *closed* surface to the triple integral of a scalar function over the region enclosed by the surface. Before stating the theorem, we need to develop some machinery.

Outward flux and divergence

Orientation of the boundary of a solid

Gauss' Theorem deals with the flux integral of a vector field through the boundary ∂K of a solid K. Before we can make sense of this, we need a rule for attaching an orientation to the surface ∂K.

The outward normal. Take a solid K, and a point P on its boundary surface ∂K. There are two unit vectors normal to ∂K at P. One points *into* the solid K, and the other points *out* of K. The one that points out of K is called the *outward normal* at P. We denote it $\vec{n}(P)$. When we integrate over the boundary surface of a solid, we use the orientation determined by the outward normal.

Example 1. Let B be the solid ball $x^2 + y^2 + z^2 \le 1$. Its boundary ∂B is the sphere $x^2 + y^2 + z^2 = 1$. At a point P of ∂B, the outward normal $\vec{n}(P)$ is the unit vector pointing directly away from the origin. In formulas,

$$\vec{n}(x, y, z) = x\,\vec{\imath} + y\,\vec{\jmath} + z\,\vec{k}.$$

Example 2. Let A be the solid defined by $2 \le x^2 + y^2 + z^2 \le 3$. Its boundary ∂A consists of two concentric spheres, S_1 and S_2, of radius 2 and 3 respectively. The solid A consists of the two spheres, together with all points inside S_2 and outside S_1. Along the *outer* boundary sphere S_2, the outward normal points directly away from the origin. Along the *inner* boundary sphere, the outward normal points *toward* the origin. A cross section of A is shown below, together with normals at an inner boundary point P and an outer boundary point Q.

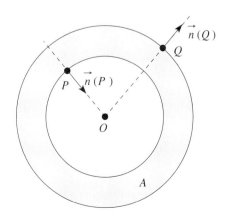

Problem

43.1. The solid K shown below is a solid cylinder with a hole drilled in it. The height is 4, the inner diameter is 1, and the outer diameter is 2.

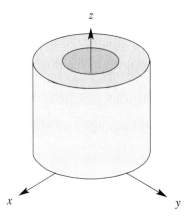

a) The point P is on the top surface of K. Find the components of the outward normal $\vec{n}(P)$.

b) The point $Q(2, 0, 1)$ is on the outer lateral surface. Find the components of the outward normal $\vec{n}(Q)$?

c) The point $R(1, 0, 1)$ is on the inner lateral surface. Find the components of the outward normal $\vec{n}(R)$?

Outward flux of a field from a solid

We start with two things:

1. A solid K;
2. A vector field $\vec{F}$ defined along the boundary ∂K.

We assume that the boundary ∂K consists of one or more surfaces, and orient each of the boundary surfaces by the outward normal vector, as discussed previously. The *outward flux* of $\vec{F}$ from K is simply the flux integral $\iint_{\partial K} \vec{F} \cdot d\vec{A}$.

Example 3. $\vec{F}$ is the velocity field of an incompressible fluid, and K is a hypothetical solid completely immersed in the fluid. Assume there are no sources of new fluid or sinks to remove fluid inside the region K. In this case, the amount of fluid inside the region K does not change, so the rate that fluid flows into K must exactly match the rate that it flows out. In other words, the flux of $\vec{F}$ out of K is 0. This is a special case of the Divergence Theorem, which will come up soon. ∎

Problems

43.2. $\vec{F}$ is a constant vector field. What is the outward flux of $\vec{F}$ from a generic solid K? You do not need to know what K is to answer this question. *Hint:* Think of $\vec{F}$ as the velocity field of a fluid, with each fluid particle moving along with the same constant velocity.

43.3. $\vec{r}$ is the radial field defined by

$$\vec{r} = x\,\vec{\imath} + y\,\vec{\jmath} + z\,\vec{k}.$$

Calculate the outward flux of $\vec{r}$ from the ball B_a defined by $x^2 + y^2 + z^2 \leq a^2$. Here a can be any positive number. Your answer may depend on a. *Hint:* What is the normal component of $\vec{r}$ along the boundary of B_a?

43.4. Let K be a rectangular box with edges parallel to the coordinate axes. The edges parallel to the x, y, and z axes have lengths Δx, Δy, and Δz, respectively. The boundary ∂K consists of the six faces of the box K. Each face is oriented by the normal pointing out of K.

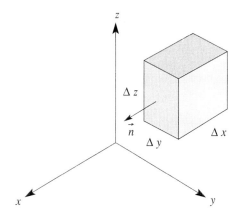

a) $\overrightarrow{F}$ is the field with components $(x, 0, 0)$. What is the flux of $\overrightarrow{F}$ across the faces parallel to the xy-plane? What about the faces parallel to the xz-plane? *Hint:* Draw a typical field vector along each of these four faces. How is it related to the outward normal vector? What is its normal component?

b) Calculate the flux of $\overrightarrow{F}$ through each of the faces parallel to the yz-plane. Be careful with signs! *Hint:* The x coordinate is constant on each of these faces. Let x_0 and x_1 denote the x coordinates on the "back" and "front," respectively. Your answers will involve x_0 and x_1.

c) Use your answers to parts a and b to calculate the outward flux of $\overrightarrow{F}$ from K.

d) $\overrightarrow{G}$ is the vector field with components $(y, 0, 0)$. Calculate the outward flux of $\overrightarrow{G}$ from K.

The divergence of a vector field in space

How is divergence defined?[1]   The *divergence* of a vector field at a point P is a scalar that measures the tendency of the field to flow away from the point P. As

---

[1]The divergence of a vector field was defined previously in Section 33, using partial derivatives of the component functions of the field. Here we give a purely geometric definition, without reference to a choice of coordinates. We will reconcile the two definitions shortly.

we have done previously with integrals, we *define* the divergence of the field $\overrightarrow{F}$ at the point P by giving a recipe for approximating it.

The divergence of $\overrightarrow{F}$ at the point P is denoted div $\overrightarrow{F}|_P$. It is the number which is approximated by the following procedure.

1. Choose a small solid K that contains the point P. For example, K may be chosen to be a small ball or box.

2. Measure the outward flux of the field $\overrightarrow{F}$ from the solid K, and divide by the volume of K. The resulting scalar approximates div $\overrightarrow{F}|_P$:

$$\text{div } \overrightarrow{F}|_P \approx \frac{1}{V(K)} \iint_{\partial K} \overrightarrow{F} \cdot d\overrightarrow{A}. \tag{43.1}$$

The precise meaning of (43.1) is that the div $\overrightarrow{F}|_P$ can be approximated to any precision we wish by making the solid K small enough. It means roughly that div $\overrightarrow{F}|_P$ measures the outward flux per unit volume at P.

Example 4. $\vec{r}$ is the radial field defined by $\vec{r} = x\,\vec{\imath} + y\,\vec{\jmath} + z\,\vec{k}$. We will calculate the divergence of $\vec{r}$ at the origin.

In Problem 43.3 you found that the outward flux of $\vec{r}$ from a ball B_a about the origin with radius a is $4\pi a^3$. Formula (43.1), therefore, gives

$$\text{div } \vec{r}|_O \approx \frac{\text{outward flux}}{\text{volume}} = \frac{4\pi a^3}{\frac{4}{3}\pi a^3} = 3.$$

Thus, the divergence of $\vec{r}$ at O is 3. ■

Problems

43.5. $\overrightarrow{F}$ is a constant field, and P is a generic point. What is div $\overrightarrow{F}|_P$? Explain your answer in one or more complete sentences. *Hint:* Your answer to Problem 43.2 may be useful.

43.6. $\overrightarrow{F}$ and $\overrightarrow{G}$ are the fields of Problem 43.4. Find the divergence of $\overrightarrow{F}$ and $\overrightarrow{G}$ at a generic point P. *Hint:* Use a small box containing P in the divergence formula (43.1).

43.7. $\overrightarrow{F}$ is the gradient of a scalar function Φ. The function Φ has a strict local maximum at the point P. This means that $\Phi(Q) < \Phi(P)$ for any point Q that is near P but not equal to P. Do you expect div $\overrightarrow{F}|_P$ to be positive, negative, or zero? Please explain. *Hint:* Let c be a constant that is slightly smaller than $\Phi(P)$, and let K be the solid defined by $\Phi \geq c$. Then K contains P. (Why?) What can you say about the sign of the flux of grad Φ out of K?

How do we calculate divergence? We work with a generic vector field $\overrightarrow{F}$ with components (F_1, F_2, F_3). *How do we calculate the divergence of $\overrightarrow{F}$ from the components of $\overrightarrow{F}$?* You can work out a formula for the divergence in two stages:

1. In the definition of divergence, we use a *small* ball. For points in such a small ball, we can safely replace the components $(F_1, F_2, F_3$ of $\overrightarrow{F}$ by their linear approximations (see Section 16). Therefore, we have to deal only with vector fields whose components are linear functions of three variables.

$$\vec{r} = x\,\vec{i} + y\,\vec{j} + z\,\vec{k}.$$

2. The divergence of a field with linear components can be calculated by measuring the outward flux from boxes. You did this yourself for some linear fields in Problem 43.6.

We will omit the details of the derivation and cut right to the chase. If $\overrightarrow{F}$ is a vector field with components (F_1, F_2, F_3), then the divergence of $\overrightarrow{F}$ is given by

$$\operatorname{div} \overrightarrow{F}|_P = \frac{\partial F_1}{\partial x}(P) + \frac{\partial F_2}{\partial y}(P) + \frac{\partial F_3}{\partial z}(P).$$

Example 5. $\vec{r} = x\,\vec{\imath} + y\,\vec{\jmath} + z\,\vec{k}$ is the vector field of Example 4. In Example 4, we worked out the divergence of $\vec{r}$ at the origin directly from the definition. Here, we will calculate the divergence at any generic point in space using the above formula. The divergence of $\vec{r}$ is

$$\operatorname{div} \vec{r} = \frac{\partial x}{\partial x} + \frac{\partial y}{\partial y} + \frac{\partial z}{\partial z} = 1 + 1 + 1 = 3.$$

Notice that for this particular field, the divergence came out to be constant. ∎

Problem _____

43.8. $\overrightarrow{E}$ is the radial field defined by

$$\overrightarrow{E} = q\,\frac{\vec{r}}{r^3}.$$

Here q is a constant, $\vec{r}$ is as in Example 5 above, and $r = |\vec{r}|$. This is the electrostatic field of a point charge of magnitude q at the origin. (See Example 1 in Section 41.) Notice that $\overrightarrow{E}$ is undefined at the origin. Calculate $\operatorname{div} \overrightarrow{E}|_P$ for a generic point P that is not the origin.

The Divergence Theorem

The Divergence Theorem in space

The divergence of a vector field measures the outward flux per unit volume. Thus, we may think of $\operatorname{div} \overrightarrow{F}$ as a "flux density" for the field $\overrightarrow{F}$. It is, therefore, reasonable to expect that integrating $\operatorname{div} \overrightarrow{F}$ over a solid K will give the outward flux from K. Gauss' Divergence Theorem asserts that this is in fact the case.

Statement of the Divergence Theorem. We consider:

- A solid K in space;
- The boundary ∂K of K, oriented by the *outward* normal;
- A vector field $\overrightarrow{F} = (F_1, F_2, F_3)$, whose components have the property that all their partial derivatives are defined and continuous at all points of K.

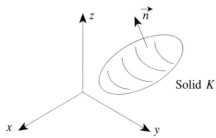

Gauss' Divergence Theorem. *We are given a solid K and a vector field $\overrightarrow{F}$, as specified previously. Then the outward flux of the vector field $\overrightarrow{F}$ across the boundary of the solid K equals the triple integral over K of the divergence of $\overrightarrow{F}$. In formulas:*

$$\iint_{\partial K} \overrightarrow{F} \cdot d\overrightarrow{A} = \iiint_K \operatorname{div}\overrightarrow{F}\, dV. \qquad (43.2)$$

To get a feeling for the meaning of the Divergence Theorem, we look at some special cases.

- If the field $\overrightarrow{F}$ is constant, then the right side of (43.2) is 0, so the Divergence Theorem asserts that the outward flux of a constant field from a solid K is 0. You had already worked this out in Problem 43.2 on page 372.

- More generally, if the divergence of $\overrightarrow{F}$ is 0 at each point of K, then the integrand on the right side of (43.2) is 0, so again the outward flux is 0. We had already made this observation in Example 3. Borrowing from the jargon of fluids, we say that the field $\overrightarrow{F}$ is *incompressible* if its divergence is 0 at every point.

- If the divergence of $\overrightarrow{F}$ is positive at each point of K, then the right side of (43.2) is positive, so the outward flux of $\overrightarrow{F}$ from K is positive. Similarly, if the divergence of $\overrightarrow{F}$ is negative at each point of K, then the outward flux is negative.

Problems _____

Hint: First, write down Gauss' Theorem in general form. Then ask: Is it OK to use the theorem? What is given? What do I have to find?

43.9. Throughout this problem, K is a solid in xyz-space. We are interested in the flux of different vector fields across ∂K.

    a) Find the flux of $\overrightarrow{F} = (x+\sin(y^2 z^{2398}), y+(xz)^4, z)$ across the boundary ∂K of K. Because K is not specified, the answer cannot be a number; push the computation as far as you can.

    b) As in part a, but for the field $\overrightarrow{F} = (yz, xz, xy)$.

    c) As in part a, but for the field $\overrightarrow{F} = \operatorname{curl}(xyz, x^2 y^2 z^2, x^3 y^3 z^3)$.

43.10. The surface S consists of the six faces of the cube K shown below. We orient S by the outside normal. Find the flux of $\overrightarrow{F} = (xy, yz, xz)$ across S.

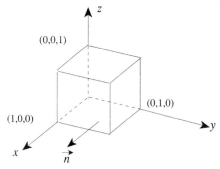

43.11. Go back to the table of integrals at the beginning of Section 42. Gauss' Theorem establishes a relation between two of them. Which are these integrals? Which concept expresses the link?

43.12. We write $\vec{r} = (x, y, z)$ and $r = |\vec{r}| = \sqrt{x^2 + y^2 + z^2}$, and let $\vec{E}$ be the vector field defined by $\vec{E} = qr^{-3}\vec{r}$. Here q is a constant. K will be the solid ball $x^2 + y^2 + z^2 \leq a^2$, and $S : x^2 + y^2 + z^2 = a^2$ is the boundary ∂K of K. We orient S by taking the outside normal.

   a) Find the flux of $\vec{E}$ across ∂K by direct computation, that is, without using Gauss' Theorem. *Hint:* (1) What is the normal component E_n on the sphere S? (2) A sphere of radius a has a surface area of $4\pi a^2$.

   b) In Problem 43.8, you calculated div $\vec{E}$ and found that it is zero. Your friend says: "OK, let's use Gauss' Theorem to compute the flux:

$$\iint_{\partial K} \vec{E} \cdot d\vec{A} = \iiint_K \operatorname{div} \vec{E}\, dV = \iiint_K 0\, dV = 0.$$

   Your computation in part a must be wrong." What do you say?

Example 6. Gauss' Law Here we exploit the germ of truth in the incorrect calculation of Problem 43.12 part b to derive the basic equation of electrostatics. We work with the field

$$\vec{E} = q\,\frac{\vec{r}}{r^3}$$

which is the electrostatic field of a charge of magnitude q situated at the origin. Let S be *any* closed surface that does not pass through the origin. Then S bounds a solid K. We orient S by the outward normal to K. We ask: *What is the flux of $\vec{E}$ across S?* We consider two cases.

Case 1: The surface S does not enclose the origin. This is illustrated at left in Figure 43.1.

Case 2: The surface S encloses the origin. This is illustrated at right in Figure 43.1

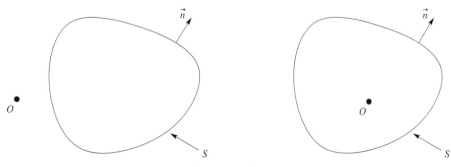

Figure 43.1

Calculation of the flux in Case 1. This is the easy case. We know that the divergence of $\vec{E}$ is 0 *except at the origin where it is undefined.* Since the origin is not part of K or its boundary, the Divergence Theorem gives

$$\iint_S \vec{E} \cdot d\vec{A} = \iiint_K \operatorname{div} \vec{E}\, dV = 0.$$

Conclusion: *The flux of $\vec{E}$ through any closed surface which does not enclose or pass through the charge at the origin is* 0.

Calculation of the flux in Case 2.   In this case, we cannot apply the Divergence Theorem directly, because the solid K enclosed by the surface S contains the origin, where the field $\vec{E}$ is undefined. We will instead apply the Divergence Theorem on a closely related region K' which excludes the origin. The region K' is obtained by removing a small spherical bubble about the origin from K, as illustrated in Figure 43.2.   Since the origin is not part of the region K' or its boundary, we can

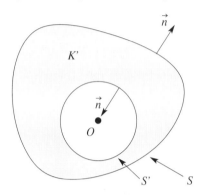

Figure 43.2

apply the Divergence Theorem to obtain

$$\iint_{\partial K'} \vec{E} \cdot d\vec{A} = \iiint_{K'} \operatorname{div} \vec{E} \, dV = 0. \tag{43.3}$$

We now examine the left side of (43.3) to see what this tells us about the flux of $\vec{E}$ across S. The boundary of K' consists of two surfaces. One is just the original surface S. Its orientation is given by the normal pointing out of K', which is the same as the one pointing out of K. The second part of the boundary of K' is a sphere S' centered at the origin O. It is oriented by the normal vector point our of K', which points *into* the sphere, toward the origin. Thus, (43.3) reads

$$\iint_{S} \vec{E} \cdot d\vec{A} + \iint_{S'} \vec{E} \cdot d\vec{A} = 0.$$

Rearranging gives

$$\iint_{S} \vec{E} \cdot d\vec{A} = -\iint_{S'} \vec{E} \cdot d\vec{A}$$

so calculating the flux of $\vec{E}$ across S is reduced to calculating the flux across the sphere S'. You worked this out in Problem 43.12, obtaining a value of $4\pi q$ for the answer. However, in that calculation you used the opposite orientation, determined by the normal that points out of the sphere. Changing the orientation reverses the sign of the flux integral, so the flux of $\vec{E}$ through S' is $-4\pi q$. Thus, we obtain

$$\iint_{S} \vec{E} \cdot d\vec{A} = -(-4\pi q) = 4\pi q.$$

Conclusion: *The outward flux of $\vec{E}$ through any surface that encloses the origin is $4\pi q$.*

In the language of electrostatics, this says the following. Let S be any surface that encloses a single point charge. The electrostatic flux across S is 4π times the magnitude of the enclosed charge. ■

Problems

43.13. $\overrightarrow{F}$ is the field $r^{-3}\vec{r}$.

    a) S is a sphere of radius 2 and center $(0,3,0)$, oriented by the normal that points away from its center. What is the flux of $\overrightarrow{F}$ across S?

    b) T is a sphere of radius 2 and center $(0,-1,0)$, oriented by the normal that points away from its center. What is the flux of $\overrightarrow{F}$ across T?

43.14. $\overrightarrow{F}$ is the vector field $r^{-2}\vec{r}$. If you compute the divergence of $\overrightarrow{F}$, you find that it equals r^{-2}, except at the origin where the field is undefined.

    a) S is a closed surface that does not enclose the origin or pass through the origin. Is the outward flux of $\overrightarrow{F}$ through S positive, negative, or zero? Explain your answer. *Hint:* What does the Divergence Theorem tell you?

    b) T is a surface that encloses the origin. Is the flux of $\overrightarrow{F}$ through T positive, negative, or zero? Explain your answer. *Hint:* Reason as in Example 6.

43.15. Repeat Problem 43.14 for the field $\overrightarrow{F} = r^{-4}\vec{r}$. In this case, div $\overrightarrow{F} = -r^{-4}$.

The Divergence Theorem in the plane

Flux of a vector field in the plane. Up to this point, we have worked the flux of a vector field $\overrightarrow{F}$ in space across a surface S in space. The flux of $\overrightarrow{F}$ across S measures the amount of fluid which crosses S per unit of time. To keep track of the direction of the flow relative to S, we have to choose a normal of the surface.

In the plane, there are no surfaces to cross, only curves. So we consider a vector field $\overrightarrow{F}$ and a curve C in the plane. The flux of $\overrightarrow{F}$ across C measures the amount of fluid which crosses C per unit of time. To keep track of the direction of the flow relative to C, we have to choose a normal of the curve.

The following figure illustrates flux in space (left) and flux in the plane (right).

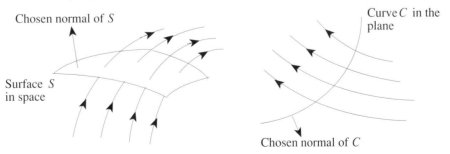

What is the flux of a vector field $\overrightarrow{F}$ in the plane across a curve C in the plane? It measures the amount of fluid which crosses C per unit of time, relative to the direction of the normal chosen for C. Graphically, the amount is represented by an area. *How do we compute the flux of $\overrightarrow{F}$ across the curve C with the normal $\vec{n}$?* We compute it as the integral of the normal component F_n of the field $\overrightarrow{F}$ over the curve

C, taken with respect to arc length along C:

$$\text{Flux of } \overrightarrow{F} \text{ across } C = \int_C F_n \, ds = \int_{t_A}^{t_B} \overrightarrow{F} \cdot \vec{n}(t) \, |\vec{r}'(t)| \, dt.$$

In this formula, $\vec{r} = r(t)$ is a parametrization of C, $\vec{n}(t)$ is an outward *unit* normal of C, and we are working with the normal component the way we are working with the tangential component in Section 35 on page 295.

Gauss' Theorem in the plane: *Let D be a (filled-in) region in the xy-plane. Let $\vec{n}$ be the outward unit normal for its boundary curve C as shown below.*

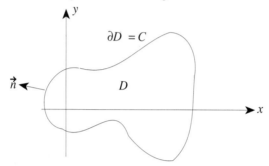

Let $\overrightarrow{F} = (P(x,y), Q(x,y))$ be a vector field such that all partial derivatives of P and Q are defined and continuous for all points of D. Then the flux across C can be computed as the double integral of $\operatorname{div} \overrightarrow{F}$ over the region D:

$$\int_C F_n \, ds = \iint_D \operatorname{div} \overrightarrow{F} \, dA = \iint_D (P_x + Q_y) \, dx \, dy.$$

Problem _____

43.16. The ellipse C given by $(x-3)^2 + 9(y-1)^2 = 9$ is shown below. We consider the vector fields $\overrightarrow{F}$ and $\overrightarrow{G}$ given by $\overrightarrow{F} = (5x, 2y+3)$, $\overrightarrow{G} = (2y+3, -5x)$. Let D be the region of which C is the boundary.

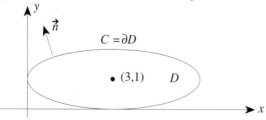

  a) Find the flux of $\overrightarrow{F}$ across C. *Hint:* The area of the filled-in ellipse $\dfrac{x^2}{a^2} + \dfrac{y^2}{b^2} = 1$ equals $\pi\,a\,b$.

  b) Find the flux of $\overrightarrow{G}$ across C.

  c) Your colleague turns toward you and says: "I don't get it. The two vector fields $\overrightarrow{F}$ and $\overrightarrow{G}$ are nearly the same—at any point (x,y), their field vectors have the same magnitude and are perpendicular to each other. Why isn't the flux across C the same?" Write down an answer that will convince your colleague.

Additional Problems

A special case of Gauss' Theorem

43.17. K is the solid rectangular box shown below. It has length $b - a$, width 1, and height 1. We orient its boundary by the outward normal.

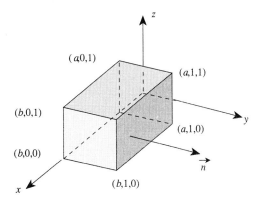

$\overrightarrow{F}$ is the vector field given by $\overrightarrow{F} = (f(x), 0, 0)$. In this formula, $f(x)$ is a generic function.

a) Find the flux of $\overrightarrow{F}$ across ∂K *without* using Gauss' theorem, that is, by finding the flux across each of the six faces which make up ∂K. Simplify your answer as much as possible.

b) Now find the flux by means of Gauss' theorem. Simplify the triple integral as much as possible, but do *not* evaluate it.

c) The quantities in parts a and b are the same. This reflects a theorem of elementary calculus. Which theorem is it?

Applying Gauss' Theorem

43.18. The surface S consists of the five faces of the triangular box shown below. As normal, we take the one which points away from the box.

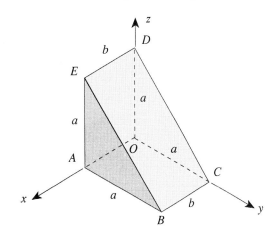

The vector field $\overrightarrow{F}$ is given by

$$\overrightarrow{F} = (2x - 3y + 6z, x - y - z + 4, 2x - 7z + 15).$$

Find the flux of $\overrightarrow{F}$ across S.

43.19. The surface S looks like a cylindrical tin can with one lid removed, as follows: The cylinder has radius a. The axis of the cylinder is the y-axis. The lid that has not been removed lies in the xz-plane. The lid in the plane $y = b$ has been removed. The surface S is shown below. The open end faces the viewer.

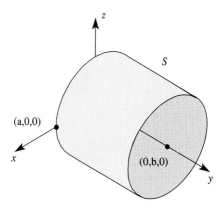

As normal of S we take the outward normal.

  a) Describe the boundary ∂S of S in words, using one or more complete sentences.

  b) Let $\overrightarrow{F}$ be the vector field given by $\overrightarrow{F} = (y, y, y)$. Find the flux of $\overrightarrow{F}$ across S. *Hint:* Consider the *solid* cylinder. What is its boundary?

43.20. Let S be the surface which consists of all six faces of the rectangular box shown below. We orient S by the normal which points away from the box.

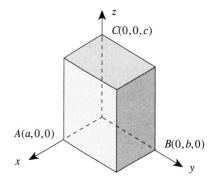

Let $\overrightarrow{F}$ be the vector field defined by

$$\overrightarrow{F}(x, y, z) = (x^2, 5y, 3).$$

Find the flux of $\overrightarrow{F}$ across S.

43.21. All through this problem, $\overrightarrow{F}$ is the vector field defined by

$$\overrightarrow{F} = (3x^2, 5y, -6xz),$$

and K is the triangular solid block shown below. We orient all faces of K by choosing the exterior normal.

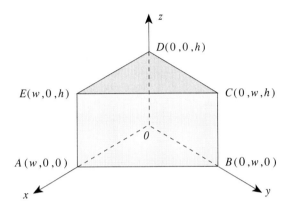

S is the surface consisting of the following faces of K: The triangles OAB and ECD, the front face $ABCE$, and the rectangle $OAED$.

    a) Describe the following boundaries in words: ∂S, ∂K, $\partial(\partial S)$.

    b) Find the flux of $\overrightarrow{F}$ across S.

    c) Find the flux of curl $\overrightarrow{F}$ across S.

    d) Find the flux of $\overrightarrow{F}$ across ∂K.

    e) Find the flux of curl $\overrightarrow{F}$ across ∂K.

43.22. Let K be the solid circular cone shown below, and and let $\overrightarrow{F}$ be the vector field given by $\overrightarrow{F} = (x^2yz, x^2 + z^2, x^2 + y^2)$.

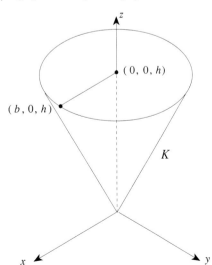

    a) Find the flux of $\overrightarrow{F}$ across ∂K. Hint: $\dfrac{d \sin^2 t}{dt} = -2 \sin t \cos t$.

    b) The flux of a vector field across a closed surface tells us something about the sources and sinks of the vector field in the interior of the closed surface. In the case of our field $\vec{F}$ and closed surface ∂K, what can you say about the sources and sinks within the solid K?

c) Your friend is confused and asks: "Wait a minute, there must be something wrong. Did the professor not say in class that the flux of constant vector field across a closed surface is zero. Our vector field $\vec{F}$ is not constant, but the flux is zero. That is a contradiction." Write down a response.

43.23. The surface S consists of the four faces QRS, QRT, RST, QST of the pyramid shown below. As normal of S, we take the one that points out of the pyramid. We are also given he vector field $\vec{F} = (y^2 + z^3, x + z^2, x^3 + z^2)$.

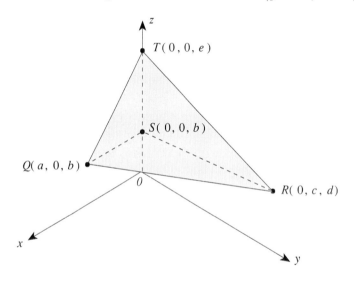

Estimate the flux of $\vec{F}$ across S. That is, find scalars m and M such that

$$m \le \iint_S \vec{F} \cdot d\vec{A} \le M$$

Hint: (1) We do not know the values of $a, b \dots$, but all these scalars are positive, and given any two of them, we can decide which one is greater. (2) The volume of the solid pyramid equals $\frac{1}{6}a(e - b)c$.

43.24. We are dealing with a vector field $\vec{G} = (G_1(x, y, z), G_2(x, y, z), G_3(x, y, z))$ and a point $P(a, b, c)$. We are interested in the flux across very small spheres centered at P. We consider such a sphere S_1 of radius $4 \cdot 10^{-2}$. Give the best estimate you can for the radius R_2 of a sphere S_2 such that the flux across S_2 is one seventh of the flux across S_1. *Hint:* Remember that

$$\frac{\iint_S \vec{G} \cdot d\vec{A}}{V(B)} \approx \operatorname{div} \vec{G}|_P.$$

In this formula, B is a solid ball, $V(B)$ its volume, and S its boundary.

43.25. Throughout this problem, $\vec{F}$ is a vector field, $f, g, \dots$ are scalar functions, K is a solid, S a surface, and C a curve.

Six statements follow. Some of the statements are true, and some are not. Also, some of the statements have terms that do not make sense, such as "the square root of a vector." Go over each statement, and write down your answer in complete sentences:

- Do all the terms make sense? If some do not make sense, explain why.
- Is the statement true? If not, explain.

1. The flux of $\vec{F}$ across the solid K equals the line integral of $\vec{F}$ over the boundary of K.

2. The triple integral of div $\vec{F}$ over the solid K equals the flux of $\vec{F}$ over the boundary of K.

3. The line integral of grad f over the boundary of the surface S equals the flux of grad f across S.

4. The flux of $\vec{F}$ across the boundary of the solid K equals the triple integral of div $\vec{F}$ over K.

5. The flux of curl $\vec{F}$ across the boundary of the solid K equals zero. *Hint:* Apply Gauss' Theorem to curl $\vec{F}$.

6. The triple integral of div $\vec{F}$ over the boundary of the solid K is zero.

44   Stokes' Theorem

Orientation of a surface and its boundary

We are given a surface S and a vector field $\vec{F}$. We want to study the flux of $\vec{F}$ across S and the line integral of $\vec{F}$ along the closed curve ∂S. Remember that such a line integral is also called the *circulation* of $\vec{F}$ along ∂S.

$$\iint_S \vec{F} \cdot d\vec{A} = ? \qquad \oint_{\partial S} \vec{F} \cdot d\vec{r} = ?$$

To integrate, we have to orient the surface and its boundary. We have to choose a normal vector for S and a direction arrow for ∂S. We choose the orientations in an interlocking way:

1. Given an orientation of S, we orient ∂S as follows: Stand on S such that the normal vector has its initial point at your feet and the endpoint at your head. Define an orientation on ∂S by walking along ∂S so that S is on your *left*.

2. Given an orientation of ∂S, we orient S as follows: Stand on the surface such that it is on your left if you walk along ∂S in the direction of orientation. Take the normal with initial point at your feet and endpoint at your head.

When the orientations are related in this way, we say that S and ∂S are *coherently* (or *consistently*) oriented.

Problem _____

44.1.   a) At left in Figure 44.1 on the next page, S is the triangle ABC, oriented by the normal which points away from the origin. Draw the arrow on its boundary so that S and ∂S are coherently oriented.

   b) At right in Figure 44.1 on the facing page, S is part of a right cylinder (a can with top and bottom removed). Its boundary ∂S consists of two circles. One of them is oriented, as shown. Choose a normal of S and orient the other circle so that S and ∂S are coherently oriented.

Statement of Stokes' Theorem

We consider:

- An oriented surface S in space;
- The boundary ∂S of S, oriented coherently with S;
- A vector field $\vec{F} = (F_1, F_2, F_3)$ whose components have the property that all their partial derivatives are defined for all points of S.

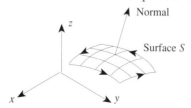

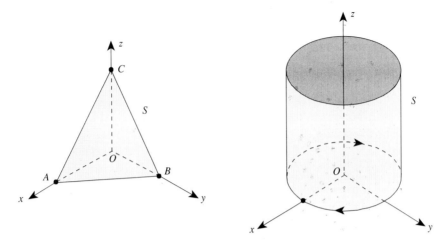

Figure 44.1

For the statement of the theorem, remember that the curl of a vector field $\overrightarrow{F} = (F_1, F_2, F_3)$ is given by

$$\operatorname{curl} \overrightarrow{F} = \nabla \times \overrightarrow{F} = \left(\frac{\partial F_3}{\partial y} - \frac{\partial F_2}{\partial z}, \frac{\partial F_1}{\partial z} - \frac{\partial F_3}{\partial x}, \frac{\partial F_2}{\partial x} - \frac{\partial F_1}{\partial y} \right).$$

Stokes' Theorem: *We are given a surface S and a vector field $\overrightarrow{F}$, as specified above. Then the line integral over the boundary ∂S of S equals the flux of* $\operatorname{curl} \overrightarrow{F}$ *across S. In formulas:*

$$\oint_{\partial S} \overrightarrow{F} \cdot d\vec{r} = \iint_S \operatorname{curl} \overrightarrow{F} \cdot d\overrightarrow{A}.$$

The meaning of Stokes' Theorem

To become familiar with the theorem, we first examine its statement. Then we work through examples and problems.

We look at Stokes' Theorem and observe the following:

- On the *left* side, we have a line integral. It represents the circulation of $\overrightarrow{F}$ around the boundary ∂S, that is, the work done by $\overrightarrow{F}$ on the closed curve ∂S.
- On the *right* side, we have the surface integral of $\operatorname{curl} \overrightarrow{F}$ over the surface S. It can be thought of representing the flow of $\operatorname{curl} \overrightarrow{F}$ across S.

It means:

(i) The circulation of $\overrightarrow{F}$ along *boundary* of S is determined by the vector field $\operatorname{curl} \overrightarrow{F}$ on the *interior* of the surface S.

(ii) Now we ask: *How* does $\operatorname{curl} \overrightarrow{F}$ determine the circulation? It does it by means of an integral: The circulation of $\overrightarrow{F}$ along ∂S is the flux integral of $\operatorname{curl} \overrightarrow{F}$ over the surface S.

Now we go to examples and problems.

Example 1. We rewrite the theorem with left and right sides interchanged:

$$\iint_S \operatorname{curl} \overrightarrow{F} \cdot d\overrightarrow{A} = \oint_{\partial S} \overrightarrow{F} \cdot d\vec{r}.$$

Let S be a *closed* surface. Then ∂S is empty, and the line integral equals zero.

Result:　*The flux of* curl $\overrightarrow{F}$ *over any closed surface S equals zero, no matter what* $\overrightarrow{F}$ *is.*

∎

Example 2. We consider two surfaces S_1 and S_2 as follows. S_1 consists of all faces of the cube shown below, except the bottom face $OABC$. The surface S_2 consists of the filled-in bottom face $OABC$.

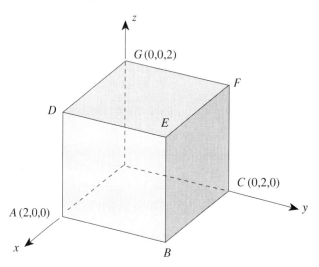

We orient S_1 by choosing the normal which points away from the origin, and we orient S_2 by $(0,0,1)$. Note that S_1 and $_2$ have the same boundary: $\partial S_1 = \partial S_2 = AB + BC + CO + OA$.

As vector field $\overrightarrow{F}$, we take one whose components have the property that all their partial derivatives are defined for all points in space. In other words, $\overrightarrow{F}$ is a vector field which can be used for Stokes' theorem everywhere.

Now we apply Stokes' theorem to S_1, rewritten as in Example 1:

$$\iint_{S_1} \text{curl}\,\overrightarrow{F} \cdot d\overrightarrow{A} = \oint_{\partial S_1} \overrightarrow{F} \cdot d\overrightarrow{r}.$$

Similarly for S_2.

$$\iint_{S_2} \text{curl}\,\overrightarrow{F} \cdot d\overrightarrow{A} = \oint_{\partial S_2} \overrightarrow{F} \cdot d\overrightarrow{r}.$$

We have $\partial S_1 = \partial S_2$. Therefore

$$\iint_{S_1} \text{curl}\,\overrightarrow{F} \cdot d\overrightarrow{A} = \iint_{S_2} \text{curl}\,\overrightarrow{F} \cdot d\overrightarrow{A},$$

and this is true despite the fact that S_1 is different from S_1. Instead of our particular S_1 and S_2 we could have taken any two surfaces S and T with the property $\partial S = \partial T$.

Result:　*We are given two surfaces S and T which have the same boundary. Then the flux of* curl $\overrightarrow{F}$ *across S is the same as across T, no matter what $\overrightarrow{F}$ is as long as the partial derivatives of its components are defined where necessary.*

∎

Problem _____

44.2. Go back to the cube of Example 2. Describe two more surfaces S_3 and S_4 and their orientations such that the following is true: $\partial S_3 = \partial S_4 = AB + BC + CO + OA$. Make sure that the orientations of S_3 and S_4 are consistent with the orientation $AB + BC + CO + OA$.

Example 3. The curve C consists of the four sides of a square $DEFG$ in space. Each side is of length b. $\vec{n} = (4, 1, 5)$ is a normal vector of the plane of the square. We orient C consistently with $\vec{n}$. The figure below is not to scale.

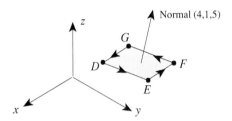

Given a vector field $\overrightarrow{F}$, we want to find the circulation along the closed curve $DEFG$. We can either evaluate the circulation directly, or we can try to apply Stokes' theorem. *Can we apply the theorem? If yes, will it be easier than to compute the circulation directly?* Problems 44.3 and 44.4 deal with these two questions. ∎

Problems _____

44.3. C is the closed curve $DEFG$ shown in the figure above, and $\overrightarrow{F}$ is the vector field defined by $\overrightarrow{F} = (x, x, x)$. The problem is to find the circulation of $\overrightarrow{F}$ along C.

  a) Can we apply Stokes' theorem? If yes, why? How?

  b) What is easier, computing the circulation directly or via Stokes' theorem?

  c) Find the circulation.

44.4. S is the disk $x^2 + y^2 \leq 1$ in the xy-plane. As normal of S, we take $\vec{k} = (0, 0, 1)$. We consider the vector field $\overrightarrow{F} = r^{-2}(y, -x, 3z)$ where $r = \sqrt{x^2 + y^2}$. The following figure shows S in the xy-plane.

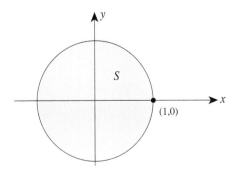

  a) Orient ∂S consistently with S. Show the arrow of the orientation of ∂S.

b) Find the tangential component $F_{\tan}$ of $\overrightarrow{F}$ along ∂S. Use $F_{\tan}$ to find the circulation along ∂S.

c) We compute curl $\overrightarrow{F}$, and find it is the zero vector. Your friend says: "OK, let's use Stokes' Theorem to compute the line integral:

$$\oint_C \overrightarrow{F} \cdot d\vec{r} = \iint_S \text{curl}\,\overrightarrow{F} \cdot d\overrightarrow{A} = \iint_S \vec{0} \cdot d\overrightarrow{A} = 0.$$

Your computation in part b must be wrong." What do you say?

Theorems of Gauss and Stokes compared

Both theorems establish a link between different kinds of integrals according to a basic template:

$$\int_{\partial(\text{something})} \overrightarrow{F} = \int_{\text{something}} \text{derivative of } \overrightarrow{F}.$$

In the case of Gauss' Theorem, the "something" is a solid K, and the "derivative" is div, so the template becomes

$$\iint_{\partial K} \overrightarrow{F} \cdot d\overrightarrow{A} = \iiint_K \text{div}\,\overrightarrow{F}\,dV.$$

In the case of Stokes' Theorem, the "something" is a surface S, and the "derivative" is curl, so the template becomes

$$\oint_{\partial S} \overrightarrow{F} \cdot d\vec{r} = \iint_S \text{curl}\,\overrightarrow{F} \cdot d\overrightarrow{A}.$$

The particulars vary, but the template is the same in both cases.

Physical meaning of curl

Up to now, we have worked with the curl in terms of the partial derivatives of the components (F_1, F_2, F_3) of the field $\overrightarrow{F}$. We now ask whether the curl $\overrightarrow{F}$ has a physical or geometric meaning that does not depend on any particular choice of a xyz coordinate system. As we did with the divergence in Section 43, we will resolve this issue by coming up with a recipe for estimating curl $\overrightarrow{F}(P)$ using geometric measurements that do not depend on the choice of coordinates. Here, Stokes' Theorem will play the role that was played by Gauss' Theorem in the discussion of divergence.

The situation for the curl is complicated by the fact that curl $\overrightarrow{F}(P)$ is a *vector* quantity. Instead of estimating the vector curl $\overrightarrow{F}(P)$ directly, we will instead estimate the *scalars* curl $\overrightarrow{F}(P) \cdot \vec{n}$ when $\vec{n}$ is a unit vector. Taking $\vec{n}$ to be $\vec{i}$, $\vec{j}$, and $\vec{k}$ will give the components (F_1, F_2, F_3) of $\overrightarrow{F}$.

As illustrated below, in the plane perpendicular to $\vec{n}$, draw a small circle C_ρ with center P and radius ρ. Let D_ρ be the filled-in circle with C_ρ as its boundary, oriented by the normal vector $\vec{n}$. We orient the circle C_ρ in the manner required by Stokes' Theorem.

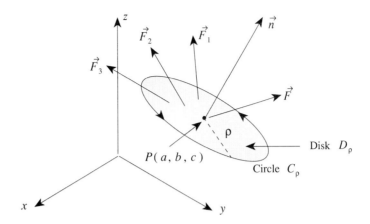

If the radius ρ is small enough, then

$$\text{curl } \overrightarrow{F}(P) \cdot \vec{n} \approx \text{curl } F(Q) \cdot \vec{n} \qquad \text{for } Q \text{ in } D_\rho$$

and the approximation can be made as close as you wish by choosing ρ small enough. This tells you that the *average* value of curl $\overrightarrow{F}$ over the disk D_ρ is a good approximation to curl $\overrightarrow{F}(P) \cdot \vec{n}$:

$$\text{curl } \overrightarrow{F} \cdot \vec{n} \approx \frac{1}{A(D_\rho)} \iint_{D_\rho} \text{curl } F \cdot \vec{n} \, dA.$$

In this formula, $A(D_\rho)$ stands for the area of the disk D_ρ. Now apply Stokes' Theorem to express the integral on the right as the circulation of $\overrightarrow{F}$ around C_ρ. The result is

$$\text{curl } \overrightarrow{F}(P) \cdot \vec{n} \approx \frac{1}{A(D_\rho)} \int_{C_\rho} \overrightarrow{F} \cdot d\vec{r}.$$

The left side of this formula is the scalar component of curl $\overrightarrow{F}(P)$ along the unit vector $\vec{n}$. The right side is the circulation of $\overrightarrow{F}$ per unit area around a small circle about P perpendicular to $\vec{n}$.

Summary: *The scalar component of* curl $\overrightarrow{F}(P)$ *along the unit vector $\vec{n}$ is approximated by the circulation of $\overrightarrow{F}$ per unit area around a circle centered at P and perpendicular to $\vec{n}$.*

It means that you can estimate the scalar component of curl $\overrightarrow{F}(P)$ in any direction, to any desired precision, by measuring the circulation around a small circle about P perpendicular to the given direction, and dividing by the area of the circle.

What does this mean physically or geometrically? To answer this question, we insert a probe into the flow of the vector field. The probe is a small paddlewheel centered at P. The paddles are attached to the small circle C_ρ. The wheel turns around the axis. The axis is normal to the plane of the circle, as shown in the following figure.

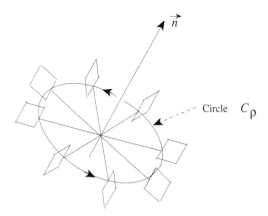

Will the paddlewheel turn or not? If yes, in which direction will it turn? What makes the paddlewheel turn is the tangential component F_{tan} of the field vector along the points of the circle C_ρ. If F_{tan} is more positive than negative along C_ρ, the paddles get more push in the direction of the orientation of C_ρ than in the opposite direction. The paddlewheel turns in the direction of the orientation of C_ρ. Here "F_{tan} is more positive than negative" means that the circulation $\oint_C F_{\text{tan}}\, ds$ of $\vec{F}$ along C_ρ is positive.

Result: *If the circulation along C_ρ is positive, the paddlewheel turns in the direction of the orientation of C_ρ.*

Similarly, if the circulation is negative, the paddlewheel turns in the direction opposite the orientation of C_ρ. The direction and intensity of the spin is determined by the scalar product $\operatorname{curl}\vec{F}\cdot\vec{n}_0$.

Summary

- For a small disk S_ρ, the circulation, per unit area, along its boundary circle $C_\rho = \partial S_\rho$ is approximately the scalar projection of $\operatorname{curl}\vec{F}|_P$ onto the normal $\vec{n}$ of the plane of the circle.

- The scalar projection of $\operatorname{curl}\vec{F}|_P$ onto $\vec{n}$ tells you how fast, and in which direction, the paddlewheel with axis $\vec{n}$ and center P will spin.

Note that at one and the same point P, the direction of spin of the paddlewheel will depend on the position of the normal $\vec{n}$.

Problem

44.5. $\vec{F}$ is the vector field with components $(z^2, x + x^2 + z^2, y^2 + x)$. We put a paddlewheel into the flow and look at the paddlewheel in the direction of $-\vec{n}$. That is, our direction of vision is opposite to the normal $\vec{n}$ of the paddlewheel. This is shown in Figure 44.2 on the next page.

  a) We put the center of the paddlewheel at $P(1,1,1)$ and align $\vec{n}$ in the direction $(-2,5,-1)$. Do you expect the paddlewheel to turn, and, if yes, clockwise or counterclockwise as seen by us?

  b) We leave the paddlewheel at $P(1,1,1)$. Can you position its normal $\vec{n}$ so that the paddlewheel will hardly spin at all? If yes, find such a vector $\vec{n}$; if no, explain.

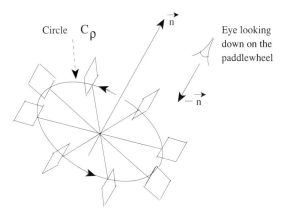

Figure 44.2

c) We keep the paddlewheel at $P(1,1,1)$. How will you position the normal $\vec{n}$ so that we see the strongest possible spin in clockwise direction?

d) Can you find a point Q for the center of the paddlewheel so that the following is true: No matter how you position the plane of the paddlewheel, the paddlewheel does not seem to spin. If yes, find such a point; if no, explain.

Stokes' Theorem in the plane: Green's Theorem

Does Stokes' Theorem mean anything for vector fields in the plane? The idea is (i) to consider a vector field $\vec{F} = (P(x,y), Q(x,y))$ in the plane as a vector field $\vec{F} = (P(x,y), Q(x,y), 0)$ in space, and (ii) to consider a region D in the xy-plane as a surface in space. We orient D by $(0,0,1)$, and we orient the boundary of D consistently, as shown below.

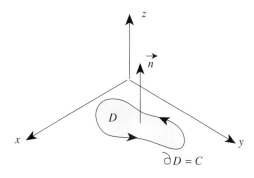

We apply Stokes' Theorem to $\vec{F} = (P(x,y), Q(x,y), 0)$ and to the surface D:

$$\oint_C \vec{F} \cdot d\vec{r} = \iint_D \operatorname{curl} \vec{F} \cdot d\vec{A}$$

On the left side, we have $\vec{F} \cdot d\vec{r} = P\,dx + Q\,dy + 0\,dz$, and on the right side $\operatorname{curl} \vec{F} = (0, 0, Q_x - P_y)$. We obtain what is usually called Green's Theorem:

Green's Theorem: *Let D be a (filled-in) region in the xy-plane. We orient its boundary curve C as shown below.*

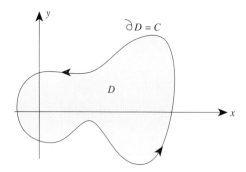

$\vec{F} = (P(x,y), Q(x,y))$ *is a vector field such that all partial derivatives of P and Q are defined for all points of D. Then we have*

$$\oint_C \vec{F} \cdot d\vec{r} = \oint_C P\,dx + Q\,dy = \iint_D (Q_x - P_y)\,dx\,dy = \iint_D \mathrm{curl}(P,Q)\,dx\,dy.$$

The theorem of Green links a line integral with a double integral.

Remember: The curl of a vector field (P,Q) in the plane is given by $\mathrm{curl}(P,Q) = Q_x - P_y$. It is a scalar.

Problem

44.6. $\vec{F}$ is the vector field in the plane defined by $\vec{F} = (2y, x)$. The ellipse C is given by $(x-3)^2 + 9(y-1)^2 = 9$, with counterclockwise orientation, as shown below.

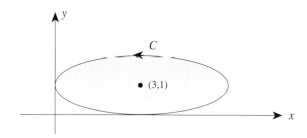

Evaluate $\displaystyle\oint_C \vec{F} \cdot d\vec{r}$. Hint: The area of the filled-in ellipse $\dfrac{x^2}{a^2} + \dfrac{y^2}{b^2} = 1$ equals $\pi\,a\,b$.

Additional Problems

44.7. Let C be the circle $x^2 + z^2 = 9$, $y = 0$ in the xz-plane, as shown in Figure 44.3 on the facing page. We orient C as follows: If you look onto the xz-plane from the point $(0, 10, 0)$, then C is oriented clockwise.

a) Indicate the orientation of C by drawing an arrow on C.

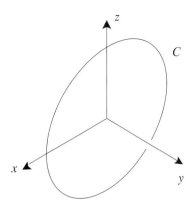

Figure 44.3

b) The vector field $\vec{F}$ is given by $\vec{F} = (2x - 3z, y + 4x, z + 2y)$. Evaluate $\oint_C \vec{F} \cdot d\vec{r}$. *Hint:* Use Stokes' Theorem to convert the line integral to a surface integral.

44.8. We consider a circular cone of height h and radius b, as shown in the following figure. We call this surface S. The cone is open at the bottom. That is, the "bottom lid" $x^2 + y^2 \leq b^2, z = 0$ is not part of S. We take the normal which points away from the z-axis.

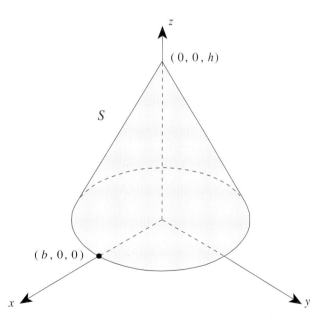

We also consider the vector field given by $\vec{F} = (3xz, 3yz, z^2)$. *In this problem, we are interested in the flux of* curl $\vec{F}$ *across different surfaces.*

a) Describe the boundary of S in words and in formulas.

b) Find the flux of curl $\vec{F}$ across S. *Hint:* Use Stokes' Theorem.

c) We write B for the bottom lid $x^2 + y^2 \leq b^2, z = 0$. Describe the boundary of B in words and in formulas.

d) Find the flux of curl $\vec{F}$ across B.

e) You find that the flux of curl $\vec{F}$ across the cone and its bottom lid is the same. Is that a coincidence, or is there a general reason for that? *Hint:* Write down the theorem of Stokes. Then ask yourself: What do we have to know about the surface S to find the flux of curl $\vec{F}$ across? Try to formulate your insight in a general statement that your fellow student would understand.

44.9. H is the half-circle in the xz-plane, of radius a and center at $(a, 0, 0)$, oriented as shown below.

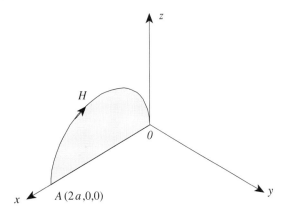

The vector field $\vec{G}$ is given by $\vec{G} = (3z, y, 2z)$.
You may or may not need the following facts: div $\vec{G} = 1$, curl $\vec{G} = (0, 3, 0)$.
Evaluate $\int_H \vec{G} \cdot d\vec{r}$. *Hint:* Add the line segment OA, use Stokes, subtract the line integral over OA.

44.10. We are dealing with a vector field $\vec{F} = (F_1(x, y, z), F_2(x, y, z), F_3(x, y, z))$ and a point $P(a, b, c)$. We have an instrument that measures the circulation of $\vec{F}$ around circles that have their center at P, and whose radius equals $R = 10^{-3}$. There are many such circles. Each of them is determined by a vector that is perpendicular to the plane in which it lies. Two such circles C_1, C_2 with normals $\vec{n}_1$, $\vec{n}_2$ are shown below.

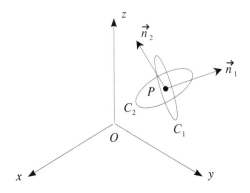

We have the results of three measurements:

$$\text{Circle } C_1: \ \overrightarrow{n}_1 = (1, 4, -7), \quad \oint_{C_1} \overrightarrow{F} \cdot d\overrightarrow{r} = 0;$$

$$\text{Circle } C_2: \ \overrightarrow{n}_2 = (0, -2, 1), \quad \oint_{C_2} \overrightarrow{F} \cdot d\overrightarrow{r} = 12 \cdot 10^{-7};$$

$$\text{Circle } C_3: \ \overrightarrow{n}_3 = (-1, 3, 0), \quad \oint_{C_1} \overrightarrow{F} \cdot d\overrightarrow{r} = 0.$$

We use these measurements to get some information on the direction and magnitude of curl $\overrightarrow{F}$ at P. To do this we have to remember that

$$\frac{\oint_C \overrightarrow{F} \cdot d\overrightarrow{r}}{A(S)} \approx (\text{curl } \overrightarrow{F})_n.$$

In this formula, S is a disk of area $A(S)$, C the boundary of S, and $(\text{curl } \overrightarrow{F})_n$ the normal component of curl $\overrightarrow{F}$ at P.

a) Give the best estimate you can of the components of a vector $\overrightarrow{g}$ which has the same direction as $\pm \text{curl } \overrightarrow{F}$. *Hint:* Suppose you knew the components of curl $\overrightarrow{F}$. For which direction of $\overrightarrow{n}$ will the circulation around C be zero?

b) Can you find a scalar s such that either $s \le |\text{curl } \overrightarrow{F}|$ or $s \ge |\text{curl } \overrightarrow{F}|$? *Hint:* You project a vector $\overrightarrow{a}$ onto a vector $\overrightarrow{m}$ and find that the projection equals 5. What can you say about the magnitudes of $\overrightarrow{a}$ and $\overrightarrow{m}$?

44.11. The vector field is defined by $\overrightarrow{F} = (y^2 + z^2, x^2 + z^2, x^2 + y^2)$.

a) Find curl $\overrightarrow{F}$.

b) We consider the circle C defined by $(x - 6)^2 + (z - 2)^2 = 10^{-16}$, $y = 5$. It is a very small circle in a plane parallel to the xz-plane. When we look onto the plane $y = 5$ from the point $(0, 10, 0)$ we get the following picture of C (the radius out of scale, of course):

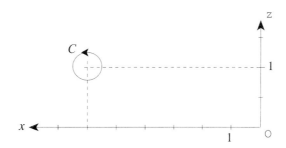

We orient C as shown. Give the best estimate you can for $\oint_C \overrightarrow{F} \cdot d\overrightarrow{r}$. *Hint:* Use the interpretation of curl as circulation per area.

44.12. $\overrightarrow{F}$ is the vector field in space given by $\overrightarrow{F} = (4y, x^2 + y, z^2)$. *You may or may not need part or all of the following information:*

$$\text{div } \overrightarrow{F} = 1 + 2z, \quad \text{curl } \overrightarrow{F} = (0, 0, 2x - 4)$$

We insert a paddlewheel as probe into the flow of the vector field, as discussed in the section on Stokes' Theorem. The center of the paddlewheel is at the point P, and $\vec{n}$ is a vector in the direction of the axis. *We look at the paddlewheel in the direction of $-\vec{n}$, as shown in Figure 44.4.*

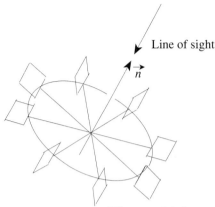

Figure 44.4

a) The center of the paddlewheel is at $P(0,0,0)$ (the origin), and its axis points in the direction of $\vec{n} = (2, 3, 6)$. Does the paddlewheel spin? If yes, clockwise or counterclockwise?

b) We leave the center of the paddlewheel at the origin, but we reposition its normal $\vec{n}$. Is there a vector $\vec{n}$ so that the paddlewheel does not seem to spin? If yes, find such a vector; if no, write "none."

c) Is there a point Q with the following property: No matter how you position the paddlewheel at Q, it does not seem to spin. If yes, find such a point Q; if no, write "none."

44.13. The vector field $\overrightarrow{F}$ is defined by $\overrightarrow{F} = (0, g(x), 0)$. In this definition, $g(x)$ is a function of one variable. We also consider the solid rectangle M in the xy-plane shown below.

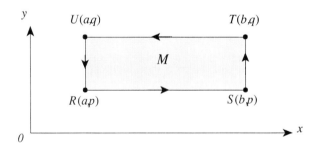

We orient the the boundary of M as shown above, and we orient the xy-plane by choosing $(0, 0, 1)$ as normal.

a) Evaluate $\oint_{\partial M} \overrightarrow{F} \cdot d\overrightarrow{r}$.

b) Set up the double integral for curl $\overrightarrow{F}$ across R. Then integrate with respect to y and stop. You end up with an expression of the form $(\ldots) \int (\ldots) dx$.

c) By virtue of Stokes' Theorem, the line integral of part a equals the expression that you got in part b. Write down the equation (answer to part a) = (answer to part b) and simplify the equation as far as you can. You end up with a fact of elementary calculus that you know very well. Describe this fact in words.

d) The fact of part c is known as the Fundamental Theorem of Calculus. What is the relation between the Fundamental Theorem of Calculus and Stokes' Theorem? *The Fundamental Theorem of Calculus is (choose one)*

> *a generalization of   a special case of*

Stokes' Theorem.

Green's Theorem

44.14. The curve C is shown below. Decide whether the line integral $\oint_C y^2\,dx - xy\,dy$ is positive, negative, or zero if C is given a counterclockwise orientation. Explain how you got your answer.

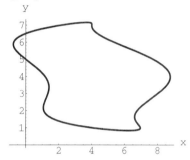

44.15. C is a circle with radius 1 centered at the origin. $\overrightarrow{F}$ is the vector field given by

$$\overrightarrow{F}(x,y) = \frac{-y}{x^2+y^2}\,\vec{i} + \frac{x}{x^2+y^2}\,\vec{j}.$$

a) Calculate $\oint_C \overrightarrow{F}\cdot d\vec{r}$. *Hint:* The tangential component of $\overrightarrow{F}$ is constant.

b) In part a, you got a positive answer. Your friend says: "Your answer must be wrong. I calculated curl $\overrightarrow{F}$ and got zero, Green's Theorem says that the circulation of $\overrightarrow{F}$ around C must be zero." What do you say?

44.16. $\overrightarrow{F}$ is the field of Problem 44.15.

a) Find the circulation of $\overrightarrow{F}$ around a counterclockwise circle of radius 2 and center $(3,0)$. *Hint:* Does the incorrect calculation of Problem 44.15 b work here? Why?

b) Find the circulation of $\overrightarrow{F}$ around a counterclockwise circle of radius 2 and center $(1,0)$. *Hint:* Let C be the circle described above, and let C' be the circle with radius $1/2$ centered at the origin. Apply Green's Theorem to the region between C and C'. What does the result tell you about the relation between the circulation around C and the circulation around C'? Can you calculate the circulation around C' directly?

Problems on both Gauss' and Stokes' Theorems

You have to figure out which theorem to use. At first, this may be confusing. Write out both theorems on a separate piece of paper so you can refer to them all the time.

44.17. Your friend says: "I am confused. Gauss' Theorem states that

$$\iint_{\partial K} \vec{F} \cdot d\vec{A} = \iiint_{K} \operatorname{div} \vec{F} \, dV$$

and Stokes' Theorem states that

$$\oint_{\partial S} \vec{F} \cdot d\vec{r} = \iint_{S} \operatorname{curl} \vec{F} \cdot d\vec{A}.$$

When I combine the two statements I obtain

$$\oint_{\partial S} \vec{F} \cdot d\vec{r} = \iint_{S} \operatorname{curl} \vec{F} \cdot d\vec{A} = \iiint_{K} \operatorname{div}(\operatorname{curl} \vec{F}) \, dV.$$

We saw earlier in the course that the $\operatorname{div}(\operatorname{curl} \vec{F})$ is always zero, no matter what $\vec{F}$ is. So the term on the right is zero. So the term on the left is zero. This means: No matter what the vector field $\vec{F}$ is, the line integral of $\vec{F}$ over the boundary of a surface is zero. For example, any circle is the boundary of a surface. So we have concluded that the line integral over a circle is zero, no matter what the field $\vec{F}$ is. That is nonsense. What is going on here?"

44.18. This is a review problem on line integrals (work), surface integrals (flux), triple integrals, boundaries, the Theorems of Gauss and Stokes, tangential and normal components of a vector field, conservative vector fields, and the curl-test. We consider the solid cube with edges of length 2 units, as shown below. The cube's corners are the points O, A, B, C, D, E, F, its edges the line segments $OA, AB, BC, AD, \ldots$, of length 2, and its faces the filled-in squares $OABC, ABDE, \ldots$.

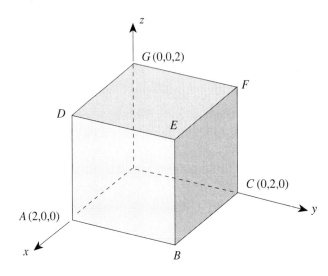

We choose the outward normal for ∂K. If we consider a single face such as $BCFE$, we will use the normal that face inherits from ∂K. In the parts a to i, we will use the same symbol $\vec{H}$ for different vector fields and S for different surfaces. *Hint for all parts:* Before you start integrating, ask yourself: Can I use Gauss' or Stokes' Theorem? In case of line integrals and surface integrals, can I say something about the tangential or normal components?

a) The vector field $\vec{F}$ is defined by $\vec{H} = (0, 0, 3z^2 + 7)$. We remove the face $DEFG$ from ∂K and call the remaining surface S. Find the flux of $\vec{H}$ across S.

b) $\vec{H}$ and S are as in part a. We orient ∂S in the direction $DEFGD$. Find $\int_{\partial S} \vec{H} \cdot d\vec{r}$.

c) $\vec{H}$ is still as in parts and b. Find the flux of $\vec{H}$ across the boundary of K.

d) $\vec{H}$ is still as in parts a to c. We consider the directed line segment OE, the diagonal of K. Find the maximum and minimum of the tangential component of $\vec{H}$ on OE.

e) Here, S is the surface that consists of the four faces $OADG$, $OABC$, $BCFE$, and $EFGD$. The surface S inherits the orientation of ∂K. Describe ∂S in a complete sentence. Orient ∂S such that you could apply Stokes' Theorem to S and ∂S if you had to.

f) $\vec{H} = (0, 0, 3z^2 + 7)$ of a)–d). Is $\vec{H}$ conservative? Or is it not? Or is it impossible to say anything? In case it is conservative, can you find a potential $\Phi(x, y, z)$ such that $\Phi(1, 1, 1) = 349$? If yes, find it; if no, explain.

g) Now $\vec{H}$ is the vector field with components $(5y, 4, -6)$. Is $\vec{H}$ conservative? Or is it not? Or is it impossible to say anything? In case it is conservative, can you find a potential $\Phi(x, y, z)$ such that $\Phi(1, 1, 1) = 691$? If yes, find it; if no, explain.

h) $\vec{H}$ is the vector field $(5y, 4, -6)$ of g). Below are given six points $P_1, \ldots, P_6$. Each of them is in one of the six faces of the boundary of K:

$$P_1(1, 1, 0), \ P_2(1, 2, 1), \ P_3(1, 1, 2), \ P_4(1, 0, 1), \ P_5(2, 1, 1), \ P_6(0, 1, 1)$$

We write $H_1, H_2, \ldots$ for the normal component of $\vec{H}$ at the points $P_1, P_2, \ldots$. Find $H_1, H_2, \ldots$.

i) $\vec{H}$ is the vector field $(5y, 4, -6)$ of parts g and h. S is the face $ABED$ of the cube. Find $\iint_S \vec{H} \cdot d\vec{A}$.

Answers to Odd Problems

In many problems, you have to work with ruler and pencil. For such problems, there is a range of correct answers simply because no two persons have the same eyesight or the same dexterity in measuring and drawing. Also, during printout and duplication, the drawings often get slightly magnified or reduced. Thus, do not worry when your answer to a ruler-and-pencil problem differs slightly from the one given here. *Do the ruler-and-pencil problems as carefully as you can and show your work.*

Answers for Section 1

A–1.1. See below; in a) and b) the point T lands at the same point U.

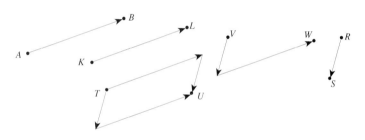

A–1.3. a)   b) It looks as if the points describe a parabola.

A–1.5. (a) and (c) are true. (b) is nonsense. (d) is not true in general (e.g., take two vectors $\vec{a}$ and $\vec{b}$ which form a right angle).

A–1.7. b) See below (the figure is at reduced scale). Note that $\vec{v} = -\vec{u}$.   c) See below.

d) $|\overrightarrow{AC}| = 5.5$ cm, $c = \pm\frac{4398}{5.5} = \pm799.6$.

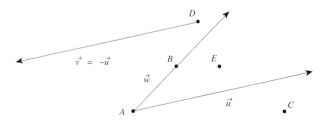

403

A–1.9.   a) $\overrightarrow{KL} = \overrightarrow{OL} - \overrightarrow{OK} = (-11, 4)$.
b) $\overrightarrow{SR} = \overrightarrow{OR} - \overrightarrow{OS} = (a - u, b - v)$.

A–1.11.   a) $(147.49)(\cos \frac{14\pi}{180}, \sin \frac{14\pi}{180}) = (147.49)(0.97, 0.24) = (143.07, 35.40)$.
b) $(x, y) = (r \cos \theta, r \sin \theta)$.

A–1.13.   a) $a = 1.55$.   b) $b = 0.64$.

A–1.15.   a) $k = 2.5$.   b) $t = 5/7$.

A–1.17.   In the xy-system, the coordinates of the points A, B, C are $(0, 0)$, $(3, 1)$, $(1, 3)$. In the uv-system, the coordinates of the same points are $(-1, -3)$, $(2, -2)$, $(0, 0)$. The components of the vectors $\overrightarrow{AB}$, $\overrightarrow{AC}$, $\overrightarrow{AB} + \overrightarrow{AC}$ are the same in both systems, namely $(3, 1)$, $(1, 3)$, $(4, 4)$.

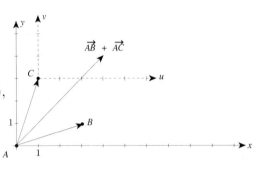

A–1.19.   a) See figure at right.
b) $|\overrightarrow{FP}|^2 = x^2 + (y - 1)^2$, $|\overrightarrow{QP}|^2 = (y + 1)^2$, $y = (1/4)x^2$. The points P lie on a parabola.

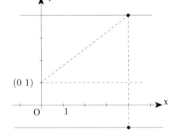

A–1.21.   a) $\vec{b} = (5.9, 2.4)$ (or any scalar multiple).   b) $Q(6.8, 2.4)$.
Please refer to Figure A1.1 on the facing page.

Answers for Section 2

A–2.1.   a) and b) See Figure A2.1 on the next page.
c) (i) Velocity: $\vec{v} = (-4, 2)$. Speed: $|\vec{v}| = \sqrt{20}$.   (ii) $t = 1/2$.

A–2.3.   Possible answers: a) $x = -1 + 3t$, $y = 1 - 2t$.   b) $x = 2t$, $y = 3t$.

A–2.5. $\vec{r}(t) = (5 - 4t, 1 + 2t)$. Velocity vector: $(-4, 2)$. Speed: $\sqrt{20}$.

A–2.7.   a) At time $t = \frac{\pi}{2}$ we have $\omega t = \frac{3\pi}{4}$. It means: At time $t = \frac{\pi}{2}$ the position vector $\vec{r}$ of P forms the angle of $\frac{3\pi}{4}$ with the positive x-axis. Velocity vector: $\vec{r}'(t) = 3\omega (-\sin \omega t, \cos \omega t)$; the magnitude of $\vec{r}'(t)$ equals 3ω at all times. See Figure A2.2 on the facing page.
b) One revolution means $\omega t = 2\pi$.
(i) Time for one revolution for $\omega = 1.5 : \frac{2\pi}{1.5} \approx 4.1888$;
(ii) Time for one revolution in general: $\frac{2\pi}{\omega}$.
c) (i) $\frac{1.5}{2\pi} \approx 0.2387$;   (ii) $\frac{\omega}{2\pi}$.

A–2.9.   a) $\displaystyle\int_{t=0}^{t=1} \sqrt{(x_1 - x_0)^2 + (y_1 - y_0)^2}\, dt$.

b) and c) I expect to get $\sqrt{(x_1 - x_0)^2 + (y_1 - y_0)^2}$ because that is the distance between (x_0, y_0) and (x_1, y_1), and I do get it when I integrate.

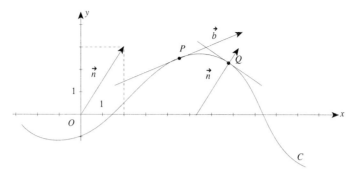

Figure A1.1

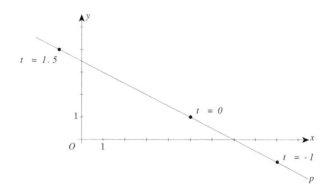

Figure A2.1

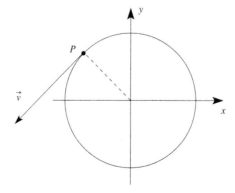

Figure A2.2

A–2.11. a) See below.   b) $\sqrt{13}$ (distance units)/(time unit).   c) $2x+3y = 8$.

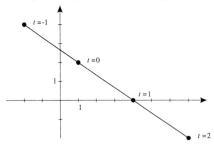

A–2.13. a) $x = t,\quad y = 2 + \frac{3}{2}t$.   b) $x = \frac{2}{3}t - \frac{4}{3},\quad y = t$.
c) $x = \frac{2}{\sqrt{13}}s,\quad y = 2 + \frac{3}{\sqrt{13}}s$.

A–2.15. $D(3, 7)$.

A–2.17. C_1: $x = \pi/2 + t,\quad y = 1 - t$. C_2: $x = 1 + t,\quad y = t$.

A–2.19. a) $\sqrt{2}(e - 1)$.   b) $s = \sqrt{2}(e^t - 1)$.
c) $t = \ln\left(\frac{s+1}{\sqrt{2}}\right),\quad x = \frac{s+1}{\sqrt{2}}\cos\left(\ln\left(\frac{s+1}{\sqrt{2}}\right)\right),\quad y = \frac{s+1}{\sqrt{2}}\sin\left(\ln\left(\frac{s+1}{\sqrt{2}}\right)\right)$.

Answers for Section 3

A–3.1.   a) The acceleration is $\vec{0}$ since the velocity is constant.   b) Since the velocity is always parallel to the line, the *change* in velocity $\Delta\vec{v}$ is always either parallel to line or zero. Thus, the acceleration is either parallel to the line or zero. Without additional information, you cannot say anything about the magnitude of the acceleration.

A–3.3.   a) $\omega = \frac{\pi}{2}$; $\vec{r}(t) = \left(3 \cos\frac{\pi}{2}t\right)\vec{i} + \left(3 \sin\frac{\pi}{2}t\right)\vec{j}$.
b)

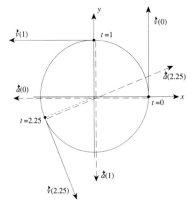

c) $v = \frac{3\pi}{2}$; $v = \rho\omega$; v is constant, i.e., independent of t.
d) $|\vec{a}| = \frac{3\pi^2}{4}$.
e) $|\vec{a}| = \frac{v^2}{\rho}$.
f) The acceleration is directed toward the center.

A–3.5. a) $x = t,\quad y = 2t - t^2/2$.   b) See below left.
c) The parabola $y = 2x - x^2/2$.
d) $x = 0,\quad y = 2t - t^2/2$.   The particle moves up and down the y-axis. The plot is shown below right.

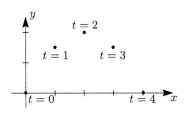 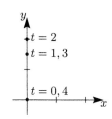

A–3.7. Trajectory: $x = 122\cos(40\deg)t$, $y = 122\sin(40\deg)t - 16t^2$. The ball reaches the wall when $t = 4.7\sec$ at a height of about 14.5 ft, clearing the fence by over 3 ft.

Answers for Section 4

A–4.1. a) (i) 1; (ii) 0; (iii) 0; (iv) −1.
b) From left to right: $|\vec{a}||\vec{b}|1 \approx 2.7$; $|\vec{a}||\vec{b}|0 = 0$;
$|\vec{a}||\vec{b}|\frac{\sqrt{2}}{2} \approx 3.4$; $|\vec{a}||\vec{b}|(-1) \approx -2.1$.

A–4.3. a) If we write out the equation $\vec{a} \cdot \vec{u} = 0$ we obtain $1 \cdot m + 6n = 0$. This is *one* equation in *two* unknowns m and n. There are infinitely many solutions (m, n), for example $(m, n) = (6, -1)$ or $(-6, 1)$ or $(15, -2.5)$ etc.
b) $\cos\theta = \frac{4+18}{\sqrt{37}\sqrt{25}}$ from which we can compute $\theta \approx 0.76215$ radians $\approx 43.7°$.
c) (i) N is a straight line that passes through the origin and is normal to $\vec{b}$.
(ii) $4x + 3y = 0$.

A–4.5. The equation of m is $2x + 3y = 6$. Therefore, $\vec{n} = (2, 3)$. For A, take $(3, 0)$. Then $d = |\overrightarrow{AP} \cdot \frac{\vec{n}}{|\vec{n}|}| = \frac{70}{\sqrt{13}}$.

A–4.7. $\vec{m} = (1.5, -0.5) + (0.5, 1.5)$.

A–4.9. (i) vector;
(ii) nonsense: the dot is used only between two vectors (or between two scalars), and not between a scalar like $\vec{a} \cdot \vec{b}$ and a vector like $\vec{c}$;
(iii) vector; (iv) scalar.

A–4.11. a) $\vec{w}$ perpendicular to $\vec{c}$.
b) $\vec{w}$ parallel to $\vec{c}$ and same direction as $\vec{c}$.
c) $\vec{w}$ parallel to $\vec{c}$ but direction opposite to $\vec{c}$.

A–4.13. (i) 12; (ii) −4; (iii) $\frac{8}{\sqrt{2}}$.

A–4.15.

Answers for Section 5

A–5.1. a) $T(p, q, 0)$ is in the xy-plane, $T(p, 0, r)$ in the xz-plane, and $T(0, q, r)$ in the yz-plane.
b) $A(-37, 4, 82)$, $B(79, -4, 1)$.

A–5.3. (x, y, z): (positive, positive, positive); (negative, negative, positive); (positive, negative, positive).

A–5.5. a) See below left. There are two answers.
b) See below right.

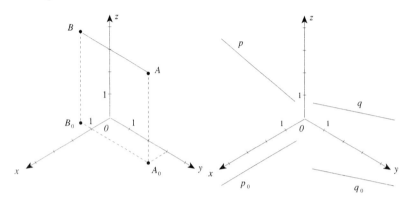

A–5.7. The vectors $\vec{q}$, $\vec{r}$, $\vec{t}$ are parallel to the xy-plane, and $\vec{r}$ is perpendicular to the yz-plane.

A–5.9. a) See Figure A5.1 on the next page.
b) $\vec{c} = (0, 0, 2) - (3.8, -2, 2.5) = (-3.8, 2, -0.5)$,
   $\vec{d} = (0, 4.5, 2) - (0, 7, 7) = (0, -2.5, -5)$.

A–5.11. a) $A(3, 2, 7)$, $B(2, 4, 4,)$.   b) See Figure A5.2 on the facing page.

Answers for Section 6

A–6.1. a) See Figure A6.1.
  b) $x = 1+t$, $y = 4-4t$, $z = 3$; often, this is written as $(x, y, z) = (1, 4, 3)+t(1, -4, 0)$.
c) Note that d has the direction of the vector $(1, -4, 0)$, and this vector is parallel to the xy-plane. Therefore, d does not intersect the xy-plane. There is no point U. To find V and W there are two methods:

Method 1—computation: At V (point of intersection with the yz-plane) we must have $x = 0$. We solve $x = 1 + t = 0$ for t and obtain $t = -1$. This corresponds to the point $(x, y, z) = (0, 8, 3)$, and this is the wanted point V. For W we operate similarly and find $W = K(2, 0, 3)$.

Method 2—looking and drawing: Look at Figure A6.1. You see that the given point K lies in the xz-plane. Therefore, $W = K$. For V you draw the shadow d_0. The point where the shadow intersects the y-axis must be the shadow V_0 of the wanted point V. You find V by going up vertically from V_0.

A–6.3.   We write $P(t)$ for P at time t. $\overrightarrow{P(5)P(8)} = (4, -9, -2)$, $\overrightarrow{OP(2)} = \overrightarrow{OP(5)} - \overrightarrow{P(5)P(8)} = (-2, 18, 3)$.

A–6.5.   One of many possible answers: $x = 5 + 2(t - 1)$, $y = t - 1$, $z = -1 + 3(t - 1)$.

A–6.7. It takes P two seconds to go from A to K. In these two seconds, Q goes from B to K. Therefore, the velocity vector of Q is $(1/2)\,\overrightarrow{CK} = (-1.5, 1, 3)$.

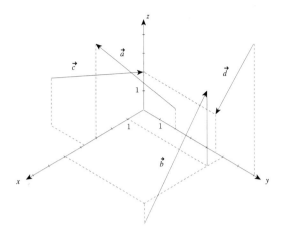

Figure A5.1

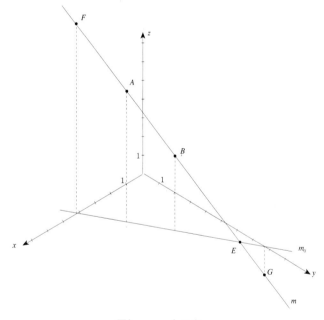

Figure A5.2

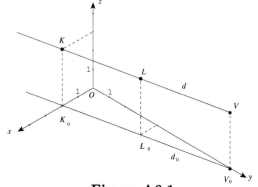

Figure A6.1

Answers for Section 7

A–7.1. a) $\overrightarrow{AB} = (1, 2, -4)$, $\overrightarrow{AC} = (-2, 2, -1)$, $\cos\theta = (21)^{-1/2} \cdot 9^{-1/2} \cdot (-2+4+4) = \ldots$, from which we get $\theta = \ldots$.

b) $\vec{p} = (u, v, w)$ at a right angle with $\overrightarrow{AB}$ means $\vec{p} \cdot \overrightarrow{AB} = 0$: $u + 2v - 4w = 0$. This is one equation for three unknowns u, v, w, and we pick two of the infinitely many solutions. For example, we can take $\vec{p} = (2, -1, 0)$ and $\vec{q} = (0, 2, 1)$.

A–7.3. a) $15/\sqrt{30}$. b) $15/\sqrt{26}$.

c) Write $\vec{m}_0$, $\vec{q}_0$ for the unit vectors in the directions of $\vec{m}$, $\vec{q}$. In a), you have to compute the dot product $\vec{m} \cdot \vec{q}_0$, and in b) the dot product $\vec{q} \cdot \vec{m}_0 = \vec{m}_0 \cdot \vec{q}$. They are products of different vectors, and *not* the same product with the factors in different order.

A–7.5. (i) and (ii): The *scalar* projections of any vector $\vec{p} = (u, v, w)$ onto $\vec{\imath}$, $\vec{\jmath}$, $\vec{k}$ are u, v, w, respectively. Consequently, the *vector* projections are $u\vec{\imath}$, $v\vec{\jmath}$, $w\vec{k}$. Therefore, we have: (i) $2\vec{\jmath}$, (ii) $3\vec{k}$.

(iii): $\vec{b} = 7\vec{\imath} + 2\vec{\jmath} + 3\vec{k} = (2\vec{\jmath}) + (7\vec{\imath} + 3\vec{k})$. The vector in the first set of parentheses is parallel to $\vec{\jmath}$, and the vector in the second set is perpendicular to $\vec{\jmath}$.

A–7.7. a) (i) **S**; (ii) **V**; (iii) **S**; (iv) **S**.

b) (i) $9 + 4(1+m)$; (ii) $(6+4+2t, 4(6+4+2t), t(6+4+2t))$; (iii) $6+4+2t+3+4m-2t$; (iv) The expression $\left((\vec{a} \cdot \vec{c})\,\vec{a}\right) \cdot \vec{c}$ up for evaluation is of the form $(k\vec{a}) \cdot \vec{c}$, where k is the scalar $\vec{a} \cdot \vec{c}$. We have $(k\vec{a}) \cdot \vec{c} = k(\vec{a} \cdot \vec{c}) = (\vec{a} \cdot \vec{c}) \cdot (\vec{a} \cdot \vec{c}) = (18 + m - 4)^2$.

A–7.9. *Remember:* The cosine of the angle between two *unit* vectors is their dot product.

a) $\cos\alpha = \vec{\imath} \cdot \vec{p} = \frac{4}{9}$, and similarly $\cos\beta = \frac{1}{9}$ and $\cos\gamma = \frac{8}{9}$. This means: The three direction cosines are the components of the unit vector $\vec{u}$.

b) 1 because $|\vec{p}| = 1$.

c) The unit vector in the direction of $\vec{v}$ has components $(\frac{a}{|\vec{v}|}, \frac{b}{|\vec{v}|}, \frac{c}{|\vec{v}|})$, and these are the three direction cosines the question asks for.

d) $\cos\alpha = \cos\gamma = \frac{1}{2}$. Therefore, $\cos^2\beta = 1 - \frac{1}{4} - \frac{1}{4} = \frac{1}{2}$, and $\cos\beta = \pm\frac{\sqrt{2}}{2}$ (two solutions). The unit vector in the direction of $\vec{m}$ has components $\frac{1}{2}(1, \pm\sqrt{2}, 1)$, and $\vec{m} = (13.5)(1, \pm\sqrt{2}, 1)$.

A–7.11. a) $|\overrightarrow{OP_0}| = \sqrt{a^2 + b^2}$. b) $\overrightarrow{OP_0} = (a, b, 0)$.

c) and d): any point on the line P_0P, for example $Q(a, b, 269.81)$. The shadow of any such point is $P_0(a, b, 0)$.

A–7.13. a) (1) no, (2) yes, (3) yes, (4) no, (5) yes, (6) no, (7) no, (8) yes, (9) no.

b) $\overrightarrow{OP} = (a, 0, 0) + (0, b, c)$.

Answers for Section 8

A–8.1. a) $\vec{\imath}$. b) $-\vec{\jmath}$. c) (pos, neg, zero).

A–8.3. a) $(-12, -13, -8)$, $(w - v, u - w, v - u)$.

b) The area equals $(1/2)|\overrightarrow{PQ} \times \overrightarrow{PR}| = (1/2)|(6, 0, 6)| = 3\sqrt{2}$.

c) $(\sqrt{2}/2)(1, 0, 1)$.

A–8.5. (pos, pos, pos).

A–8.7. a) It is impossible to decide whether $\vec{a}$ is perpendicular to $\vec{b}$.
b) $\vec{a}$ is not parallel to $\vec{b}$.
c) $\vec{a}$ is not perpendicular to $\vec{c}$.
d) $\vec{a}$ is parallel to $\vec{c}$.
e) It is impossible to decide whether $\vec{a} \cdot \vec{b}$ is zero.
f) $\vec{a} \cdot \vec{c}$ is different from zero because $\vec{a}$ and $\vec{c}$ are parallel (remember that both vectors are different from zero).

A–8.9. a) (i): $(0, 3, 6)$,   (ii): $(-3, 6, 6)$.
b) (i) = (ii): 9.
c) -3.

A–8.11. a) $\sqrt{160}$.   b) 0.
c) $t = 4$ is the only zero of $F(t)$, for the following reason: $F(t)$ is the area of the parallelogram spanned by the vectors $(3, 1, 4)$ and $(3, 1, t)$. For any $t \neq 4$, the vectors $(3, 1, 4)$ and $(3, 1, t)$ are not parallel, and the area of the parallelogram will be different from zero.

Answers for Section 9

A–9.1. a) The equation of α must be of the form $5x + 3y - 2z = K$ where K is a constant. To find K, evaluate $5x + 3y - 2z$ for the coordinates $(x, y, z) = (1, 3, 4)$ of A. You find $K = 6$.
b) Set $(x, y) = (0, 0), (0, 1), (1, 0), (1, 1)$ and determine the corresponding z. Result: $Q_1(0, 0, -3)$, $Q_2(0, 1, -1.5)$, $Q_3(1, 0, -0.5)$, $Q_4(1, 1, 1)$.
c) $R_1 = (0, 0, 0)$ is not in α because the constant term of the equation of α is different from zero. For three more, use the (x, y)-coordinates of $Q_2, \ldots$ and pick a different z. Result: $R_2(0, 1, 0)$, $R_3(1, 0, 0)$, $R_4(1, 1, 0)$.

A–9.3. a) B, C are the points of m corresponding to $t = 0$, $t = 1$. Equation of the plane: $9x + 10y + 6z = 48$.
b) The direction of k must be perpendicular to the normal $(9, 10, 6)$ of α. For example, $(-6, 0, 9)$ is normal to $(9, 10, 6)$. Equation of k: $(x, y, z) = (2, 3, 0) + t(-6, 0, 9)$.

A–9.5.

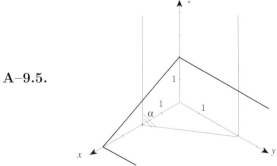

A–9.7. m_1: The direction $(1, 2, -1)$ is perpendicular to the normal $(1, 2, 5)$. Therefore, m_1 is either parallel to α or m_1 lies in α. The point $(0, 4, 0)$ lies in α. Therefore, all of m_1 lies in α.
m_2: The direction $(1, 2, 1)$ is *not* perpendicular to the normal $(1, 2, 5)$. Therefore, m_2 must intersect the plane. Point of intersection D: $(2+t)+2(3+2t)+5(4+t) = 8$

yields $t = -2$ and $D(0, -1, 2)$.

m_3 is parallel to α, but does not lie in α.

A–9.9. $\vec{n} = (4, -1, 2)$ is a normal of α. Equation of α: $4x - y + 2z = 7$.

A–9.11. The normal of α has the direction of the cross product of the directions of p and q. Equation of α: $-4x + 13y + 3z = 14$.

A–9.13. See below. The line s is parallel to the z-axis.

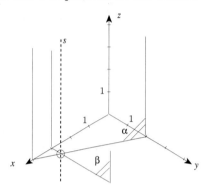

A–9.15. a) See below.

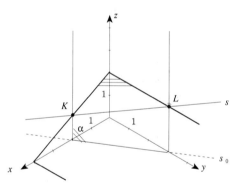

b) Method (i): $K(2, 0, 1)$, $L(0, 3, 2)$.

Line of intersection s: $(x, y, z) = (2, 0, 1) + t(-2, 3, 1)$.

Answers for Section 10

A–10.1. a) $(x, y, z) = (\cos t, \sin t, \frac{1}{4}t)$.

b) $\vec{v}(t) = (-\sin t, \cos t, \frac{1}{4})$.

c) The velocity vector $\vec{v}(t)$ is parallel to the xz-plane when its y-component is zero. For example, this happens at time $t = \frac{\pi}{2}$ sec.

d) Point of C for $t = \frac{\pi}{6}$: $(\frac{\sqrt{3}}{2}, \frac{1}{2}, \frac{\pi}{24})$. Velocity vector: $(-\frac{1}{2}, \frac{\sqrt{3}}{2}, \frac{1}{4})$. For the direction vector of the tangent line we take $\vec{b} = (-2, 2\sqrt{3}, 1)$ obtained by multiplying the velocity vector by 4. For the parameter of the tangent line we use u:

$$x = \frac{\sqrt{3}}{2} - 2u, \quad y = \frac{1}{2} + 2\sqrt{3}\,u, \quad z = \frac{\pi}{24} + u.$$

A–10.3. a) No. b) It is below the xy-plane when $1 < t < 3$. c) No. d) Yes, when $t = 2$. e) velocity $\approx 2\vec{i} + \vec{j} - 1.2\vec{k}$, speed ≈ 2.5 dist. units/sec.

A–10.5.   a) $(0, 1, 0)$.   b) $x = 2t$, $y = 1 - t$, $z = t$.

A–10.7.   a) For $t = 1$ we have $(x, y, z) = (e, 5, \tan 1)$ and $(x', y', z') = (e, 7, 1 + \tan^2 1)$.
(i) $\vec{r}(1 + h) \approx \vec{r}(1) + h\vec{r}'(1) = (e, 5, \tan 1) + (h)(e, 7, 1 + \sec^2 1)$;
(ii) $\vec{r}(t) \approx \vec{r}(1) + (t - 1)\vec{r}'(1) = (e, 5, \tan 1) + (t - 1)(e, 7, 1 + \sec^2 1)$.
b) For $t = 0$ we have $(x, y, z) = (1, 0, 0)$ and $(x', y', z') = (1, 4, 1)$.
$\vec{r}(t) \approx \vec{r}(0) + t\vec{r}'(0) = (1, 0, 0) + t(1, 4, 1)$.
c) $\vec{r}(-0.14) \approx \vec{r}(0) + (-0.14)\vec{r}'(0) = (1, 0, 0) - (0.14)(1, 4, 1) = (0.86, -0.56, -0.14)$.

A–10.9.   a) $\vec{r}'(8) \approx \frac{\Delta \vec{r}}{\Delta t} = (1.25, -0.75, 0.5)$.
b) $|\vec{r}'(8)| \approx |(1.25, -0.75, 0.5)| = \sqrt{(1.25)^2 + (-0.75)^2 + (0.5)^2} \approx 1.54$.

Answers for Section 11

A–11.1. $P : (x, y) = (2.1, 1)$, and $\vec{r}'(1) = (1.2, 0)$. Thus, $\vec{b} = (1, 0)$ is a tangent vector at P, and $\vec{n} = (0, 1)$ is a normal vector at P.
a) Tangent line: (i) $(x, y) = (2.1, 1) + u(1, 0)$ (ii) $y = 1$.
b) Normal line: (i) $(x, y) = (2.1, 1) + v(0, 1)$ (ii) $x = 2.1$.
c) "normal vector $= (c, 0)$" means "tangent vector $= (0, c)$, and this happens for $t = -5$. Result: $(x, y) = ((0.1)(-5)^2 - 5 + 1, \sin(-(2.5)\pi))$.

A–11.3.   a) The tangent line of a line is the line itself. Equation of m: $(x, y, z) = (1, 3, 2) + t(4, 1, 4)$.
b) The normal of the normal plane is the direction vector $(4, 1, 4)$. Equation of the normal plane at A: $4x + y + 4z = 15$.

A–11.5.   a) Slope of the tangent: $\cos x$. Slope of the normal: $-\frac{1}{\cos x}$.
(i) "Slope of the normal equals $\frac{2}{3}$" means "$\cos x = -1.5$," and this is impossible. There is no such point.
(ii) $\cos x = -0.5$ is true for $x = \frac{4}{3}\pi$ (many answers).
b) Tangent vector at the generic point: $(6t + 5, 2t + 4)$; normal vector at the same point: $(-2t - 4, 6t + 5)$.
(i) "Normal line parallel to $(-2, 3)$" means that the following system of two equations in two unknowns t and c must have a solution (t, c):

$$-2t - 4 = -2c$$
$$6t + 5 = 3c.$$

The system has the solution $(t, c) = (\frac{1}{3}, \frac{7}{3})$. Coordinates of wanted point: Evaluate $(3t^2 + 5t, t^2 + 4t)$ for $t = \frac{1}{3}$.
(ii) The system
$$6t + 5 = 3c$$
$$2t + 4 = 1c$$

has no solution. There is no such point.
c) Tangent vector at the generic point: $(3t^2, 5, 4t)$.
(i) "Normal plane parallel to $12x + 5y + 8z = 24$" means "tangent vector parallel to $(12, 5, 8)$." We have to find (t, c) such that $(12, 5, 8) = c(3t^2, 5, 4t)$. You find $t = 2$. Coordinates of wanted point: $(8, 10, 8)$.

(ii) "Tangent line parallel to plane $5x + 5y + 5z = 15$" means "tangent vector perpendicular to $(5, 5, 5)$." Work out the equation $(3t^2, 5, 4t) \cdot (5, 5, 5) = 0$. You get a quadratic equation for t. It has no real solution t. There is no such point.

Answers for Section 12

A–12.1.   a) P increases.   b) Cannot tell.   c) P decreases.   d) P decreases.

A–12.3.

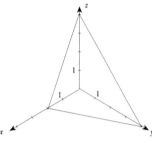

A–12.5.   a) $-2x - y + 4 = 0$, $-2x - y = 0$,
$-2x - y + 5 = 0$.
b) See figure at right.

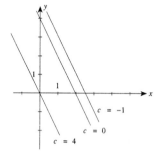

A–12.7.   a) \$127   b) The monthly payment will decrease by \$6.   c) It will take you 13 years and 9 months.

A–12.9.   a) $f(1, 3) = 0$, $f(3a + 6, 22) = -9a - 34$.   b) For example, $(x_1, y_1) = (-4, 1)$ and $(x_2, y_2) = (-5, 4)$ will work. There are many answers; take any point on the level curve of level 17.

A–12.11.   See Figure A12.1.

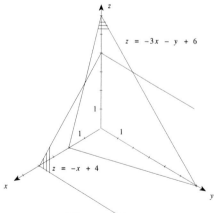

Figure A12.1

A–12.13.   a) Find a solution (x, y) of $2x - 7y + 1 = 0$, e.g., $(-\frac{1}{2}, 0)$.
b) $2x - 7y + 1 = 0$: a line.

c) It is a circle of radius 64 with center at the origin.

d) $w(t) = 1$ for $t = 7.9$. The level curve $h(x, y) = w(t) = w(x^2 + x + y + 2) = 1$ is given by $x^2 + x + y + 2 = 7.9$. This is a parabola.

A–12.15. a) $f(x, y) = -\frac{4}{5}x - \frac{8}{5}y - \frac{3}{5}$.
b) See figure at right.
c) They point in the same direction, perpendicular to the level lines.
d) growth rate $\approx 4/2.9 \approx 1.38$, and $|\vec{v}| = \frac{4\sqrt{3}}{5} \approx 1.38$. They're equal!
e) The vector $A\vec{i} + B\vec{j}$ points in the direction of steepest ascent, and its magnitude is the rate of increase of f in that direction.

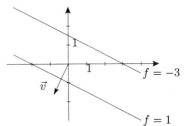

A–12.17. $f(x, y) = 5 - 5(x - 1) + 3(y - 2) = -5x + 3y + 4$.

A–12.19. a) 6. b) -1. c) For example $(3, 4)$ and $(5, 4)$. There are many answers. Any point on the level curve of level -4 will do.

A–12.21. a) $f(0, -3) = 1$, $f(4, 0) = -1$, and $f(5, 0) = 2$.
b)

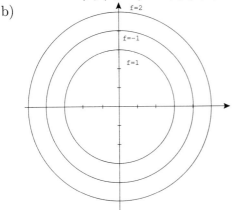

c) ... from $P(x, y)$ to *the origin*.

A–12.23. a)

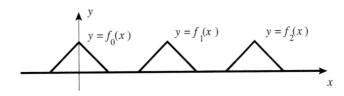

b) A "wave" in the shape of the given function $f(x)$ travels from left to right with speed c.

Answers for Section 13

A–13.1. a) Decreases. b) Cannot tell (depends on angle at which you start).
c) Cannot tell.

A–13.3. $c = 0$: The level surface is reduced to the point $(0, 0, 0)$.
$c = -4$: The level surface for that level is the empty set.

$c = 4$: The level surface is a spherical shell of radius 4 and center at the origin.

A–13.5.   a) $e^{2x-7y-z+1} = 2$ means $2x - 7y - z + 1 = \ln 2$. There are many points $P(x, y, z)$ whose coordinates satisfy $2x - 7y - z + 1 = \ln 2$, for example $P(0, 0, 1 - \ln 2)$.
b) It is the plane $2x - 7y - z + 1 = \ln 2$.
c) There are many points $Q(x, y, z)$ whose coordinates satisfy $g(x, y, z) = 3$, for example $Q(3, 0, 0)$.
d) $h(t) = 3$ for $t = 5.2$; equation of the level surface: $x^2 + xz + yz^3 = 5.2$.

A–13.7.

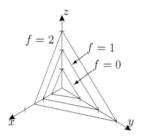

A–13.9.   a) Spheres centered at the origin with radii 2, 1, and $\frac{1}{2}$, respectively.
b) $1/|\vec{r}|$.
c) $\varphi(x, y, z)$ is one over the distance from the point (x, y, z) to the origin.
d) $1/\sqrt{(x-1)^2 + (y-2)^2 + (z-3)^2}$.

A–13.11.   a) Question 1: The points in the xz-plane which satisfy $x^2 + z^2 = 4$ lie on a circle of radius 2 and center at the origin. Question 2: Take any point (x, y, z) on the straight cylinder S whose axis is the y-axis and whose basis is C. Such a point will still satisfy $x^2 + z^2 = 4$ no matter where on the cylinder it is. Thus, the level surface of level 2 is the cylinder S. It extends infinitely in both directions of the y-axis. See below left.
b) $z^2 = 9$ means $z = 3$ and $z = -3$. The level surface $v(x, y, z) = 9$ consists of the two planes $z = 3$ and $z = -3$. See below right.

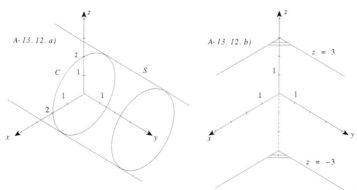

A–13.13.   a) The domain of $w = f(x, y, z)$ consists of all triples (x, y, z) for which $f(x, y, z)$ is defined. The range of $w = f(x, y, z)$ consists of all scalars w for which the equation $w = f(x, y, z)$ has a solution (x, y, z).
b) Scalars.
c) Points in xyz-space.
d) The domain of the function $f(x, y, z)$ in question consists of all points (x, y, z) which satisfy $x^2 + y^2 + z^2 \geq 4$. These are all the points on the sphere $x^2 + y^2 + z^2 = 4$

and on the *outside* of that sphere. The range consists of all scalars which are greater than one.

Answers for Section 14

A–14.1. a) and b): See figure below.

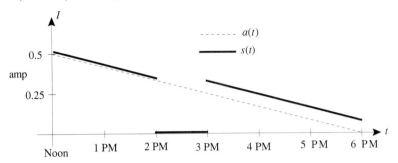

A–14.3. a) $f(0,0) = 1$; at any point near to $(x,y) = (0,0)$, $f(x,y)$ takes the value 1. The function is continuous at $(0,0)$.

b) $f(1,0) = 1$; the value of $f(x,y)$ for (x,y) near to $(1,0)$ depends where (x,y) lies. If (x,y) lies on the unit circle $x^2 + y^2 = 1$ or in its interior, then $f(x,y) = 1$. If (x,y) lies outside the unit circle, even by a tiny little bit, $f(x,y) = 0$. The function is discontinuous at $(1,0)$.

c) $f(x,y)$ is discontinuous for all points on the unit circle.

A–14.5. See figure below (not to scale): $\frac{\epsilon}{\delta}$ is the slope of $y = f(x) = 6x + 5$. Therefore we have $\delta = \frac{4}{6} 10^{-6}$.

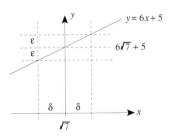

A–14.7. a) 1.   b) 0.   c) The limit does not exist.

A–14.9. Your friend thinks that for a function to be continuous it is sufficient to find a δ for just one ϵ. That is a misunderstanding. For continuity it must be possible to find a δ for any value of ϵ whatsoever, no matter what you choose.

A–14.11. $\phi(x,y)$ is not continuous at $(0,0)$ because $\ln 0$ is not defined.

A–14.13. $f(x) = \ln(1 + \sin x)$ is discontinuous at all points x where $\sin x = -1$. That is, $x = \frac{3}{2}\pi + 2\pi n$ where n is any integer.

A–14.15. We have to say what the density $f(P)$ at a point on the surface of the block of ice is. We consider P as a point of the ice, and the density at P is that of ice.

Now we check continuity at a point P on the surface of the ice. The density of water w is higher than the density of ice i. Take for ϵ a number between w and i, for example $\frac{1}{2}(w + i)$. Can we find a k so that at all points in the solid ball with radius k and center P the density is within $\epsilon = \frac{1}{2}(w + i)$ of the density of ice at P? The schematic figure below shows the surface as a curve and the ball as a circle. It illustrates the fact that *any* solid ball centered at P contains points of water, and there the density is w.

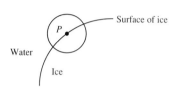

In other words: We *cannot* find a k so that at all points in the solid ball with radius k and center P the density is within our ϵ of the density of ice at P. Therefore the density function is discontinuous on the surface of the block of ice.

A–14.17. a) The function is discontinuous at $x = 0$.
b) The limit of $\frac{1}{x}$ for x tending to zero does not exist.
c) $\lim_{x \to 10^{-67}} \frac{1}{x} = \frac{1}{10^{-67}} = 10^{67}$.
d) Your friend erroneously thinks that the definition of continuity runs as follows: "$f(x)$ is continuous at x_0 if $f(x)$ is continuous for all x sufficiently close to x_0, that is, for all x such that $|x - x_0| < \epsilon$ where ϵ is any positive number, no matter how small." You have to tell your friend: "I think you misread the definition of continuity. For $f(x)$ to be continuous at x_0 it must satisfy two conditions. One of them is that $f(x_0)$ must be defined. If that condition is not satisfied, $f(x)$ is not continuous at x_0. In our case of $f(x) = \frac{1}{x}$, $f(0)$ is not defined. Therefore $\frac{1}{x}$ is not continuous at $x = 0$."

A–14.19. At $x = 0$ the function is not continuous, for the following reason. Take $\epsilon = 0.6$. No matter what δ is, the interval $-c < x < c$ contains points x where $H(x) = 0$. These are points where the values of the function are *not* within $\epsilon = 0.6$ of $H(0) = 1$. That is, $H(x)$ is not continuous at $x = 0$.
At all other points the functions is continuous, for the following reason. For each of the four points in question I can find an interval centered at the point on which the function $H(x)$ is constant. As we we found in the subsection *Constant functions* on page 113, a constant function is continuous.

A–14.21. a) $g(24.7, 8) = g(-24, 8.5) = g(24, 367) = 24.$ $g(-23.99999, 0) = 23.$
b) For example, $g(x, y)$ is continuous at $(2.6, 0)$, $(2.6, 1)$, $(2.6, 2)$ and discontinuous at $(2, 0)$, $(2, 1)$, $(2, 2)$.

A–14.23. Your friend is wrong about $x = 0$ because there the function is continuous: First, $g(0)$ is defined and equal to zero. Second, given any positive ϵ we take the interval $-\sqrt{\epsilon} < x < \sqrt{\epsilon}$. For all x in this interval the values of $g(x)$ are within ϵ of zero. For $x = 2$ your friend is right: Take $\epsilon = 0.7$. Every interval centered at $x = 2$ has points where $g(x) = 5$. That is, we cannot find a positive number δ such that at all points of the interval $2 - \delta < x < 2 + \delta$ the values of $g(x)$ are within 0.7 of $g(2) = 4$.

Answers for Section 15

A–15.1. a) At $P_0(5,2)$, the function increases in the direction of $\vec{\imath}$.
b) At $P_0(5,2)$, the function decreases in the direction of $\vec{\jmath}$.

A–15.3. a) On the line $x = 5$, the function $g(x, y) = (x - 2)^2 - xy + 7$ becomes a function $k(y) = (5 - 2)^2 - 5y + 7$ of one variable y. We take the ordinary derivative of $k(y)$: $k'(y) = -5$. This means: In the direction of $\vec{\jmath}$, the function decreases at P_0 at the rate of 5 z-units per y-unit (or it increases at the rate of -5).
b) We would have arrived at the same result by doing the following:

- In the formula $g(x, y) = (x - 2)^2 - xy + 7$, we treat x as a constant and take the derivative with respect to y.
- Then we set $(x, y) = (5, 2)$.

A–15.5. a) To find the partial derivative f_x of $f(x, y) = e^{xy}$ you treat y as a constant and take the ordinary derivative of e^{xy} with respect to x. For this, you have to use the chain rule of elementary calculus. For the partial derivative with respect to y you treat x as a constant. Result: $f_x = ye^{xy}$, $\frac{\partial f}{\partial y} = xe^{xy}$.
b) $\frac{\partial f}{\partial x}|_{(1,2)} = 2e^2$, $z_y(3, 0) = 3$.

A–15.7. $f_x = -0.35$, $f_y = 0.25$.

A–15.9. $w_z = xy$, $w_x|_{(1,2,3)} = (2x + yz)|_{(1,2,3)} = 8$.

A–15.11. $f(x, y, z) = -11x + 2y + 23z - 839$. (The -839 has been put there to make a point. What is the point?)

A–15.13. a) $u_x - u_t = f'(x - t) - (-1)f'(x - t) = 2f'(x - t)$. Answer: $2f'(1)$.
b) Remember that $u_{tx} = u_{xt}$. Result: $u_{xx} + 2u_{xt} + u_{tt} = f''(x - t) - 2f''(x - t) + f''(x - t) = 0$.

A–15.15. $(2xy + 3y^5)|_{(1,1)} = 5$; $(x^2 + 15xy^4)|_{(2,2)} = 2^2 + 15 \cdot 2 \cdot 2^4$.

A–15.17. (i) $\sqrt{d} + v$; (ii) u; (iii) $u \frac{1}{2\sqrt{d}}$; (iv) 0.

A–15.19. a) $f_x = u'(x)v(y)$, $f_y = u(x)v'(y)$.
b) $g_x = u'(3x - 5y) \cdot 3 + v'(xy) \cdot y$, $g_y = u'(3x - 5y) \cdot (-5) + v'(xy) \cdot x$.
c) $h_x = \frac{u'(x)}{v(y)}$, $h_y = u(x)\frac{-v'(y)}{(v(y))^2}$.
d) $i_x = \cos(u(x^2)v(y^3)) \cdot u'(x^2)v(y^3) \cdot 2x$, $i_y = \cos(u(x^2)v(y^3)) \cdot u(x^2)v'(y^3) \cdot 3y^2$.

A–15.21. a) $f_u = t + v$, $f_v = u + w$, $f_w = v + z$, $f_t = u$, $f_x = 0$, $f_z = w$.
b) $t_0 + v_0 = 17$. For example $(t_0, u_0, v_0, w_0, z_0) = (17, 0, 0, 0, 0)$. There are many answers.

A–15.23. a) $f_{xx} = f_{xy} = f_{yy} = 0$.
b) 0.
c) Let f be a linear function of any number of variables. Then the second-order partial derivatives of f are all zero.
d) $h_{zu} = 1$. Response: $h(x, y, z, u) = 2x - 4y + zu$ is not a linear function of the four variables $x, \ldots u$. Therefore the statement of c) does not apply to $h(x, y, z, u)$.

A–15.25. 1, 0.

A–15.27. a) Positive. b) Negative.

c) $\dfrac{\partial f}{\partial N} = \dfrac{Ar}{12} \dfrac{\ln(1 + \frac{r}{12})\,(1 + \frac{r}{12})^{-N}}{[1 - (1 + \frac{r}{12})^{-N}]^2}$. d) $\dfrac{\partial f}{\partial A} = \dfrac{\frac{r}{12}}{1 - (1 + \frac{r}{12})^{-N}}$.

A–15.29. $f_x(a, b)$ is the slope of the curve in which the plane $y = b$ intersects the graph $z = f(x, y)$. In our case, the graph is a plane α. If you intersect α with a plane $y = b$ you get a line. The slope of this line is always the same, no matter what a and b are (see right; the plane α is not drawn to scale). The same kind of argument can be used for f_y.

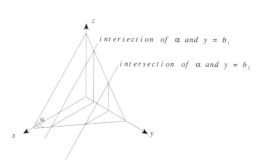

A–15.31. a) $\frac{\partial u}{\partial x} > 0$ since the slope of the center curve is positive at $x = x_0$. $\frac{\partial u}{\partial t} < 0$ since $u(x_0, t)$ is decreasing when $t = t_1$.

b) $\frac{\partial^2 u}{\partial x^2} < 0$, since the center curve is concave down at $x = x_0$. $\frac{\partial^2 u}{\partial t^2} < 0$ since the velocity $\frac{\partial u}{\partial t}(x_0, t)$ of the point $x = x_0$ is decreasing. $\frac{\partial^2 u}{\partial t \partial x} < 0$ since the slope $\frac{\partial u}{\partial x}(x_0, t)$ decreases as t increases.

Answers for Section 16

A–16.1. a) (i) $L(2, 1) = e^2 + 1$; (ii) $e^2 + 1 + e^2(-1) + (2e^2 + 1)(-1) = -2e^2$.

b) (i) = (ii) : $L_x = e^2$, $L_y = 2e^2 + 1$; since $L(x, y)$ is a linear function, its partials are the same at all points.

c) $L(x, y) = e^2 x + (2e^2 + 1)y - 3e^2$.

A–16.3. a) $L(a, b) = f(a, b)$.

b) The partials of L are constant: $L_x = f_x(a, b)$ and $L_y = f_y(a, b)$.

A–16.5. a) $2e^2 x + e^2 y - z = 2e^2$. b) $L(x, y) = 2e^2 x + e^2 y - 2e^2$.

A–16.7. $L(x, y) = 1 - 3\Delta x + \Delta y = 1 - 3(x + 1) + (y - 2)$.

A–16.9. a) $L(x, y) = f(a, b) + df$. b) 0.

c) $L(x, y)$ and df have the same partials *everywhere*, and they agree with the partials of f at $(x, y) = (a, b)$.

A–16.11. a) $\dfrac{\Delta M}{237^2} - \dfrac{(2)(87)\,\Delta s}{237^3}$. b) approx. $-5.28 \cdot 10^{-5}$. c) approx. $4.9 \cdot 10^{-5}$.

A–16.13. a) 5.3%. b) The percentage change gives a sense of the significance of the change that does not depend on the size of the quantity that is changing.

A–16.15. a) $L(r, h) = 2\sqrt{29} + \dfrac{33}{\sqrt{29}}(r - 2) + \dfrac{10}{\sqrt{29}}(h - 5)$.

b) $\dfrac{33}{\sqrt{29}}\Delta r + \dfrac{10}{\sqrt{29}}\Delta h$.

c) $\dfrac{(2r^2 + h^2)\,dr + rh\,dh}{\sqrt{r^2 + h^2}}$.

A–16.17. a) This is difficult to say. If a alone increases, the torus gets slimmer, and the surface area decreases. If b alone increases, the torus gets fatter, and the surface area increases. If a and b increase at the same time, we cannot predict without calculation which way the surface area goes unless one of a, b increases much more than the other.

b) Increases by about 0.434 m^2.

c) $|dA| \leq 2\pi^2[2.5 \cdot 0.05 + 1.6 \cdot 0.024]$, $A = \pi^2 \cdot 3.69$, $\dfrac{|dA|}{A} \leq \dfrac{0.3268}{3.69} = 0.0856$, 8.9%.

A–16.19. $x \approx 0.846$, $y \approx .536$.

Answers for Section 17

In order to avoid confusion between dot products, evaluation of a function at a point, etc., we give some answers with more parentheses than necessary.

A–17.1. $d(x,y) = \sqrt{(x-3)^2 + (y+5)^2}$;
$D(t) = d(e^t, t^3) = \sqrt{(e^t - 3)^2 + (t^3 + 5)^2}$.

A–17.3. a) $F(1) = \left[(t^2 + 1)(-t^3) + (t^2 + 1)\right]_{t=1} = 0$.

b) $F(t) = -t^5 - t^3 + t^2 + 1$, $F'(1) = -5 - 3 + 2 = -6$.

c) $F'(t) = (y + 1, x) \cdot (2t, -3t^2) = (-t^3 + 1, t^2 + 1) \cdot (2t, -3t^2)$, $F'(1) = -6$.

A–17.5. $(2(x+y), 2(x+y)) \cdot (0, 2u)|_{(u,v)=(1,1)} = 12$.

A–17.7. Note that $(a_u, b_u) = (v, 1)$, and $(a_v, b_v) = (u, 1)$.

(i) $(f_x(uv, u+v), f_y(uv, u+v)) \cdot (v, 1)|_{(u,v)=(2,3)} = f_x(6,5)\,(3) + f_y(6,5)\,(1)$;

(ii) $(f_x(uv, u+v), f_y(uv, u+v)) \cdot (u, 1)|_{(u,v)=(2,3)} = f_x(6,5)\,(2) + f_y(6,5)\,(1)$.

A–17.9. a) grad $f = (yz^3, xz^3, 3xyz^2)$; (i) $(1 \cdot 4^3, 0, 0)$; (ii) $(2 \cdot 3^3, 1 \cdot 3^3, 3 \cdot 1 \cdot 2 \cdot 3^2)$.

b) (i) $f(4, 2, 1) = 8$; (ii) $(yz^3, xz^3, 3xyz^2) \cdot (3t^2, 2t+1, 1) = (2, 4, 24) \cdot (3, 3, 1) = 42$.

A–17.11. a) $F(2) = 3(a(2) - b(2))^2 = 3 \cdot 1^2 = 3$.

b) $F(t) = \frac{1}{2}$ means $3(a(t) - b(t))^2 = \frac{1}{2}$. We find $a(t) - b(t) = \frac{\sqrt{6}}{6} = \pm 0.4$. Now we use ruler and pencil: $t = 0, 0.8, 5.3, 6.6$.

A–17.13. grad $f = (2y, 2x - y^2)$.   a) $(4, 0)$.   b) See below.

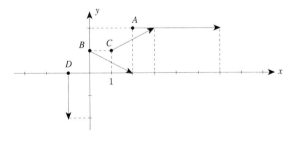

A–17.15. grad f has components $(4, -2, 7)$ at *all* points.

A–17.17. $\dfrac{d}{dt}(f(t^2, \sqrt{t}))\Big|_{t=4} = (3 + (\sqrt{t})^2, 1 + 2t^2\sqrt{t}) \cdot (2t, \frac{1}{2\sqrt{t}})\Big|_{t=4} = 56 + \frac{65}{4}$.

A–17.19. $\pm\left(\dfrac{\sqrt{2}}{2}, \dfrac{\sqrt{2}}{2}\right)$.

A–17.21. $[p_x(2u + 3v, u^2)]\,(2) + [p_y(2u + 3v, u^2)]\,(2u)$.

A–17.23. $e^{\sin x \cos y}(\sin x)(-\sin y)$.

A–17.25. $2\left[\sin(p(u,v))\right]\left[\cos(p(u,v))\right]\left[p_u(u,v)\right]$.

A–17.27. Increasing at a rate of 100π cm^3/min.

Answers for Section 18

A–18.1.   a) f_x is positive, and f_y is negative.
b) At P_0, the function $f(x, y)$ is decreasing in the direction of $\vec{u}$.
c) At P_0, the directional derivative in the direction of $\vec{v}$ is positive.
d) Along the level curves the function stays constant. Therefore, the directional derivative in a direction parallel to a level curve is zero. Take as $\vec{w}$ a vector parallel to the level curves.

A–18.3.   a) The directional derivative in the direction of $\vec{\imath}$ is nothing but $\frac{\partial f}{\partial x}$. Similarly, $\frac{\partial f}{\partial y}$ is nothing but the directional derivative in the direction of $\vec{\jmath}$.
b) Parts b and d.

A–18.5.   The directional derivative is lowest in the direction of $-\operatorname{grad} f$. Therefore, $\operatorname{grad} f$ is parallel to $-(-4, 2)$. The lowest directional derivative is $-|\operatorname{grad} f|$. Therefore, $|\operatorname{grad} f| = 5.9$. Thus, $\operatorname{grad} f = (5.9)\frac{(4, -2)}{\sqrt{4^2 + (-2)^2}}$,   and f_x, f_y are the components of $\operatorname{grad} f$.

A–18.7.   (i) 0;   (ii) $|\operatorname{grad} f| = 1.7$ (with 1 cm as unit).

A–18.9. $\operatorname{grad} f = (7, 2)$; note that $\operatorname{grad} f$ is constant because $f(x, y)$ is a linear function.
a) $(7, 2) \cdot \frac{(4, -3)}{5} = \frac{22}{5}$.
b) $\frac{22}{5}$.
c) $\vec{u}$ parallel to the level curves, i.e., $\vec{u}$ perpendicular to $\operatorname{grad} f$. E.g., $\vec{u} = (-2, 7)$.
d) "All directional derivatives are zero" is only possible if $\operatorname{grad} f$ is the zero vector. Yet $\operatorname{grad} f = (7, 2)$ at all points. Therefore there is no such point P_0.
e) The largest directional derivative equals $\sqrt{7^2 + 2^2}$, and it occurs in the direction of $\operatorname{grad} f = (7, 2)$. The smallest directional derivative equals $-\sqrt{7^2 + 2^2}$, and it occurs in the direction of $-\operatorname{grad} f = (-7, -2)$.
f) The directional derivative is the scalar projection of $\operatorname{grad} f$ onto $\vec{u}$. The absolute value of the scalar projection cannot be bigger than the magnitude of the vector $\operatorname{grad} f$ which you project. Yet $|\operatorname{grad} f| = \sqrt{7^2 + 2^2}$ is less than 8. Therefore, there is no such vector $\vec{u}$.
g) $f_y = 2$, i.e., $\vec{u} = \vec{\jmath}$

A–18.11. $\sqrt{2}$ meters.

A–18.13.   *Remember:* $(\vec{a} + \vec{b}) \cdot \vec{c} = \vec{a} \cdot \vec{c} + \vec{b} \cdot \vec{c}$, and similarly $\vec{c} \cdot (\vec{a} + \vec{b}) = \dots$.
a) $D_{\vec{u}}(f + g) = \operatorname{grad}(f + g) \cdot \frac{\vec{u}}{|\vec{u}|}$.
What is $\operatorname{grad}(f + g)$? $\operatorname{grad}(f + g) = ((f + g)_x, (f + g)_y) = (f_x + g_x, f_y + g_y) = (f_x, f_y) + (g_x, g_y) = \operatorname{grad} f + \operatorname{grad} g$. Therefore:

$$D_{\vec{u}}(f + g) = \operatorname{grad}(f + g) \cdot \frac{\vec{u}}{|\vec{u}|} = (\operatorname{grad} f + \operatorname{grad} g) \cdot \frac{\vec{u}}{|\vec{u}|}$$

$$= \operatorname{grad} f \cdot \frac{\vec{u}}{|\vec{u}|} + \operatorname{grad} g \cdot \frac{\vec{u}}{|\vec{u}|} = D_{\vec{u}}f + D_{\vec{u}}g.$$

Statement a) is true.

b) First, we work out the left side of statement b):

$$D_{\vec{u}+\vec{v}}f = \operatorname{grad} f \cdot \frac{\vec{u}+\vec{v}}{|\vec{u}+\vec{v}|} = \operatorname{grad} f \cdot \left(\frac{\vec{u}}{|\vec{u}+\vec{v}|} + \frac{\vec{v}}{|\vec{u}+\vec{v}|} \right).$$

Now we rewrite the right side:

$$D_{\vec{u}}f + D_{\vec{v}}f = \operatorname{grad} f \cdot \frac{\vec{u}}{|\vec{u}|} + \operatorname{grad} f \cdot \frac{\vec{v}}{|\vec{v}|} = \operatorname{grad} f \cdot \left(\frac{\vec{u}}{|\vec{u}|} + \frac{\vec{v}}{|\vec{v}|} \right).$$

Look at the last terms of the displayed equations. For example, take $\vec{u} = \vec{\imath}$ and $\vec{v} = \vec{\jmath}$. The statement b) is not true. Why? "The statement is true" means "the statement is true for *all* vectors $\vec{u}$ and $\vec{v}$." We have found vectors $\vec{u}$ and $\vec{v}$ for which the statement is false.

A–18.15.   a) 3.   b) $\frac{3\sqrt{2}}{2}$.   c) $3\cos\theta$.

Answers for Section 19

A–19.1.   a) $(2,1,1) \cdot \frac{1}{\sqrt{18}}(1,1,-4) = -\frac{1}{\sqrt{18}}$.
b) It is $-f_z$ at $(1,1,1)$ and equals -1.
c) $(18,1,9) \cdot \frac{1}{\sqrt{3}}(1,1,1) = \frac{1}{\sqrt{3}}28$.

A–19.3.   a) $|\operatorname{grad} g| = |(-2,-7,2)| = \sqrt{57}$.
b) Such a vector $\vec{u}$ has to be perpendicular to $\operatorname{grad} g = (-2,-7,2)$. There are infinitely many such vectors, for example $(2,0,2)$ and $(0,2,7)$.

A–19.5.   a) $f(Q) - f(P_0) = f(8,2,5) - f(3,1,1) = 31$, distance P_0Q equals $\sqrt{42}$, average rate of change: $(\sqrt{42})^{-1}31$.
b) $\operatorname{grad} f = (4,-1,3)$, $\overrightarrow{P_0Q} = (5,1,4)$.
Directional derivative: $(4,-1,3) \cdot (\sqrt{42})^{-1}(5,1,4) = (\sqrt{42})^{-1}31$.
c) Write k for the line P_0Q and $\vec{v} = (v_1,v_2,v_3)$ for the unit vector in the direction $\overrightarrow{P_0Q}$. Parametric representation of k: $(x,y,z) = (3,1,1) + t(v_1,v_2,v_3)$. Value of $f(x,y,z)$ on points of k: $F(t) = f(3 + tv_1, \dots)$. In the notation $F(t)$, the average rate of change between P_0 and Q is $\dfrac{F(1) - F(0)}{1 - 0}$, and the directional derivative is $F'(0)$.
Now we observe that $F(t)$ is a linear function of t. Why? $f(x,y,z)$ is a linear function of x, y, z, and each of the latter three is a linear function of t. Now we recall from elementary calculus: For a linear function $g(x)$, the average rate of change $\dfrac{g(x_2) - g(x_1)}{x_2 - x_1}$ equals the constant derivative $g'(x)$, no matter what x_1 and x_2 are, or geometrically "slope of secant = slope of tangent." Applied to our linear function $F(t)$: The average rate of change $\dfrac{F(1) - F(0)}{1 - 0}$ equals the constant $F'(0)$.

A–19.7.   a) $\operatorname{grad} f = g'(r)\frac{\vec{r}}{r} = g'\left(\sqrt{x^2 + y^2 + z^2}\right) \frac{x\vec{\imath}+y\vec{\jmath}+z\vec{k}}{\sqrt{x^2+y^2+z^2}}$.

b) $\operatorname{grad}\varphi = -\frac{\vec{r}}{r^3} = -\frac{x\vec{\imath}+y\vec{\jmath}+z\vec{k}}{(x^2+y^2+z^2)^{3/2}}$.

A–19.9.   a) The only information we have is: (i) a level surface passes through $P(4,3,1)$, and (ii): some directional derivatives are zero. (i) is not much; it says only that $P(4,3,1)$ is part of the domain of definition of $g(x,y,z)$. (ii) says nothing about the value of the function, but only something about directional derivatives. There is not enough information to determine $g(4,3,1)$.

b) $\vec{a}$ and $\vec{b}$ are parallel to the tangent plane to the level surface because the directional derivatives in these directions is zero. Therefore, $\vec{n} = \vec{a} \times \vec{b} = (2,4,-5)$ is a normal vector of the tangent plane. Equation of the tangent plane: $2x + 4y - 5z = 15$.

A–19.11. $\frac{3a}{2}$.

A–19.13. a) (i) $dW = q^3 r^7 dp + 3pq^2 r^7 dq + 7pq^3 r^6 dr$;
(ii) $dA = b \sin \theta \, da + a \sin \theta \, db + a \, b \cos \theta \, d\theta$.

b) (i) $\dfrac{dp}{p} + 3\dfrac{dq}{q} + 7\dfrac{dr}{r}$;   (ii) $\dfrac{da}{a} + \dfrac{db}{b} + \dfrac{\theta \cos \theta}{\sin \theta}\dfrac{d\theta}{\theta}$.

Answers for Section 20

A–20.1. $T_1(x) = x - 1$; $T_2(x) = (x-1) - \frac{1}{2}(x-1)^2$.
$T_1(1.1) = 0.1$; $T_2(1.1) = 0.095$.

A–20.3. a)

$$T_3(x,y) = f(x_0,y_0) + f_x(x_0,y_0)(x-x_0) + f_y(x_0,y_0)(y-y_0)$$
$$+\frac{1}{2}f_{xx}(x_0,y_0)(x-x_0)^2 + f_{xy}(x_0,y_0)(x-x_0)(y-y_0) + \frac{1}{2}(f_{yy}(x_0,y_0)(y-y_0)^2$$
$$+\frac{1}{6}f_{xxx}(x_0,y_0)(x-x_0)^3 + \frac{1}{2}f_{xxy}(x_0,y_0)(x-x_0)^2(y-y_0)$$
$$+\frac{1}{2}f_{xyy}(x_0,y_0)(x-x_0)(y-y_0)^2 + \frac{1}{6}f_{yyy}(x_0,y_0)(y-y_0)^3.$$

b) $T_3(x,y) = 1 + y + (x-1)y + \dfrac{1}{2}y^2 + (x-1)y^2$.

A–20.5.

$$T_2(x,y,z) = f(x_0,y_0,z_0)$$
$$+f_x(x_0,y_0,z_0)(x-x_0) + f_y(x_0,y_0,z_0)(y-y_0) + f_z(x_0,y_0,z_0)(z-z_0)$$
$$+\frac{1}{2}\Big(f_{xx}(x_0,y_0,z_0)(x-x_0)^2 + f_{yy}(x_0,y_0,z_0)(y-y_0)^2 + f_{zz}(x_0,y_0,z_0)(z-z_0)^2\Big)$$
$$+f_{xy}(x_0,y_0,z_0)(x-x_0)(y-y_0) + f_{xz}(x_0,y_0,z_0)(x-x_0)(z-z_0)$$
$$+f_{yz}(x_0,y_0,z_0)(y-y_0)(z-z_0).$$

A–20.7. a) $L_1(x,y) = 1$.
b) $Q_1(x,y) = 1 + (x-1)y$.
c) $L_2(x,y) = 2 + (x-2) + 2\ln 2(y-1)$.
d) $Q_2(x,y) = 2 + (x-2) + 2\ln 2(y-1) + (\ln 2)^2(y-1)^2 + (1+\ln 2)(x-2)(y-1)$.
e) $Q_2(2.2, 0.9)$ should give the best approximation. To get a good approximation, (x,y) should be near the basepoint. Since $(2.2, 0.9)$ is nearer to the basepoint $(2,1)$ than to $(1,0)$, we expect Q_2 to give a better approximation than Q_1. Also, we expect the quadratic Q_2 to give a better approximation than the linear L_2.

Answers for Section 21

A–21.1.   a) The point P where the lines $f(x, y) = 4$ intersect is a critical point.
b) If the gradient of $f(x, y)$ is different from zero, then it must be perpendicular to the level curve. At P, the gradient would have to be perpendicular to two different straight lines at the same time. This is impossible. Therefore, the gradient must be zero at P.
c) The point $g(x, y) = 3$ is a critical point.
d) For points (x, y) different from the point "3" we have $g(x, y) > 3$. If $\operatorname{grad} g$ were different from zero at "3," then $g(x, y)$ would decrease in the direction of $-\operatorname{grad} g$. But $g(x, y)$ is not less than 3. Therefore, we must have $\operatorname{grad} g = (0, 0)$ at "3."

A–21.3.   We have to find all points (x, y) that satisfy the two equations $f_x = -y + 2x = 0$ and $f_y = 3y^2 - x = 0$ at the same time. Note that $(0, 0)$ is a solution. From the first equation we get $y = 2x$. We go with that into the second equation and get $12x^2 - x = 0$, $x = 0$ and $x = \frac{1}{12}$. There are two critical points $(0, 0)$ and $(\frac{1}{12}, \frac{1}{6})$.

A–21.5.   a) The level curves are parallel lines. This excludes both $f(x, y)$ and $g(x, y)$ (see the level curves in Problem 21.1). The level curves of $h(x, y)$ are also circles. Therefore, the surface cannot be any of f, g, h.
b) It is very likely the graph $z = g(x, y)$. Why? The upward sloping surface looks like having concentric circles as level curves, and the graph shows that the function has a minimum. It is unlikely to be the graph of $h(x, y)$ because h has a maximum.
c) The surface has a saddle point—in two directions the surface slopes up, in two directions it slopes down. It is likely to be the graph $z = f(x, y)$. Neither of the graphs of $g(x, y)$ and $h(x, y)$ have a saddle point.
d) The surface is a plane. It is clear from the level curves of $f(x, y)$, $g(x, y)$, and $h(x, y)$ that neither of them is a linear function. The surface cannot be the graph of any of the three functions.

A–21.7. a) only critical point $(0, 0)$; $D(x, y) = -4y - 4x^2$; Second Derivative Test is inconclusive.
b) See figure.
c) See figure. The function cannot have a local extremum at $(0, 0)$. Why? Draw a circle of radius r around the origin. No matter how small or how big r is, the circle will always contain points where $f(x, y)$ is greater than $f(0, 0) = 0$ and points where $f(x, y)$ is less than $f(0, 0) = 0$. $f(0, 0) = 0$ cannot be a local extremum.

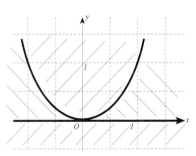

A–21.9. a) $(0, 0)$ is the only critical point. The Second Derivative Test fails.
b) See figure.
c) Saddle point.

A–21.11.   Critical points: $(0, 0)$ and $(6, 6)$; $D(x, y) = 36xy - 324$; saddle point at $(0, 0)$, local maximum at $(6, 6)$.

Answers for Section 22

A–22.1. a) See Figure A22.1 on the next page.

b) "Curve: max, min:" $J : 5.9, 0$; $K : -2, -12$; $L : 8, 4$; $M : 11, -5$; $N : 13, 0.5$;
$Q : 6, -4$.

A–22.3. $G(x)$: $\max G(2) = \frac{5}{3}$, $\min G(1) = \frac{1}{3}$;
$g(x, y)$: $\max g(2, \frac{4}{3}) = \frac{5}{3}$, $\min g(1, \frac{1}{3}) = \frac{1}{3}$.

A–22.5. See Figure A22.2 on the facing page. The critical points along C are P, Q, and R. We read off: Max 3.7 at $Q(-1.7, -0.6)$, min -1.7 at $P(1.7, -0.6)$.

A–22.7. a) There are no endpoints.

b) There are no corners.

c) See Q, R, S, T, U shown in Figure A22.3 on the next page.

d) There are no points where $\operatorname{grad} g$ is zero.

e) Q: maximum on C; $\quad$ R: no extremum; $\quad$ S: minimum on C; $\quad$ T: local maximum; $\quad$ U: local minimum.

A–22.9. a) $(0, 0)$, $(2, \frac{4}{3})$; $\quad$ b) None. $\quad$ c) $(1, \frac{1}{3})$; $\quad$ d) None.

A–22.11.

a)

b) Left end pt to A: decreasing
$\quad$ A to B: decreasing
$\quad$ B to C: increasing
$\quad$ C to right end pt: decreasing.

c) Maximum: C $\quad$ Minimum: B.

A–22.13. Critical points: $(\pm 1, 0)$, $(0, \pm 1)$, $(2/3, \pm\sqrt{5}/3)$;
Maximizers: $(0, \pm 1)$, $(1, 0)$; $\quad$ Minimizer: $(-1, 0)$.

A–22.15. $(4/3, \pm\sqrt{5}/3)$.

A–22.17. a) For $\mathcal{A}$ the Lagrange method is better.

Why: Solving $H(x, y) = 1$ yields unwieldy solutions $x = \ldots$ or $y = \ldots$.
The work is similar to the work for Problem 22.12. We have to solve the system

$$
\begin{aligned}
1 &= \lambda \cdot 2(x - 6) \\
-2 &= \lambda \cdot 2(y - 1) \\
(x - 6)^2 + (y - 1)^2 &= 1.
\end{aligned}
$$

We want to eliminate λ. $x - 6 = 0$ and $y - 1 = 0$ are both impossible (look at first two equations). Therefore, we can divide by $x - 6$ and $y - 1$. This allows us to eliminate λ from the first two equations: $2(x - 6) = -(y - 1)$. We go with that into the constraint and obtain two critical points

$$
(\frac{1}{\sqrt{5}} + 6, -\frac{2}{\sqrt{5}} + 1) \quad \text{and} \quad (-\frac{1}{\sqrt{5}} + 6, \frac{2}{\sqrt{5}} + 1).
$$

b) We use substitution because the constraint can be solved easily for x: $x = 2y$. If we go with that into $H(x, y)$ we obtain a function $G(y)$ of one variable:

$$
G(y) = H(2y, y) = (2y - 6)^2 + (y - 1)^2.
$$

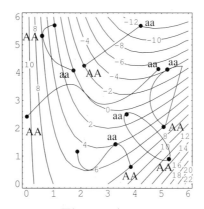

Figure A22.1

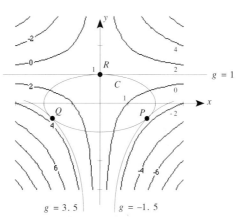

Figure A22.2

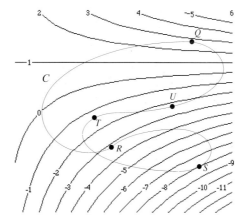

Figure A22.3

The usual routine $G'(y) = 0$ yields the critical point $(x, y) = (5.2, 2.6)$.

c) See below. The constraint is the circle shown. The level curves of $p(x, y)$ are parallel lines of slope $\frac{1}{2}$. The points on the circle where the tangent has slope $1/2$ are the critical points along the circle. The function has the minimum at A and the maximum at B. See a) for the exact coordinates of A and B.

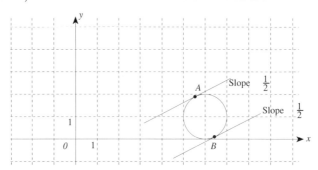

d) See below. The constraint is the line k: $y = (1/2)x$. Draw k. We have to find a point E on k so that the level curve of $H(x, y)$ is parallel to k at E. The level curves are circles centered at $(6, 1)$. At E the line through E and $(6, 1)$ is perpendicular to k. E is the critical point on k. At E the function $H(x, y)$ has the minimum on k. See b) for the exact coordinates of E.

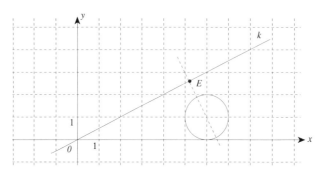

Answers for Section 23

A–23.1.   All that we can say is this: The global minimum of $f(x, y)$ is approximately -1.1, and the global maximum is likely to be greater than 9.1.

A–23.3.   The function $g(x, y)$ has the global minimum on the boundary of T and the global maximum in the interior of T.

A–23.5.   See Figure A23.1 on the facing page; $f_1(x, y) = r$: Maximum 9.2 at T, minimum 1.7 at U; $f_2(x, y) = \theta$: Maximum 1.43 (radian measure) at V, minimum 0.21 at W

A–23.7.   a) Critical point of $f(x, y)$ in the interior of OAB: $(2, 1)$.
On OA: $f(x, y) = f(x, 0) = x^3 - 12x + 5 = G(x)$, $G'(x) = 3x^2 - 12$, candidates: $(2, 0)$, endpoints $(0, 0)$ and A.
On OB: $f(x, y) = f(x, x) = 2x^3 - 15x + 5 = H(x)$, $H'(x) = 6x^2 - 15$, candidates: $(\sqrt{2.5}, \sqrt{2.5})$, endpoints $(0, 0)$ and B.
On AB: $f(x, y) = f(3, y) = y^3 - 3y - 4 = J(y)$, $J'(y) = 3y^2 - 3$, candidates: $(3, 1)$, endpoints A and B.

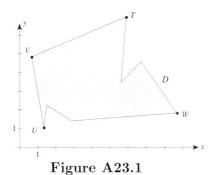

Figure A23.1

Corners: already among the candidates on the boundary.

Evaluation of $f(x, y)$: $f(2, 1) = -13$, $f(2, 0) = -11$, $f(\sqrt{2.5}, \sqrt{2.5}) = -10.81$, $f(3, 1) = -6$, $f(0, 0) = 5$, $f(3, 0) = -4$, $f(3, 3) = 14$.

Maximum 14 at $(3, 3)$, minimum -13 at $(2, 1)$.

b) None of the critical points of $f(x, y)$ lie in the interior of $OEDC$.

On OE: $f(x, y) = f(x, 0) = x^3 - 12x + 5 = G(x)$, $G'(x) \neq 0$ on OE, candidates: endpoints $(0, 0)$ and E.

On ED: $f(x, y) = f(-1, y) = y^3 - 3y + 16 = L(y)$, $L'(y) = 3y^2 - 3$, candidates: $(-1, -1)$, endpoints E and D.

On DC: $f(x, y) = f(x, -2) = x^3 - 12x + 3 = M(x)$, $M'(x) \neq 0$ on DC, candidates: endpoints D and C.

On CO: $f(x, y) = f(0, y) = y^3 - 3y + 5 = K(y)$, $K'(y) = 3y^2 - 3$, candidates: $(0, -1)$, endpoints C and $(0, 0)$.

Corners: already among the candidates on the boundary.

Evaluation of $f(x, y)$: $f(-1, -1) = 18$, $f(0, -1) = 7$, $f(0, 0) = 5$, $f(-1, 0) = 16$, $f(-1, -2) = 14$, $f(0, -2) = 3$.

Maximum 18 at $(-1, -1)$, minimum 3 at $(0, -2)$.

A–23.9.   Min ≈ 0.26,   Max ≈ 1.43.

A–23.11.   a) Revenue $= 1000p_1 - 2.3p_1^2 + 1400p_2 - 4.1p_2^2$.

b) Cost $= 218000 - 161p_1 - 287p_2$.

c) Profit is maximized with $p_1 = 252.39$ and $p_2 = 205.73$. At these prices, the demand in market 1 is 419.5 units, and the demand in market 2 is 556.5 units.

Answers for Section 24

A–24.1. $47.75 \leq$ mass in metric tons ≤ 87.75.

A–24.3. If function values are sampled at the lower left vertices of the small regions, the estimate obtained is 0.494. Sampling at other points gives different estimates.

A–24.5. We integrate the constant function $f(x) = 1$ over an interval $[a, b]$ which has length $b - a$:

$$\int_a^b 1 \, dx = \text{area} = (b - a) \cdot 1 = b - a.$$

A definite integral is defined as area. In this special case the integral is the length $b - a$ of the interval. *As a rule, a definite integral should be thought of as representing an area or, in applications, an area-like quantity.* The integral of $f(x) = 1$ is written $\int_a^b dx$.

A–24.7. a) Since $f(x, y) \leq b$, the Comparison Principle gives $\int\int_D f(x, y)\, dA \leq \int\int_D b\, dA = b \cdot A(D)$.
b) Since $f(x, y) \geq a$, the Comparison Principle gives $\int\int_D f(x, y)\, dA \geq \int\int_D a\, dA = a \cdot A(D)$.

A–24.9. Double integral: $7(-\frac{64}{3} + 32)$ (see A–23.3). Area of region R: $(7)(4) = 28$.
a) $f^* = \frac{1}{28} 7(-\frac{64}{3} + 32) = \frac{8}{3}$ b) $d = f^* = \frac{8}{3}$

A–24.11. a) 10.
b) Write p^* for the average of $p(x, y)$ over T. Then $p^* \cdot 6 = 10$, $p^* = \frac{5}{3}$.
c) (i) $10k$; (ii) The second statement is true.

A–24.13. The level curves of $q(x, y) = x - y$ are lines of slope 1. The level curves passing through C and B are those which give the extrema of $q(x, y)$ in T, and they are shown below as broken lines. We determine their level and find that $q(x, y)$ has the global minimum $\min q = -7.1$ and the global maximum $\max q = 7.5$. Estimate: $(-7.1) \cdot \frac{1}{2}(7 \cdot 9) \leq \int\int \leq (7.5) \cdot \frac{1}{2}(7 \cdot 9)$.

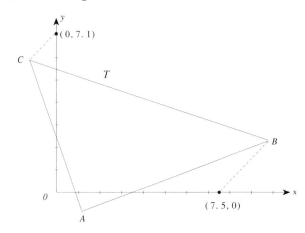

A–24.15. If the plate covers the square S in the xy-plane defined by $0 \leq x \leq 1$ and $0 \leq y \leq 1$, and if the zero density edge is along the y-axis, then the mass density is $\rho(x, y) = 2x$, and the total mass is given by $\int\int_S 2x\, dA$. Interpretting this integral as the volume of a prism gives a total mass of 1 kg.

A–24.17. In an xy-system with the origin at the center of town, the town covers a disk D of radius 10 km. The population density is $500 - r(x, y)/10$, where $r(x, y) = \sqrt{(x^2 + y^2)}$ is the distance from the center of town. The total population is given by $\int\int_D(500 - r(x, y)/10)\, dA$. The integral may be interpretted as the volume of a cone height 500 and radius 10, so the total population is $\frac{1}{3}\pi(10)^2 500 \approx 52360$.

Answers for Section 25

A–25.1. a) $A(y)$ and $t(y)$ increase, $s(y)$ decreases
b) $A(1) = 0$, $s(3.5) = 2$, $t(6) = 5$.

A–25.3. $\displaystyle\int_{y=1}^{y=6}\left(\int_{x=-0.4y+3.4}^{x=0.4y+2.6}xy\,dx\right)dy.$

A–25.5. D_1: $\iint f(x,y)\,dydx$; we use x-slices because for y-slices you need two integrals with respect to x.

D_2: $\iint f(x,y)\,dydx$ or $\iint f(x,y)\,dxdy$. It makes no difference because either way you have to split the region into two.

D_3: $\iint f(x,y)\,dxdy$; we use y-slices because for x-slices you need two integrals with respect to y.

D_4: It makes no difference which slices we use.

A–25.7. a) $y^2(\frac{3^2}{2}-\frac{1^2}{2})$.　　b) $x\left(\frac{(x+1)^3}{3}-\frac{(x-1)^3}{3}\right)$.　　c) $xy^2(q^2-(3p+4))$.
d) $(\frac{y^2}{2}-\frac{(e^y)^2}{2})+2y(y-e^y)$.

A–25.9. Equation of circle of center $M(4,2)$ and radius 3: $(x-4)^2+(y-2)^2=9$, therefore $y-2=\pm\sqrt{9-(x-4)^2}$.

a)　$\displaystyle\int_{x=4}^{x=7}\left(\int_{y=-\sqrt{9-(x-4)^2}+2}^{y=\sqrt{9-(x-4)^2}+2}xy\,dy\right)dx.$

b)　$\displaystyle\int_{x=4}^{x=7}\frac{x}{2}\left(\left(\sqrt{9-(x-4)^2}+2\right)^2-\left(-\sqrt{9-(x-4)^2}+2\right)^2\right)dx.$

A–25.11. $\displaystyle\iint_S 2x\,dA=\int_{x=0}^{x=1}\int_{y=0}^{y=1}2x\,dy\,dx=\int_0^1 2x\,dx=1$ kg.

Answers for Section 26

A–26.1. a) $Q_1:(x,y)=(0,-2)$;　　$Q_2:(x,y)=(3,0)$;　　$Q_3:(x,y)=(-2\sqrt{2},-2\sqrt{2})$;　　$Q_4=Q_5:(x,y)=(-1,0)$.
b) Polar coordinates (r,θ): $T_1:(r,\theta)=(\sqrt{242},\frac{7}{4}\pi)$;　　$T_2:(r,\theta)=(2,\frac{\pi}{3})$;
$T_3:(r,\theta)=(328,\pi)$.

A–26.3.

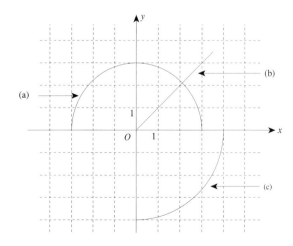

A–26.5. a) $r^2=16$.　　b) $2r\cos\theta+r\sin\theta=0$.
c) Equation of m: $-x+3y-2=0$, $-r\cos\theta+3r\sin\theta-2=0$.

A–26.7. a) See Figure A26.1 on the next page.

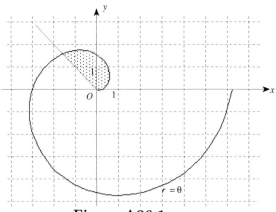

Figure A26.1

b) $(x, y) = (\theta \cos \theta, \theta \sin \theta)$.

c) The area is given by $\iint_D dA$ where D is the shaded area in Figure A26.1.

$$\iint_D dA = \iint_D r \, dr \, d\theta = \int_{\theta=0}^{\theta=0.75\pi} \left(\int_{r=0}^{r=\theta} r \, dr \right) d\theta = \frac{1}{6} \left(\frac{3\pi}{4} \right)^3.$$

A–26.9. $\iint_{\text{sector}} dA = \iint_{\text{sector}} r \, dr = \int_{\theta=0}^{\theta=\theta_0} \left(\int_{r=0}^{r=r_0} r \, dr \right) d\theta = \frac{1}{2} \theta_0 \, r_0^2,$

$f(r, \theta) = \frac{1}{2} \theta \, r^2$.

A–26.11.

a)   (i) $\int_{\theta=0}^{\theta=\alpha} \int_{r=0}^{r=2m \cos \theta} r \, dr \, d\theta;$ (ii) $m^2(\alpha + \sin \alpha \cos \alpha)$.

b)   (i) $\int_{y=0}^{y=1} \int_{x=1-y}^{x=\sqrt{1-y^2}} (x^2 + y) \, dx \, dy;$

(ii) $\int_{\theta=0}^{\theta=\frac{\pi}{2}} \int_{r=\frac{1}{\cos \theta + \sin \theta}}^{r=1} (r^2 \cos^2 \theta + r \sin \theta) \, r \, dr \, d\theta.$

Answers for Section 27

A–27.1.   400 kg.

A–27.3.   Lower bound: $\frac{1}{64}$;   Upper bound: $\frac{27}{64}$.

A–27.5. $\frac{4}{3}$.

A–27.7.   a) $\int_{z=0}^{z=3} \int_{y=0}^{y=2-(2/3)z} \int_{x=0}^{x=1-(1/2)y-(1/3)z} z \, dx \, dy \, dz.$ b) $\frac{3}{4}$.

A–27.9. $\iiint_K (z + 3) \, dV = \iiint_K z \, dV + 3 \frac{a^2 b}{2} = \frac{a^3 b}{6} + 3 \frac{a^2 b}{2}.$

A–27.11.

a)

b) $$\iiint_K y\, dV \;=\; \int_{x=0}^{x=1}\int_{y=0}^{y=1}\int_{z=x}^{z=x+b} y\, dz\, dy\, dx$$

$$= \int_{x=0}^{x=1}\int_{y=0}^{y=1} y\big((x+b)-x\big) dy\, dx = \int_{x=0}^{x=1}\frac{b}{2}\, dx = \frac{b}{2}.$$

A–27.13. a) $W(z)$ is a filled in circle (a disk) of radius $b(1-\frac{c}{h})$.

b) $\displaystyle\int_{\theta=0}^{\theta=2\pi}\int_{r=0}^{r=(b/h)(h-z)} (r\cos\theta)\,(r\sin\theta)\, z\, r dr d\theta.$

c) $\displaystyle\int_{z=0}^{z=h}\int_{\theta=0}^{\theta=2\pi}\int_{r=0}^{r=(b/h)(h-z)} (r\cos\theta)\,(r\sin\theta)\, z\, r\, dr\, d\theta\, dz.$

A–27.15.

a) $\displaystyle\int_{y=0}^{y=b}\int_{z=0}^{z=(a/b)y}\int_{x=0}^{x=(a/b)y-z} H(x,y,z)\, dx\, dz\, dy.$ b) y and z only.

Answers for Section 28

A–28.1. See Figure A28.1 on the following page.

A–28.3. See Figure A28.2 on the next page.

A–28.5.

a) $\displaystyle\iiint_K (4x^2+4y^2+z^2)\, dV = \int_{z=0}^{z=h}\int_{\theta=\frac{\pi}{2}}^{\theta=\frac{3\pi}{2}}\int_{r=a}^{r=b} (4r^2+z^2)\, r\, dr\, d\theta\, dz.$

b) $\displaystyle\iiint_T f(P)\, dV = \int_{z=-b}^{z=b}\int_{\theta=0}^{\theta=2\pi}\int_{r=m-\sqrt{b^2-z^2}}^{r=m+\sqrt{b^2-z^2}} (600b-z)\, r\, dr\, d\theta\, dz.$

Answers for Section 29

A–29.1. a) See Figure A29.1 on the following page.

b) x is positive, y and z are negative.

c) (i) $\rho=\sqrt{38}$;   (ii) $0<\theta<\frac{\pi}{2}$;   (iii) $\frac{\pi}{2}<\phi<\pi$.

A–29.3. $z=\frac{\sqrt{3}}{3}\sqrt{x^2+y^2}.$

A–29.5. In both cases, the equation is $z=0$.

A–29.7. A solid cone with a hemispherical bubble removed.

A–29.9. a) $\displaystyle\iiint_S (x+y+z)\, dV =$

$\displaystyle\int_{\theta=0}^{\theta=2\pi}\int_{\phi=0}^{\phi=\frac{\pi}{4}}\int_{\rho=b\frac{\sqrt{2}/2}{\cos\phi}}^{\rho=b} (\rho\sin\phi\cos\theta + \rho\sin\phi\sin\theta + \rho\cos\phi)\,\rho^2\sin\phi\, d\rho\, d\phi\, d\theta.$

b) $\displaystyle V = \int_{\theta=0}^{\theta=2\pi}\int_{\phi=0}^{\phi=\arccos(\frac{b-h}{b})}\int_{\rho=\frac{b-h}{\cos\phi}}^{\rho=b} \rho^2\sin\phi\, d\rho\, d\phi\, d\theta.$

c) The volume V of the ball is twice the volume of the upper half. To get the volume of the upper half, set $h=b$ in b):

$$V = 2\int_{\theta=0}^{\theta=2\pi}\int_{\phi=0}^{\phi=\arccos 0}\int_{\rho=0}^{\rho=b} \rho^2\sin\phi\, d\rho\, d\phi\, d\theta$$

$$= 2\int_{\theta=0}^{\theta=2\pi}\int_{\phi=0}^{\phi=\frac{\pi}{2}}\frac{b^3}{3}\sin\phi\, d\phi\, d\theta = \frac{4\pi b^3}{3}.$$

A–29.11. Mass $= \dfrac{4\pi\delta_0}{a}\left(\rho_0^2+\dfrac{2\rho_0}{a}+\dfrac{2}{a^2}\right)\approx 6.15\times 10^{18}$ kg. Here ρ_0 denotes the radius of the earth.

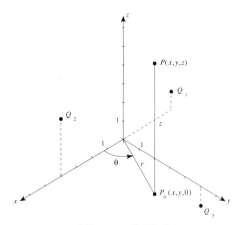

Figure A28.1

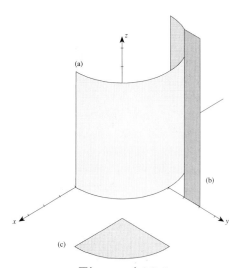

Figure A28.2

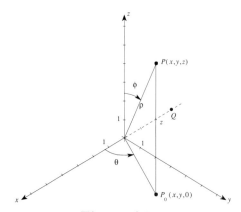

Figure A29.1

Answers for Section 30

A–30.1.   a) and b) see Figure A30.1 on the next page. Note that the figure is reduced in size.

c) $|\overrightarrow{M}|$ is a scalar. Therefore assigning $|\overrightarrow{M}|$ to the point M defines a scalar function (of two variables), and not a vector field.

A–30.3. $\overrightarrow{F}$ is the constant vector field $\overrightarrow{F} = \vec{b}$. The vector $\vec{b}$ is perpendicular to the line g and points towards the top of the page, and its magnitude is 3 units.

A–30.5.   a) $\vec{v}(P) = (-8.4, 3.6)$.   b) $Q(2, -5)$.

A–30.7.   a) $r = \sqrt{x^2 + y^2}$.

b) Then $(F_1, F_2) = (\frac{x}{r}, \frac{y}{r}) = \frac{\vec{r}}{r}$. The magnitude is 1.

c) For every point P on the line UO *other than the origin* the field vector is a unit vector in the direction of $\overrightarrow{OU}$. This is true no matter how close P is to the origin. At the origin, the field vector is not defined.

A–30.9.   a) See Figure A30.2 on the following page.

b) $F_1(x, y) = (1/2)\sqrt{x^2 + y^2}$ (3.1), $F_2(x, y) = 0$.

A–30.11.   a) $\operatorname{grad} xy = (y, x)$.   b) $|\operatorname{grad} xy| = \sqrt{y^2 + x^2} = |\overrightarrow{OP}|$.

c) See Figure A30.3 on the next page.

A–30.13.   a) $(-10, 1)$.

b) (i) K is a curve in the plane. The equation of K is $-x^2 + y + 4 = -13$. This is the parabola $y = x^2$ with the vertex moved to $(0, -17)$.

(ii) Equation of tangent line of K at $P(5, 8)$: $-10x + y = -42$.

(iii) $(x, y) = (5 - 10t, 8 + t)$.

c) (i) $-2/5$;   (ii) $\vec{v} = (2, -1)$, $-\sqrt{5}$.

A–30.15.   The level curve is a line which is parallel to g and which is 4/3 cm above g.

Answers for Section 31

A–31.1.   a) The flow lines are circles with center at the point O.

b) The flow lines are straight lines which are parallel to the vector $\vec{a}$.

c) The flow lines are halflines issuing from the origin. The origin itself is not part of any flow line because the vector field is not defined at the origin.

A–31.3.   a) $x'(t) = \frac{x(t)}{\sqrt{[x(t)]^2 + [y(t)]^2}}$, $y'(t) = \frac{y(t)}{\sqrt{[x(t)]^2 + [y(t)]^2}}$.

b) $(x, y) = (7, 48) + t\left(\frac{7}{\sqrt{7^2 + 48^2}}, \frac{48}{\sqrt{7^2 + 48^2}}\right)$.

A–31.5.   a) $(5, -11)$.

b) (i) $(a(3), b(3)) = (4, 6)$.

(ii) The velocity vector $(a'(3), b'(3))$ at the point $(4, 6)$ must be the field vector at $(4, 6)$, and at this point the components of the field vector are $(10, -18)$.

c) To find out whether K is a flow line we have to check whether the functions $p(t)$ and $q(t)$ are solutions of the following system of ODE:

$$\begin{aligned} x'(t) &= x(t) + y(t) \\ y'(t) &= -3\,x(t) - y(t). \end{aligned}$$

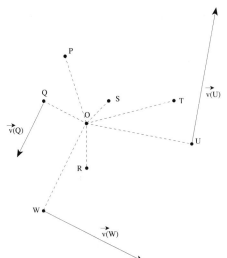

Figure A30.1

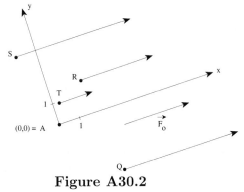

Figure A30.2

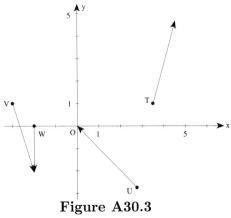

Figure A30.3

If you work out the expressions on the left and the right side, you see that the function $p(t)$ and $q(t)$ are solutions of the system. That is, K must be a flow line of $\vec{F}$.

A–31.7. The field vector at $(5,3)$ is the velocity vector of the flow line at the same point. We compute the velocity vector: $\vec{v} = (x'(0), y'(0)) = (1, 10)$. That is, the field vector at $(5,3)$ has components $(1, 10)$.

A–31.9. a) See below.
b) The speck will move along a line parallel to $\vec{a} = (1, 1)$ towards the upper right. Its speed increases.
c) $(x, y) = (e^t + 2, e^t)$

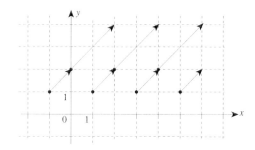

Answers for Section 32

A–32.1. $(F_1(x, y, z), F_2(x, y, z), F_3(x, y, z) = (0, -\omega z, \omega y)$.

A–32.3. a) $(6, 3, 2)$.
b) $6x + 3y + 2z = 18$, $\vec{b} = (6, 3, 2)$.
c) The following is a standard fact (discussed in Section 19): *The gradient of $f(x, y, z)$ at a point P is perpendicular to the level surface passing through P.* Why is that true? The directional derivative in the direction of $\vec{u}$ is computed as the scalar projection of $\operatorname{grad} f$ onto $\vec{u}$. The directional derivative in a direction $\vec{u}$ parallel to the tangent plane is zero. That is, $\operatorname{grad} f \cdot \vec{u} = 0$. Therefore, $\operatorname{grad} f$ is perpendicular to the tangent plane at P.

A–32.5. a) The gravity field has components $(0, 0, -g)$. Is there a function $h(x, y, z)$ so that $h_x = h_y = 0$ and $h_z = -g$? Yes: Take $h(x, y, z) = -gz$. Therefore, the field is a gradient field.
b) Let (a_1, a_2) be the components of $\vec{a}$. Take $f(x, y) = a_1 x + a_2 y$. Then $\operatorname{grad} f = a_1 x + a_2 y = \vec{a}$. Therefore, $\vec{F} = \vec{a}$ is a gradient field.
c) The field $\operatorname{grad}(x^2 + y^3 + z^4)$ is not constant.

A–32.7. a) $(F_1(x, y, x), F_2(x, y, z), F_3(x, y, z)) = \overrightarrow{AB} = (-2, 2, -3)$.
b) The flow lines are straight lines parallel to $\overrightarrow{AB}$.
c) $x'(t) = -2$, $y'(t) = 2$, $z'(t) = -3$.
d) $(-734, 745, -1088)$.
e) $g(x, y, z) = -2x + 2y - 3z + 79$.

A–32.9. Consider a circle C of radius b which lies in a plane parallel to the yz-plane, and which is centered on the x-axis. Let (x, y, z) be a point on C. The vector $\vec{t} = (0, -z, y)$ is tangent to C at (x, y, z). We have $|\vec{t}| = \sqrt{y^2 + z^2} = b$, which is the distance of (x, y, z) from the x-axis. We therefore take

$$\overrightarrow{F} = \frac{k}{b^2}\overrightarrow{t} = \frac{k}{y^2 + z^2}\,(0, -z, y).$$

A–32.11.   a) $(-2, -4, 1)$.   b) See below.

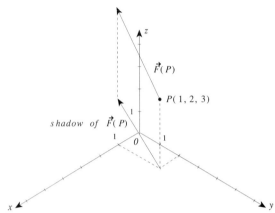

c) System of equations: $x'(t) = -2x(t)$, $y'(t) = -2y(t)$, $z'(t) = 1$.
Solution: $(x, y, z) = (e^{14-2t}, 2e^{14-2t}, t - 4)$.

A–32.13.   a) $(\dfrac{1}{\sqrt{17}}, \dfrac{4}{\sqrt{17}}, 0)$.

b) The flow line is a ray (half-line) issuing from P^* and passing through P. It points away from the z-axis.

c) $(a(t), b(t), c(t)) = (4 + \frac{4}{5}t, 3 + \frac{3}{5}t, 7)$.

Answers for Section 33

A–33.1.   a) $\operatorname{grad} g = (3x^2, 1, 2z)$.

b) $\operatorname{div} g$: Bad question! The divergence is defined only for vector fields, and not for scalar functions.

c) $\operatorname{div}\overrightarrow{F} = 2z + \cos y + 0$.   d) $\nabla \times \overrightarrow{F} = (e^y, 2x, 0)$.   e) $\nabla \cdot (\nabla g) = 6x + 0 + 2$.

f) $\operatorname{curl}(\nabla g) = (0, 0, 0)$.   g) $\operatorname{div}(\nabla \times \overrightarrow{F}) = 0$.

A–33.3. $\operatorname{curl}(\operatorname{grad} g) = \dfrac{\partial g_y}{\partial x} - \dfrac{\partial g_x}{\partial y} = g_{yx} - g_{xy} = 0$, provided that $g(x, y)$ has continuous second order partial derivatives.

Answers for Section 34

A–34.1.   If I double the distance, the work should double, and if I double the force, the work should also double. For this to be true, I have to choose Fs or $-Fs$. All the other formulas would violate the requirement "If I double s....."

A–34.3.   Write $r = \sqrt{x^2 + y^2 + z^2}$. Unit vector in direction of $\overrightarrow{F}$: $(-1/r)(x, y, z)$. Magnitude of field vector: $k(1/r^2)$. We get $\overrightarrow{F} = -\frac{k}{r^3}(x, y, z)$. Note that the constant k must be positive.

A–34.5.   a) $W(K, L) = \overrightarrow{KL} \cdot \overrightarrow{F}$;   b) $=$

c) By definition of vector addition we have $\overrightarrow{KM} = \overrightarrow{KL} + \overrightarrow{LM}$. We use this in the formula for the work we found in a): $W(K, M) = \overrightarrow{KM} \cdot \vec{F} = (\overrightarrow{KL} + \overrightarrow{LM}) \cdot \vec{F} = \overrightarrow{KL} \cdot \vec{F} + \overrightarrow{LM} \cdot \vec{F} = W(K, L) + W(L, M)$. Note that we used a general property of dot products, namely the fact $(\vec{a} + \vec{b}) \cdot \vec{c} = \vec{a} \cdot \vec{c} + \vec{b} \cdot \vec{c}$.

A–34.7. At all points of C, the field vector is perpendicular to the tangent vector of C. It means: For any short stretch along C the work is practically zero. Therefore, we expect the work to be zero.

Answers for Section 35

A–35.1. $\overrightarrow{AB} \cdot \vec{H} = -20$.

A–35.3. A typical term of a Riemann sum looks like $\vec{F} \cdot \Delta\vec{r}$. Here $\Delta\vec{r}$ points in the direction of the orientation of C. If the orientation of C is reversed, then each $\Delta\vec{r}$ must be changed to $-\Delta\vec{r}$. Then all Riemann sums are multiplied by -1. Therefore the line integral is multiplied by -1.

A–35.5. (i) $\int_a^b [f(t) + g(t)]dt = \int_a^b f(t)dt + \int_a^b g(t)dt$.

(ii) Let k be a constant. Then $\int_a^b k f(t)dt = k \int_a^b f(t)dt$.

(iii) $\int_a^b f(t)dt + \int_b^c f(t)dt = \int_a^c f(t)dt$ where a, b, and c are three points on the t-axis as shown below.

A–35.7.

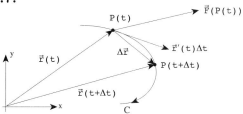

A–35.9. a) $\int_{t=0}^{t=\pi} (\cos t \sin t, \cos t) \cdot (-\sin t, \cos t)\, dt = \int_{t=0}^{t=\pi} (-\cos t \sin^2 t + \cos^2 t)\, dt$.

b) $\int_{t=-1}^{t=1} (0, -t) \cdot (-1, 0)\, dt = \int_{t=-1}^{t=1} 0\, dt = 0$.

A–35.11. a) $\int_C \overrightarrow{G} \cdot d\vec{r} = \int_0^1 (16t^2, 16t, 2t) \cdot (4, 2, 16)\, dt = (64/3) + 32$.

b) $\int_K \overrightarrow{G} \cdot d\vec{r} = \int_0^2 (t^4, 2t^3, t) \cdot (2t, 1, 6t^2)\, dt = (64/3) + 32$.

A–35.13. Switching the orientation of C reverses the direction of $\overrightarrow{T}$ and therefore changes the sign of $F_{\tan} = \overrightarrow{F} \cdot \overrightarrow{T}$.

A–35.15. The field vector at $P(a, b, c)$ is parallel to the xy-plane. It is tangent to the circle that lies in the plane $z = c$, which is centered at $(0, 0, c)$, and which has radius $\sqrt{a^2 + b^2}$, as shown below.

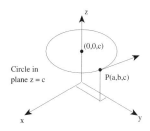

a) (i) $(1/2)\pi$;   (ii) $=$ (iii) 0.   b) (i) 0;   (ii) 12;   (iii) $(12)(24\pi)$.

A–35.17. a) (i) 9.8 nt m;   (ii) 4.43 m/sec.   c)–d) 4.43 m/sec.

A–35.19. a) 5ω.   b) 5ω.

A–35.21. a) Maximum of F_{tan} at A: 4. Minimum at B: -4. See Figure A35.1.
b) $e = (-4)(3\pi)$, $E = (4)(3\pi)$.

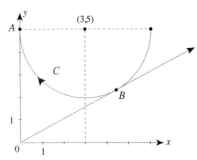

Figure A35.1

A–35.23. a) $=$ b): 6.5.

A–35.25. a) $x = t, y = 0, t_A = a, t_B = b$.   b) $\displaystyle\int_{t=a}^{t=b} f(t)dt$.

c) (i) The ordinary integral is a special case of the line integral. (ii) The line integral is a generalization of the ordinary integral.

Answers for Section 36

A–36.1. a) $\displaystyle\int_{t=t_A}^{t=t_B} [ma'(t) + nb'(t)]dt$.

b) $\displaystyle\int_C \overrightarrow{F} \cdot d\vec{r} = m\int_{t=t_A}^{t=t_B} a'(t)dt + n\int_{t=t_A}^{t=t_B} b'(t)dt = m(u - p) + n(v - q)$.

A–36.3. Write $\Phi(x, y) = x^2y + y^3 + 5$. Then $\int_C \overrightarrow{F} \cdot d\vec{r} = \Phi(1, 1) - \Phi(0, 0) = 2$.

A–36.5. We know that the line integrals of $\overrightarrow{F}$ are not path-independent. Therefore, the functions F_1 and F_2 cannot both be constant. If they were, the line integrals would have to be path-independent.

A–36.7. a) With the y-axis pointed straight up, $\Phi = -9.8y$.
b) 5.102 m. At peak, pot. energy $=$ initial kin. energy.
c) 10 m/sec. Potential energy is same as at launch, so same for kinetic energy.

A–36.9. a) We pick two points on the closed curve C. Then C equals C_1 followed by the opposite of C_2, as shown below.

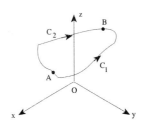

We write $-C_2$ for the opposite of C_2. Then

$$\oint_C \vec{F} \cdot d\vec{r} = \int_{C_1} \vec{F} \cdot d\vec{r} + \int_{-C_2} \vec{F} \cdot d\vec{r}$$

The line integral over the opposite of C_2 equals the negative of the integral over C_2. Therefore, we can write the equation in the form

$$\oint_C \vec{F} \cdot d\vec{r} = \int_{C_1} \vec{F} \cdot d\vec{r} - \int_{C_2} \vec{F} \cdot d\vec{r}$$

The line integral is path independent. This means that the integral over C_1 equals the integral over C_2. Thus, the right side equals zero. Therefore, the integral over the closed curve C on the left side must be zero.

b) We use the notation and figure of part a. C_1 and C_2 are two curves from A to B. Then C_1 followed by the opposite of C_2 is a closed curve C. We integrate over this closed curve C:

$$\begin{aligned} \oint_C \vec{F} \cdot d\vec{r} &= \int_{C_1} \vec{F} \cdot d\vec{r} + \int_{-C_2} \vec{F} \cdot d\vec{r} \\ &= \int_{C_1} \vec{F} \cdot d\vec{r} - \int_{C_2} \vec{F} \cdot d\vec{r} \end{aligned}$$

We know that the integrals over a closed curve are zero. That is, the left side is zero. Then the right side must be zero. Therefore, the integrals over C_1 and C_2 are the same.

c) The assertion in part a has the form "If A is true, then B is true," while part b asserts the converse "If B is true, then A is true." These assertions are logically different. For example, the statement "If Ed is a horse then Ed has four legs" does not have the same meaning as the statement "If Ed has four legs then Ed is a horse."

A–36.11. a) 0.
b) We know that if the integrals over *all* closed curves are zero, then the field is conservative. We have integrated the field only around *one* closed closed curve, namely, the circle of radius 2. We cannot conclude from the computation of the integral over this *one* closed curve that the integrals over *all* closed curves are zero. At this point, we do not know whether the field is conservative or not.
c) $-8/3$.
d) Now we know for sure that the field is *not* conservative. Why? If it were, the integral over the triangle would have to be zero, and the integral is not zero.

A–36.13. a) (i) $e^a \sin b$; (ii) $f(x,y) = e^x \sin y$;
(iii) grad $f(x,y) = (e^x \sin y, e^x \cos y)$; (iv) F T F T.

b) (i) $\displaystyle\int_{OS} \vec{F} \cdot d\vec{r} = \int_{t=0}^{t=1} \left(ae^{at} \sin(bt) + be^{at} \cos(bt) \right) dt.$
(ii) The integral must be equal to $f(a,b)$ because $f(x,y)$ is a potential for our vector field $\vec{F}$.

A–36.15. If the circle shown in the illustration below is given a counterclockwise orientation, then the tangential component of the field is positive at each point of the circle, and so the circulation about the circle is positive. Since a conservative field has zero circulation around *any* closed curve, it follows that the field is *not* conservative.

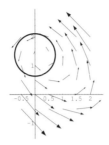

A–36.17. $\dfrac{\partial h}{\partial x} = p'(e^{xy})\, y\, e^{xy}, \qquad \dfrac{\partial h}{\partial y} = p'(e^{xy})\, x\, e^{xy}.$

A–36.19. $\nabla \phi = -3(x^2 + y^2 + z^2)^{-5/2}(x, y, z).$

Answers for Section 37

A–37.1.   The true statements are: "If the field is conservative, then its curl must be zero"; "If the curl is different from zero, then the field is guaranteed not to be conservative."

A–37.3.   The curl-test shows: $\vec{u}$ is not conservative; $\vec{v}$ is conservative; for $\vec{w}$ the test is inconclusive.

A–37.5.   We compute the curl and find $\operatorname{curl} \vec{u} = \operatorname{curl} \vec{w} = (0,0,0)$. Therefore both $\vec{u}$ and $\vec{w}$ are possibly conservative. $\operatorname{curl} \vec{v} \neq (0,0,0)$ and, therefore, $\vec{v}$ is not conservative.

A–37.7.   Write r for $\sqrt{x^2 + y^2 + z^2}$. Then $\vec{F} = -k\, r^{-3}(x, y, z)$, $F_1 = -k\, r^{-3}\, x$, and
$$\frac{\partial F_1}{\partial x} = (-k)\frac{r^3 - x\, 3r^2 \frac{x}{r}}{r^6}.$$
It is defined at all points except $(0,0,0)$. The other eight partial derivatives also have a negative power of r and are defined at all points except $(0,0,0)$. By the curl-test for vector fields in *space*, the field is conservative.

A–37.9.   The partial derivative of $\vec{F}_1$ with respect to x is the same as in 9.11, but this time, we must consider this partial derivative as a function of (x, y, z). It is not defined for all (x, y, z) such that $x^2 + y^2 = 0$. The collection of all these points (x, y, z) is the z-axis. The z-axis contains infinitely many points. The curl-test for vector fields *in space* is inconclusive.

A–37.11.   Are the fields conservative? $\operatorname{curl} \vec{u} = (0,0,0)$, and $\vec{u}$ is conservative. $\operatorname{curl} \vec{v} \neq (0,0,0)$, and $\vec{v}$ is not conservative. $\vec{w}$ is conservative because it is a constant field.
Potential for $\vec{u}$: $\Phi(x, y, z) = xyz + k$. Potential for $\vec{w}$: $\Phi(x, y, z) = 3x - y + 7z + k$.

A–37.13.   Is $\vec{u}$ conservative? $\Phi(x, y, z) = xyz + x + 2y + 3z + k$. Did you check $\operatorname{grad} \Phi$?

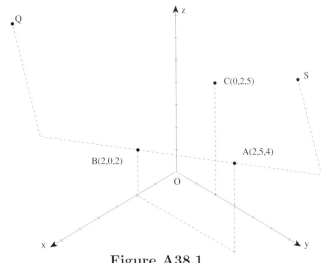

Figure A38.1

A–37.15. For a change we take $P_0 = (1,1)$. Curve C from $(1,1)$ to (a,b):
$(x,y) = (1 + t(a-1), 1 + t(b-1))$.

$$\Phi(a,b) = \int_0^1 (1 + t(b-1))e^{1+t(a-1)}, e^{1+t(a-1)}) \cdot (a-1, b-1)dt$$

$$= \int_0^1 \left([1 + t(b-1))e^{1+t(a-1)}](a-1) + [e^{1+t(a-1)}](b-1) \right) dt.$$

A–37.17. The field is conservative, and $\phi(x,y,z) = xy + z^2 + 376$.

A–37.19. $\operatorname{curl} \vec{H} = (0, -1, 0)$. $\vec{H}$ is not conservative.

A–37.21. a) The curl-test tells is inconclusive.
b) $h_x = \dfrac{x}{\sqrt{x^2 + y^2}}$, $h_y = \dfrac{y}{\sqrt{x^2 + y^2}}$.
c) $\vec{F}$ is conservative because we found in part b that $\operatorname{grad} h = \vec{F}$.

Answers for Section 38

A–38.1. a) See Figure A38.1.
b) Q is a point in α because we reach Q from A by first moving on the dotted line AB that is in α and then on the dotted line parallel to AC that is also in α. Therefore we always remain in α, that is, Q is a point in α.
c) $B : (u,v) = (1,0)$; $A : (u,v) = (0,0)$; $C : (u,v) = (0,1)$.
d) $(u,v) = (-0.9, 1.2)$ (see dotted lines extending to S in Figure A38.1).

A–38.3. a) $\overrightarrow{OS} = \overrightarrow{OA} - 0.9\,\overrightarrow{AB} + 1.2\,\overrightarrow{AC} = (2,5,4) - 0.9\,(0,-5,-2) + 1.2\,(-2,-3,1)$
$= (-0.4,\ 5.9,\ 7)$.
b) Set $(x,y) = (1,1)$: $u = v = 0.5$; $z = 3.5$.

A–38.5. Pick three points R, S, T in the plane: $R(6,0,0)$, $S(0,2,0)$, $T(0,0-3)$. (How do you check that they do not lie on a line?) Then $\overrightarrow{OP} = \overrightarrow{OR} + u\,\overrightarrow{RS} + v\,\overrightarrow{RT}$, that is, $(x,y,z) = (6,0,0) + u\,(-6,2,0) + v\,(-6,0,-3)$.

A–38.7. See Figure A38.2 on the following page.

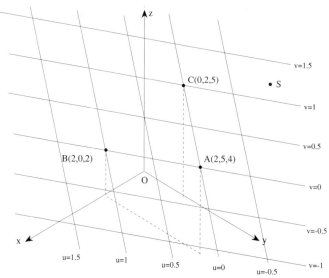

Figure A38.2

A–38.9. Usually, it is easier to remember and to recognize one formula with a variation than two seemingly unrelated formulas.

A–38.11. a) There many correct answers. If we take $K(2,0,0)$, $L(0,0,3)$, $M(2,1,0)$ we obtain:

$(x, y, x) = (2, 0, 0) + u(-2, 0, 3) + v(0, 1, 0)$.

b) If we take $K(0, 2, 0)$, $L(1, 2, 0)$, $M(0, 2, 1)$ we obtain:

$(x, y, x) = (0, 2, 0) + u(1, 0, 0) + v(0, 0, 1)$.

A–38.13. a) (i) The three points $K(2,0,0), L(0,-4,0), M(0,0,2)$ lie in α. We use these points for the parametric representation $(x, y, z) = (2, 0, 0) + u(-2, -4, 0) + v(-2, 0, 2)$.

(ii) There are infinitely many different parametric representations for α. For example, the point $N(0, 1, 2.5)$ lies in α. With K, L and N we get the parametric representation $(x, y, z) = (2, 0, 0) + u(-2, -4, 0) + v(-2, 1, 2.5)$.

b) (i) The vector $\vec{n} = (3, -1, 0) \times (-1, 4, 1) = (-1, -3, 11)$ is normal to β. Therefore, $-x - 3y + 11z = 50$ is an equation for β.

(ii) If we multiply $-x - 3y + 11z = 50$ by 2 the equation $-2z - 6y + 22z = 100$ is still an equation for β. Why?

A–38.15. a) You draw a parallel to KM through Q and intersect it with the line defined by KL. You get a point R, as shown below.

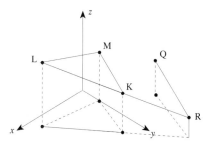

Now $\overrightarrow{KR} = -(0.8)\overrightarrow{KL}$ and $\overrightarrow{RM} = (1.4)\overrightarrow{KM}$. Therefore, $u = -0.8$ and $v = 1.4$. To find the shadow of Q you repeat for the shadows in the xy-plane what what you just did for the points themselves.
b) *(negative, positive, positive)*.

Answers for Section 39

A–39.1. a) $(x, y, z) = (h \cos u, h \sin u, v)$.
b) See below.
c) $u = k_1 = 3\pi/2 : (x, y, z) = (0, -h, 0) + v(0, 0, 1)$. $v = k_2 = m : (x, y, z) = (h \cos u, h \sin u, m)$.
d) T^0 is the rectangle in the uv-plane with corners $(0, -1)$, $(2\pi, -1)$, $(2\pi, 2)$, $(0, 2)$.

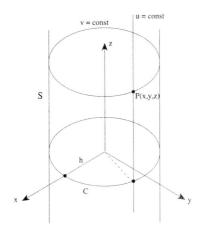

A–39.3.   On the curve $u = k$, the parameter is v, and we have the parametric representation $(x, y, z) = (k, v, f(k, v))$. Similarly, for $v = l$, the parameter is u, and we have $(x, y, z) = (u, l, f(u, l))$.

A–39.5.   a) $u =$ constant: half-circles in planes parallel to the yz-plane; $v =$ constant: half-circles in planes parallel to the xz-plane.
b) $u = 1 : (x, y, z) = (1, v, \sqrt{16 - 1 - v^2})$, $v = 2 : (x, y, z) = (u, 2, \sqrt{16 - u^2 - 4})$.
c) Tangent vector to $u = 1 : (x, y, z) = (0, 1, \frac{-2}{\sqrt{11}})$. Tangent vector to $v = 2 :$ $(x, y, z) = (1, 0, \frac{-1}{\sqrt{11}})$. Cosine of the angle between these two tangent vectors: $\frac{2}{\sqrt{180}} \neq 0$. That is, the two coordinate curves do not intersect at a right angle.

A–39.7.   a) $f_x = y - 1 = -\frac{1}{2}$, $f_y = x - 1 = \frac{1}{2}$. Normal vector of tangent plane: $(-\frac{1}{2}, \frac{1}{2}, -1)$. Equation of tangent plane: $-\frac{1}{2}x + \frac{1}{2}y - z = -\frac{5}{4}$.
b) $\vec{r}_u = (1, 0, -\frac{1}{2})$, $\vec{r}_v = (0, 1, \frac{1}{2})$. Normal: $(\frac{1}{2}, -\frac{1}{2}, 1)$.

A–39.9.   a) $(x, y, z) = (u, v, g(v))$.
b) S^0 is the rectangle $0 \le x \le a, 0 \le v \le b$ in the uv-plane.
c) The coordinate curves $u = const$ are curves parallel to the curve $z = g(y)$ in the yz-plane. The coordinate curves $v = const$ are line segments parallel to the line segment DB.
d) $\vec{n} = (0, -g'(v), 1)$. The standard normal points toward you.
e) $-g'(v)y + z = -g'(v)v + g(v)$.

A–39.11. a) (i) $Q(b\cos u, b\sin u, 0)$.

(ii) $\dfrac{(-b\cos u, -b\sin u, h)}{\sqrt{b^2 + h^2}}$.

(iii) $\dfrac{v\,(-b\cos u, -b\sin u, h)}{\sqrt{b^2 + h^2}}$.

(iv) $(x, y, z) = (b\cos u, b\sin u, 0) + \dfrac{v\,(-b\cos u, -b\sin u, h)}{\sqrt{b^2 + h^2}}$

$\qquad = \left(b\cos u \left(1 - \dfrac{v}{\sqrt{b^2 + h^2}}\right), b\sin u \left(1 - \dfrac{v}{\sqrt{b^2 + h^2}}\right), \dfrac{hv}{\sqrt{b^2 + h^2}} \right).$

b) The coordinate curves $u = constant$ are line segments that connect the top point H with the base circle in the xy-plane. The coordinate curves $v = constant$ are circles which are parallel to the xy-plane.

c) S^0 is the rectangle $0 \le u \le 2\pi, 0 \le v \le \sqrt{b^2 + h^2}$.

d) $(x, y, z) = \left(b\,\dfrac{\sqrt{2}}{2}, b\,\dfrac{\sqrt{2}}{2}, 0\right) + s\left(-\dfrac{\sqrt{2}}{2}, \dfrac{\sqrt{2}}{2}, 0\right) + t\left(-b\,\dfrac{\sqrt{2}}{2}, -b\,\dfrac{\sqrt{2}}{2}, h\right).$

Answers for Section 40

A–40.1. a) For all points, the field vector has a negative x-component. Therefore the flux is negative.

b) For all points of S, the field vector is parallel to S. Therefore, no fluid flows across S. The flux is zero.

c) $(e^{-x-y-z}, 0, 0)$: The field vector always points in the positive x-direction, and the flux is positive. $(0, \cos(-x - y - z), \sin(-x - y - z))$: The flux is zero for the reasons given in part b. The total flux is positive. *Note: The reasoning is so easy only because the surface is part of the yz-plane.*

A–40.3. a) $\vec{n} = (-1, -2, -\sqrt{11})$.   b) Negative.

A–40.5. a) $\overrightarrow{AB} = (-1, 0, 2)$, $\overrightarrow{AD} = (3, -2, 0)$.
The area of $ABCD$ equals $|\overrightarrow{AB} \times \overrightarrow{AD}| = |(4, 6, 2)| = \sqrt{56}$.

b) A normal of magnitude one has components $\dfrac{1}{\sqrt{56}}(4, 6, 2)$.

c) 34.

d) Any constant vector field $\vec{G}$ which is perpendicular to the normal $(4, 6, 2)$ has flux zero across $ABCD$. For example $\vec{G} = (-6, 4, 0)$ is perpendicular to $(4, 6, 2)$.

e) Let the components of such a vector field be (u, v, w). "Flux across $ABCD$ equals 8" means $(u, v, w) \cdot (4, 6, 2) = 8$. There ar many such (u, v, w), for example $(u, v, w) = (2, 0, 0)$.

Answers for Section 41

A–41.1. a) Positive, since the flow is in the same direction as the normal.

b) Negative, since the flow is in the opposite direction to the normal.

A–41.3. a) Both are $-12/\sqrt{237}$.   b) -6.

A–41.5.

$$\vec{r}_\phi \times \vec{r}_\theta = (\cos\phi\cos\theta\,\vec{i} + \cos\phi\sin\theta\,\vec{j} - \sin\phi\,\vec{k}) \times (-\sin\phi\sin\theta\,\vec{i} + \sin\phi\cos\theta\,\vec{j})$$

$$
= \begin{vmatrix} \vec{\imath} & \vec{\jmath} & \vec{k} \\ \cos\phi\cos\theta & \cos\phi\sin\theta & -\sin\phi \\ -\sin\phi\sin\theta & \sin\phi\cos\theta & 0 \end{vmatrix}
$$

$$
= \sin^2\phi\cos\theta\,\vec{\imath} + \sin^2\phi\sin\theta\,\vec{\jmath} + (\cos\phi\sin\phi\cos^2\theta + \cos\phi\sin\phi\sin^2\theta)\,\vec{k}
$$

$$
= \sin^2\phi\cos\theta\,\vec{\imath} + \sin^2\phi\sin\theta\,\vec{\jmath} + \cos\phi\sin\phi\,\vec{k}.
$$

A–41.7. a) 0.010. b) 0.479. c) $k = |\vec{r}_\phi \times \vec{r}_\theta|$.

A–41.9. a) $\vec{n} = (v, 1, 0) \times (u, 1, 1) = (1, -v, v - u)$.

b) $F_n = (v, v, 0) \cdot \dfrac{(1, -v, v - u)}{\sqrt{1^2 + v^2 + (v - u)^2}} = \dfrac{v - v^2}{\sqrt{1^2 + v^2 + (v - u)^2}}$.

A–41.11. Let P be a point on S, and let P_0 be the shadow of P in the xy-plane. The line $0P_0$ forms an angle with the x-axis, and we write α for this angle.

a) $F_n = (a\cos\alpha, a\sin\alpha, \ldots) \cdot (\cos\alpha, \sin\alpha, 0) = a$, $A(S) = 2\pi a h$. Flux : $2\pi a^2 h$.

c) $F_n = 0$. Flux: 0.

A–41.13. a) $(x, y, z) = (2 - 2u - 2v, 2u, 2v)$. b) $\displaystyle\int_{v=0}^{v=1}\int_{u=0}^{u=1} 4(2 - 2u - 2v)\,du\,dv = 0$.

A–41.15. a) In S, all the field vectors are parallel to $(0, 0, 1)$ (but are of different magnitude). That is, in S the normal component of $\vec{F}$ is positive. Therefore, the flux is positive.

b) $\omega\, q$.

c) $\displaystyle\int_{y=0}^{y=c}\int_{x=a}^{x=b} \omega\, y\, dx\, dy = \omega\,(b - a)\,\dfrac{c^2}{2}$.

A–41.17. $\frac{\pi}{6}\left(5^{3/2} - 1\right)$.

Answers for Section 42

A–42.1. List of integrals for reference: **(1):** $\int f(x)\,dx$ **(2):** $\int\int g(x, y)\,dA$ **(3):** $\int\int\int h(x, y, z)\,dV$ **(4):** $\int \vec{F} \cdot d\vec{r}$ **(5):** $\int \vec{F} \cdot d\vec{A}$
Common features of the integrals:

(1) through (5): In the course of the evaluation of integrals **(2)**–**(5)** you have to repeatedly perform integrations of type **(1)**.

(1) through (3): If you integrate the function which has the constant value 1 ("one"), you get a measure for the "size" of the domain of integration (length of interval, area of region, volume of solid).

(1) and (2): The two integrals have a similar interpretation as "amount of bulk under graph" (area under curve, volume under surface).

(4) and (5): The domain of integration (curve, surface) has to be oriented.

(1) and (4): The domain of integration is "string-like" (interval, curve as "interval which has been bent").

(2) and (5): The domain of integration is "sheet-like" (region, surface as "region which has been bent").

(1) through (5): Each integral is based on Riemann sums. You take Riemann sums based on finer and finer subdivisions of the domain of integration. Then the integral is the value at which the Riemann sums stabilize (i.e., is the limit of the Riemann sums). As for the Riemann sums, they all look alike. A typical term of a Riemann sum is of the form

$$\mathcal{F}(P) \cdot \Delta \mathcal{X}$$

In this formula, $\mathcal{F}(P)$ is the value of a function at a point P. P lies in one of the pieces into which the domain of integration has been subdivided. $\Delta \mathcal{X}$ is a measure for the "bulk" (length, area, volume) of the piece in which P lies.

A–42.3. a) ∂K is a closed surface. It is formed by the four faces of the pyramid.
b) ∂C consists of two points. The two points are Q and U.
c) ∂S is a closed curve. It consists of the edges QR, RV, VU, UQ joined end to end.
d) Take as surface S_1 the face (filled-in triangle) QRV.
e) Take as S_2 the surface that you obtain by removing the face QRV from the boundary of the solid pyramid.

A–42.5.
a) C: Circle of radius b, center O in the xy-plane.
    D: Disk of radius b, center O in the xy-plane.
    S: Half sphere of radius b, center O above the xy-plane.
    K: Solid half ball of radius b, center O above the xy-plane.

b) 1, 2, 2, and 3 for C, D, S, and K, respectively.

c) C.   d) D, S.   e) (D, S).   f) dim $= 2$; $x^2 + y^2 = 3b^2/4$, $z = b/2$.

g) None.   h) K.

A–42.7.   a) ∂K has two parts: A surface like a cone-shaped paper hat (such as a dunce hat) and a disk which closes off the paper hat at the bottom.
b) ∂S is circle of radius b; it is only the curve, not the disk. It lies in the xy-plane, and its center is at the origin.

Answers for Section 43

A–43.1. a) $\vec{n}(P) = \vec{k}$ b) $\vec{n}(Q) = \vec{\imath}$.   c) $\vec{n}(R) = -\vec{\imath}$.

A–43.3. $4\pi a^3$.

A–43.5. Since the outward flux of $\overrightarrow{F}$ from any solid is 0 (by Problem 43.2), the divergence of $\overrightarrow{F}$ is 0 at any point.

A–43.7. Let c be a number which is slightly less than $\Phi(P)$, and let K be the solid defined by $\Phi \geq c$. Then P is in K, since $\Phi(P) > c$. The solid K might consist of several nonoverlapping pieces. Let K' be the one that contains P. The surface ∂K is (part of) the level surface $\Phi = c$, so grad Φ is normal to ∂K. Moreover, since Φ is less than c outside of K and greater than c inside of K, it follows that grad Φ

(if it is not zero) points in the direction of the *inward* normal. Thus the *outward* normal component of K is either negative or zero, so the outward flux from K' is less than or equal to 0. If the constant c is taken to be very close to $\Phi(P)$, then the solid K' will be a small solid containing P, so the flux per unit volume from a small solid containing P will be less than or equal to 0. Thus $\operatorname{div}\operatorname{grad}\Phi|_P$ can be approximated to any desired precision by a number which is less than or equal to 0, so $\operatorname{div}\operatorname{grad}\Phi(P) \le 0$.

A–43.9. a) $\operatorname{div}\overrightarrow{F} = 3$, $\iint_{\partial K}\overrightarrow{F}\cdot d\overrightarrow{A} = \iiint_K 3\,dV = 3\,V(K)$ where $V(K)$ is the volume of K.

b) 0 because $\operatorname{div}\overrightarrow{F} = 0$.

c) $\operatorname{div}(\operatorname{curl}\overrightarrow{F}) = 0$ for any vector field $\overrightarrow{F}$. Therefore, $\iint_{\partial K}\overrightarrow{F}\cdot d\overrightarrow{A} = 0$.

A–43.11. The theorem of Gauss establishes a relation between surface integrals and triple integrals. The link is constructed by means of the notion of divergence.

A–43.13. a) 0.   b) 4π.

A–43.15. a) Let K be the solid bounded by S. By the Divergence Theorem,

$$\iint_S \overrightarrow{F}\cdot d\overrightarrow{A} = \iiint_K \operatorname{div}\overrightarrow{F}\,dV = -\iiint_K r^{-4}\,dV < 0.$$

b) Let S be a large sphere of radius a about the origin, and let K' be the solid consisting of all points outside T and inside of S. Applying the Divergence Theorem to K' gives

$$-\iiint_{K'} r^{-4}\,dV = \iint_{\partial K'}\overrightarrow{F}\cdot d\overrightarrow{A} = -\iint_T \overrightarrow{F}\cdot d\overrightarrow{A} + \frac{4\pi}{a}.$$

Rearranging gives

$$\iint_T \overrightarrow{F}\cdot d\overrightarrow{A} = \iiint_{K'} r^{-4}\,dV + \frac{4\pi}{a}.$$

Since both terms on the right are positive, the outward flux through T is positive.

A–43.17. a) The flux through the faces parallel to the xy-plane and to the xz-plane vanishes. For the face in plane $x = b$, the normal component F_n equals $f(b)$, and the flux is $f(b)(1^2) = f(b)$. For the face in the plane $x = a$, $F_n = -f(a)$ (the normal is $-\vec{\imath}$), and the flux is $-f(a)$. Adding the flux across all six faces gives a total flux of $f(b) - f(a)$.

b) $\iiint_K \operatorname{div}\overrightarrow{F}\,dV = \iiint_K f'(x)\,dV = (1)(1)\int_a^b f'(x)dx = \int_a^b f'(x)\,dx$.

c) The Fundamental Theorem of Calculus. Gauss' Theorem is a generalization of the Fundamental Theorem of Calculus.

A–43.19. a) It is the circle given by $x^2 + z^2 = a^2, y = b$.

b) We write: $A = $ flux across boundary of solid cylinder; $B = $ flux across S (the desired quantity); $C = $ flux across lid $y = b$. Then we have $A = B + C$. Also, $A = \pi a^2 b$ (Gauss), $C = [(b, b, b)\cdot\vec{\jmath}]\pi a^2 = \pi a^2 b$, therefore, $B = 0$.

A–43.21. Note that $\operatorname{curl}\overrightarrow{F} = (0, 6z, 0)$ and $\operatorname{div}\overrightarrow{F} = 5$.

a) ∂S consists of the line segments OB, BC, CD, DO. ∂K consists of S plus the filled-in rectangle $OBCD$. $\partial(\partial S)$ is empty.

b) The flux of $\overrightarrow{F}$ across S plus the flux across rectangle $OBCD$ is the flux across ∂K. By Gauss, the flux across ∂K equals $\iiint_K 5dV = 5\cdot h\frac{w^2}{2}$. Flux across rectangle

$OBCD$: The normal is $(-1, 0, 0)$. On the rectangle we have $x = 0$ and therefore $\overrightarrow{F} = (0, 5y, 0)$. We conclude: The flux across $OBCD$ is zero. Therefore, the flux across S is $5 \cdot h\frac{w^2}{2}$.

c) By Gauss, the flux of curl $\overrightarrow{F}$ across ∂K equals the triple integral over K of $\text{div}(\text{curl } \overrightarrow{F})$. No matter what the vector field is, the divergence of the curl is zero. That is, the flux of curl $\overrightarrow{F}$ across ∂K vanishes. In consequence, the flux across S is the negative of the flux across $OBCD$. The normal of $OBCD$ is $(-1, 0, 0)$. We know that curl $\overrightarrow{F} = (0, 6z, 0)$ and, therefore, the flux of curl $\overrightarrow{F}$ across $OBCD$ is zero. We see: The flux of curl $\overrightarrow{F}$ across S is zero, too.

d) The flux of $\overrightarrow{F}$ across ∂K has been computed in part b.

e) The flux of curl $\overrightarrow{F}$ across ∂K has been computed in part c.

A–43.23. Write K for the solid pyramid. We have to estimate $\int\int\int_K 2z \, dV$. The maximum of $2z$ on K equals $2e$, and the minimum $2b$. Therefore, $m = (2b)\frac{1}{6}a(e-b)c$ and $M = (2e)\frac{1}{6}a(e-b)c$.

A–43.25. (1) "Flux across a solid" does not make sense. You find the flux across a surface. Similarly, "line integral over ∂K does not make sense because ∂K is a surface, and not a curve.

(2) This is Gauss' Theorem, and it is true.

(3) The line integral of grad f over the boundary of the surface S is always zero. But the flux of grad f across S is not necessarily zero. For example, if S is the sphere of radius 1 about the origin, oriented by the outward normal, and if $f(x, y, z) = x^2 + y^2 + z^2$, then the flux of grad f across S is 8π.

(4) This is again Gauss' Theorem, stated in the direction opposite to (2).

(5) Gauss' Theorem tells us $\int\int_{\partial K} \text{curl } \overrightarrow{F} \cdot d\overrightarrow{A} = \int\int\int_K \text{div}(\text{curl } \overrightarrow{F}) \, dV$. Write out $\text{div}(\text{curl } \overrightarrow{F}): \nabla \cdot (\nabla \times \overrightarrow{F})$, and the latter expression is the scalar zero. Therefore, the statement is true.

(6) "Triple integral over a surface" does not make sense.

Answers for Section 44

A–44.1. a) See below.

b) Normal pointing towards the z-axis ("inside" normal). See below.

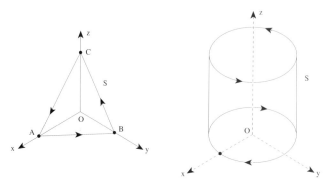

A–44.3. a) The partial derivatives of the components are defined and continuous for all points in space. We can apply Stokes' Theorem to convert the line integral to a surface integral over the filled in square.

b) S is the filled-in square $DEFG$. We compute the curl: curl $\overrightarrow{F} = (0, -1, 1)$. It is constant. The normal of S is constant because S is part of a plane. Therefore the normal component is constant, and the surface integral is curl $\overrightarrow{F})_n A(S_n)$. This is easier to evaluate than the line integral.

c) The normal component of curl $\overrightarrow{F}$ is curl $\overrightarrow{F}_n = 4/\sqrt{42}$. The area of S is $A(S) = b^2$. The surface integral is $(4/\sqrt{42}) b^2$, and that is the value of the line integral.

A–44.5. curl $\overrightarrow{F} = (2y - 2z, 2z - 1, 2x + 1)$, curl $\overrightarrow{F}$ at $(1, 1, 1)$: $(0, 1, 3)$.

a) Counterclockwise.

b) Take $\vec{n}$ perpendicular to curl $\overrightarrow{F}$ at $(1, 1, 1)$, for example $\vec{n} = (0, 3, -1)$.

c) Take $\vec{n}$ parallel to curl $\overrightarrow{F}$ at $(1, 1, 1)$, but in the opposite direction. For example, $\vec{n} = (0, -1, -3)$.

d) Take Q such that curl $\overrightarrow{F} = (0, 0, 0)$. This is the case for $Q(-0.5, 0.5, 0.5)$.

A–44.7. a) The arrow points from $(3, 0, 0)$ to $(0, 0, 3)$.

b) Stokes' Theorem says $\oint_{\partial S} \overrightarrow{F} \cdot d\vec{s} \stackrel{\text{\tiny ?}}{=} \int\int_S$ curl $\overrightarrow{F} \cdot d\overrightarrow{A}$. We can consider our given C as the boundary of the solid circle $x^2 + z^2 \leq 9$, $y = 0$ in the xz-plane with $-\vec{j}$ as normal. The curl of $\overrightarrow{F}$ is $(2, -3, 4)$. It will be easy to find the flux for a constant vector field. Therefore we compute the circulation along C by the flux of $\overrightarrow{F} = (2, -3, 4)$ across D: $\oint_C \overrightarrow{F} \cdot d\vec{r} = \int\int_D$ curl $\overrightarrow{F} \cdot d\overrightarrow{A} = (3)(3^2\pi)$.

A–44.9. Write C for the closed curve "first OA, then H," and write D for the half disk. We orient D by means of $-\vec{j}$. Then $(\text{curl } \overrightarrow{F})_n = -3$, and $\int_C \overrightarrow{G} \cdot d\vec{r} = \int\int_D (-3) d\overrightarrow{A} = -(3/2)\pi a^2$. We have to subtract $\int_{OA} \overrightarrow{G} \cdot d\vec{r}$. On the x-axis the field $\overrightarrow{G}$ is zero. Therefore $\oint_H \overrightarrow{G} \cdot d\vec{r} = -(3/2)\pi a^2$.

A–44.11. a) $(2y - 2z, 2z - 2x, 2x - 2y)$.

b) At the point $(6, 5, 2)$ we have curl $\overrightarrow{F} = (6, -8, 2)$. We take $\vec{j}$ as the normal of the plane $y = 5$. Then the orientations of the circle and the plane $y = 5$ are right for Stokes' Theorem. The scalar projection of $(6, -8, 2)$ onto the normal $(0, 1, 0)$ equals -8. Then we have $\int_C \overrightarrow{F} \cdot d\vec{r} \approx$ curl $\overrightarrow{F} \cdot \vec{n}(10^{-16})\pi = (-8) \cdot 10^{-32}\pi$.

A–44.13. a) You have four line integrals. From R to S: 0; from S to T: $(q-p)g(b)$; from T to U: 0; from U to R: $-(q - p)g(a)$. Total: $(q - p)g(b) - (q - p)g(a)$.

b) curl $\overrightarrow{F} = (0, 0, g'(x))$. The required expression is

$$\int\int_R (0, 0, g'(x)) \cdot d\overrightarrow{A} = (q - p) \int_{x=a}^{x=b} g'(x)dx.$$

c) $g(b) - g(a) = \int_{x=a}^{x=b} g'(x)dx$. To get the definite integral of a function $f(x)$ from $x = a$ to $x = b$ you take an antiderivative $g(x)$ and compute the difference $g(b) - g(a)$.

d) The Fundamental Theorem of Calculus is a special case of Stokes' Theorem.

A–44.15. a) 2π.

b) Green's Theorem cannot be applied, since the field is undefined at the origin, which is inside the curve.

A–44.17. Tell your friend the following: Take a line integral over a circle. The circle is the boundary of a surface S. The surface S cannot be closed; if it were, its boundary would have to be empty. A surface S which is not closed cannot be the boundary of a solid. Therefore you cannot apply the theorem of Gauss. Why? Because it is about the flux across a surface which is the boundary of a solid.

Index

Notes

Notes

Notes

Notes

Notes